The Fast Track to Your Extra Class Ham Radio License

Michael Burnette, AF7KB with Kerry Burnette, KC7YL

Covers all FCC Amateur Extra Class Exam Questions July 1, 2020 through June 30, 2024

*The Fast Track
to Your
Extra Class Ham Radio License*
is now available as an unabridged audiobook.
Over 20 hours of instruction.
Available from Amazon.com, Audible.com, and iTunes

Dedications

For Darrell Cutter, John Hanley, Sr., and John Hanley, Jr.

Passionate teachers, brilliant minds, loyal friends, living demands for excellence, It has been a privilege to have known them.

Thanks to all those who have served as sounding boards or sources of information that got me through some of the tough spots of creating this program – and there were plenty. Many thanks to Eric Nordin, AD7BF, Curt Green, N7OBI, and Tim Pozar, KC6GNJ, who threw in some much appreciated tech support. Dennis Bradford, W7DEB, is my Guru of Grounding. Dennis Carpenter, KB7DY, graciously and generously got this train back on track numerous times by catching errors. Dan Agun, KF7HJ, and Bill Thomassen, N6NBN, dragged me kicking and screaming into the arcane world of LaTeXtypesetting, vastly improving the look of our books.

Once again I have the honor of thanking my friend and "personal physics coach", David Cornell, PhD, W9LD, for contributing so much to my work.

Some incredibly special thanks to Colleen Edwards, KI7DRS, who has taken on the giant task of turning the Fast Track programs – graphics and all – into material that visually impaired people can use to get their licenses through the Courage Kenny Rehabilitation Institute's HandiHam program. That project has included Colleen training with the National Federation of The Blind to be certified as a Braille transcriber. That's one amazing example of serving the community.

And especially to my wife, Kerry, KC7YL, who, with boundless good humor, has patiently endured countless muddle-headed rants about obscure topics while I worked out concepts and ways of teaching them during the course of creating these programs. Then she bravely served as one of the "test pilots" for the program by going and taking (and passing with flying colors!) the exams. She also edits and proofs all of our books and audio programs and is a radiant presence at our booth at major hamfests. She well deserves her co-author credit; her contribution to the quality of our programs is enormous.

Contents

I believe that in every person is a kind of circuit which resonates to intellectual discovery—and the idea is to make that resonance work.

- Carl Sagan

CAPACITY TO LEARN
IS A *GIFT*;
ABILITY TO LEARN
IS A *SKILL*;
WILLINGNESS TO LEARN
IS A
CHOICE.

Introduction

Welcome, and congratulations on your decision to pursue your Extra Class ham license. We have quite a journey ahead of us, so get your commitment and curiosity dialed up to 11.

The Extra Class license – officially the "Amateur Extra Class License" – brings with it privileges to operate in the "VIP rooms" of the ham bands, where the bands are less crowded, and the propagation is often better. I think for most of us, too, there's a certain pride in having accomplished getting that ticket.

Those benefits aside, my hope is that you'll take the knowledge you gain from this program and assist other hams and your community with your expertise.

I think it is safe to say that unless you already have a broad and deep education in electronics, ham radio operations, and math, there are parts of this journey you will find challenging, and that this will not be a weekend-long jaunt, nor even a week-long excursion. While I've done my best to get everything boiled down to plain language, there will no doubt be a few concepts that just don't quite click for you the first time through. I hope you won't get discouraged when that happens – I doubt anybody gets all of this the first time through.

My end goal here is to get you your Extra Class license, not to give you a high-level academic style electronics education. That's far beyond the scope of the book, not to mention my capabilities. I've simplified a lot of concepts here to make them understandable, sometimes at the expense of absolute scientific rigor, so you're not going to be a PhD Electrical Engineer when you finish this course. This course is a starting point, not an end point. I hope you'll take your learning here and springboard off it into deeper explorations of the topics that particularly interest you. The good news about that is there are some great resources in the ARRL collection and elsewhere to do just that.

I'm also starting from the assumption that you know nothing more than is covered at the Technician and General Class levels, and I'll even review a lot of that stuff.

Fair warning: I'm not above using history, science trivia, stories, and corny jokes and illustrations to help stuff stick in your head.

Plan on spending at least a month, probably two, with this material. There's an accompanying audio program (available from Amazon.com, Audible.com, or iTunes) so you can also be gaining knowledge and reviewing while you commute or exercise. I think that audio program really helps with the memorization parts of the exam – and there are a lot of things to memorize. The audio program consists of over 22 hours of instruction, so you can see we have a lot of ground to cover.

There are about 600 questions in the Extra Class question pool. We'll go over the right and wrong answers for each and every one.

The challenge of the Extra exam lies not just in the number of questions, nor in the breadth and depth of the topics covered. In addition to those factors, the answer sets often contain four answers that all sound fairly reasonable. Let's face it, many of the questions on the Technician exam had one right answer and at least two, often three, that were easily eliminated with a little common sense and some basic electronic skills. (I don't know about you, but some of those wrong answers were so wildly unlikely they made me laugh out loud.) The General exam stepped it up a bit, often giving you one or two wrong answers that looked pretty likely. Still, test skills, Ohm's Law, and some common sense would get you a long way with that exam.

The philosophy of the Fast Track series has been, from the beginning, that the best way to pass an exam is to actually know and understand the material. (Shocking, right?) On the Extra exam, I doubt you'll pass by being a clever test taker, nor by being a memorizer of answers. I'll point out a few useful answer patterns and give you some memorization aids, but they won't get you through. This exam really does demand you not only know the material covered, but that you master many of the skills behind that material.

Previous question banks have contained material that I think most Extra Class candidates found more than a little irrelevant. I'm happy to report that the question pool committee for this 2020 edition of the questions put a lot of work into modernizing the question pool and eliminating a lot of the less relevant material. On the other hand, they also did a lot of work on the wrong answers, and made a lot of the choices a bit tougher.

Before I drive you off into another hobby with this part, let me point out that for the curious mind, there's some joy in learning this stuff. My hope is that by the end of this, aside from your new ticket from the FCC, and maybe a new Extra class call sign, you'll have a richer, deeper appreciation for the people and concepts of the science behind our hobby, and, indeed, behind much of our modern world. Once you really get the science behind how it works, there's an elegance and beauty to even a simple circuit like a regulated power supply. The average person looks at a tool like the Smith chart and thinks, "Huh. Funny lines on paper. Looks complicated." You're going to learn about it and in the process get a glimpse into the mind of a man who created a work of genius. Personally, I find the history and science of the things that make all the magic around us endlessly fascinating, and I suggest you open yourself up to that fascination as well. I think if you do, you'll find this an exciting and rewarding effort.

Before we get going on the questions, there's one tool you really need as you prepare for your exam, and that you will definitely want with you at the exam. One good, scientific calculator. I've shopped lots of calculators, and recommend the TI-30XS as your choice. If you catch a sale, you can pick one up at Wal-Mart or Amazon.com for about fifteen bucks. It has all the functions you'll need, is easy to use, and lets you see all the steps you have done in a calculation.

Figure 1: TI-30XS Calculator

On the TI-30XS, you can enter stuff like...

$$\frac{1}{2\pi\sqrt{0.03 \times (3 \times 10^{-6})}}$$

...directly, without having to convert 3 x 10-6 to 0.0000006, or break the formula down into smaller steps. When I teach you specific keys to press on a calculator for a particular problem, I'll be referring to the TI-30XS.

That said, if you already have a scientific calculator and are comfortable with its basic functions as well as log and trig functions, that calculator will certainly work.

Yes, there will be math. I'll explain each formula and even tell you what calculator buttons to push to make the answer appear, but you're going to need to really practice with those formulas so you can use them when they come up on the test. I'll also use English and math to explain some of the concepts – like pictures, math can say a lot in a little bit of space. If you are math-phobic, congratulations, you get to have a breakthrough in the area of math. Please know that even though we will even touch briefly on a couple of trigonometry functions, there's no real algebra, trigonometry, or calculus. Every single formula is what my math teacher wife calls "a plug 'n' chug." So long as you know the formula, you just plug in the values and the calculator chugs out the answer. If you solved Ohm's Law problems, you can solve these.

There are about twenty key formulas that are useful for passing the exam. One of the keys to remembering those formulas is practice, practice, practice. For me, a lot of it is getting what I call the "sense" of the formula, the logic that drives it, and how the outcomes change as the inputs change. Make up problems for yourself; take yourself through those formulas over and over until they stick. Flashcards are a great help for some people.

If you could use a boost in the math department, we do offer a separate workbook with lots of additional practice problems; *The Fast Track to Mastering Extra Class Ham Radio Math*. It is available from online book retailers and covers every math-based problem in the question pool.

There are quite a few schematic diagrams in the book. I've kept them simplified and very few are actually on the exam, but I've included schematics to explain how certain components and circuits function so that you'll understand them in some sort of context, rather than as just a collection of arbitrary trivia. If reading schematics is new for you – I mean really reading, as opposed to identifying schematic symbols for the Technician and General exams – take a few minutes and study those schematics. It's a valuable skill in our hobby, and once you get the hang of it, a lot of things get much more clear. As they say, a picture is worth a thousand words, and it's a lot easier to read a simple schematic than it is to understand a bunch of words on the same topic.

You're preparing for a 50 question, multiple choice exam. A passing score is 37 correct answers – 74%. Most likely, you'll get about one question from each chapter of the book. There's actually nothing in the law that says the testing organization where you take your exam couldn't just pick the first 50 questions from the pool, but most do the "one per group" system.

To find a place and time to take your Extra exam, you can visit the American Radio Relay League's web site at http://www.arrl.org. You'll find a handy search page at

http://www.arrl.org/find-an-amateur-radio-license-exam-session.

Type in your zip code, enter how many miles you're willing to travel to the session, and it will give you a list of upcoming exams and the contact information for the examiners.

As always, I *strongly* suggest you find the time and place of an upcoming exam right now, and *commit* to taking and passing your exam on that date. Let's face it; you're not going to get your Extra ticket without a considerable amount of work. I hope you'll find the work – or at least the learning – enjoyable, but you and I know it is the nature of human beings to avoid challenges that we don't perceive as necessary. So locate that test session, make your commitment, pretend that it is necessary, and you'll be a big step closer to your Extra. I guarantee, the biggest single obstacle between you and your upgrade is that commitment so make it today before you figure out an excuse to get out of it!

Chapter 1

Preparing for the Exam

In this book, each question is identified with its official question name, in case you want to look it up elsewhere. The heading also includes the letter of the correct answer, such as "E2C03" (A). That one translates to "question 03 of subelement 2, group C. The correct answer is answer A." The "E" is for "Extra." Correct answers are also in boldface, but I include the letter of the correct answer in the question heading just in case I boldfaced a wrong answer somewhere.

Some question headers include a pointer to a particular section of Part 97, the FCC rules for amateur radio, such as [97.301, 97.305.]

How to Study; Being an Active Learner

Over the years, we've developed and tested a very specific, research-based active learning plan for this *Fast Track* course. By "research-based," we mean "based on real peer-reviewed research by real PhD neuro-scientists with lab coats and clipboards."

Everyone uses the best learning strategy they have found to date. That doesn't mean it is the best it could be, and we'd urge you to seriously consider using this system rigorously. It is very likely to accelerate your learning.

Step one: *Stay off the practice exam sites and apps.* At this point, they'll just slow down your progress.

Start by reading or listening to the whole *Fast Track* Extra course, casually. If you just can't wait to get started, at least make your first read-through of a chapter casual. Don't try to memorize anything, don't work the problems, just read or listen to the overall subject matter and start letting your mind get accustomed to the terminology. Let this stuff start rattling around in there. If most of the material in the course is new to you, don't get overwhelmed. I know it is a lot of material, but we are going to take it step by step. Each chapter ends with "Key Concepts in This Chapter", which is a recap of all the correct answers. Don't skip those; that re-exposure to the material helps solidify the memory.

Step two: Start working your way through the book, in order. Start right away to develop the habit of taking written notes, even on the "easy" stuff. If you are keeping proper notes, they will be short versions of the Key Concepts sections.

Part of what builds memories is engaging as many senses as possible with the material, and that brain-hand-eye-brain connection is a powerful way to drive the material home.

As you are taking notes, consider what sort of knowledge you're recording. The Extra exam, as well as most any other topic, contains three broad categories of knowledge, and each can be attacked with a different strategy.

1

- **Factoids.** These are the answers to questions like "Which HF amateur bands have frequencies authorized for space stations?" If there is some simple reason behind why certain bands are authorized and others are not, it has not become clear to me! It's just a semi-random factoid. These are best attacked with flash cards; the very act of creating them for yourself will also help drive them into memory.

- **Principles.** Answers that would fall into this category would be like the answer to, "What is vestigial sideband modulation?" Once you understand vestigial sideband modulation, you'll be able to identify the correct answer, no matter how it is worded, so long as it correctly describes that form of modulation. For principles, then, the strategy is to fully understand what's behind the answer.

- **Methods.** "What is the resonant frequency of an RLC circuit if R is 33 ohms, L is 50 microhenries and C is 10 picofarads?" To answer that question, you honestly don't need to know a thing about electronics. What you *do* need is a formula, and the way to learn a formula is practice, practice, practice. Once you know the formula, the test could throw any set of numbers at you and you'll still be able to come up with the answer.

Of course, those are flexible categories, and some questions require a little of each. For instance; I'll show you a way to solve some coaxial cable stub problems using a particular type of chart. This falls into both **Principles** – you need to understand what the chart represents – and **Methods** – "how to use the chart."

The truth is, by the time you've done all the practice exams we urge you to do, the answers to many of the **Method** and **Principle** problems are probably going to have become **Factoids.** That's fine; you'll still have gained some knowledge and understanding by learning the underlying methods and principles.

If you have a "study buddy" (a great idea, by the way, so long as your buddy is dependable) you can talk this stuff through and quiz each other verbally. Another sense engaged!

If something is not quite clicking for you, do this: Focus on that formula or topic for 25 minutes straight. Not 24 minutes, not 26 minutes; 25. Don't worry about whether you're "getting it" or not, just engage 100% with that item. If it's a formula, write it a bunch of times or plug in made-up numbers and work the formula for 25 minutes straight. Then set that aside and move onto something else. Neuroscience suggests that the time of intense focus and the time of setting it aside are equally important for generating insight. (If you're interested, let me suggest a marvelous book called *A Mind for Numbers: How to Excel at Math and Science (Even If You Flunked Algebra)* by Barbara Oakley, PhD, who's not only an engineering professor and student of how people learn, but also a *Fast Track* fan.)

Step three: Practice tests. The fasttrackham.com site has a practice exam system specifically created to match with the *Fast Track* programs.

https://www.fasttrackham.com

When you finish Chapter 2, go to the web site, make yourself an account, and take the Practice Exam for Chapter 2. When you finish Chapter 3, go take the Practice Exam for Chapter 3, which will also review Chapter 2's questions. As you work through the Practice Exams, you'll always be reviewing the last few chapters. Treat every practice exam you take as an open-book exam. Never guess at an answer. If you don't know it, look it up. There's an index of questions in the back of the print edition of this book, and e-readers have a search function. We suggest you set a standard for yourself of scoring at least 85% before moving on to the next chapter.

Every few chapters, you'll have the opportunity to take a Progress Check. This is a 50-question sample of everything covered up to that point. At the end, the grading will give you some indication of groups, such as E1A, where you need to focus. The site has Group Drills that cover only the questions in a particular group; if you're not doing great on group E1A, go work on the E1A Group Drill. When you're pressed for time, it also has Chapter Quizzes that cover only one chapter's questions.

If you do well on the practice exam for a chapter, meaning you scored 85% or better (closed-book, of course) that's great. It's time to move on to the next chapter.

At the end, you'll have the opportunity to take Final Practice Exams that pull one question from each group, just like the real exam.

Fifteen to thirty minute *daily* study sessions are far more effective than weekly two-hour sessions. *Once you have read a chapter once or twice, the majority of your study time should be devoted to doing the practice exams.* "Hitting the books really hard," in other words, reading the same stuff over and over, is a dismally ineffective study method. You build memory by actively challenging your brain to recall what you have learned. As a side benefit, daily practice exams are a great treatment for test anxiety.

Take a look at the "forgetting curves" in Figure 1.1.

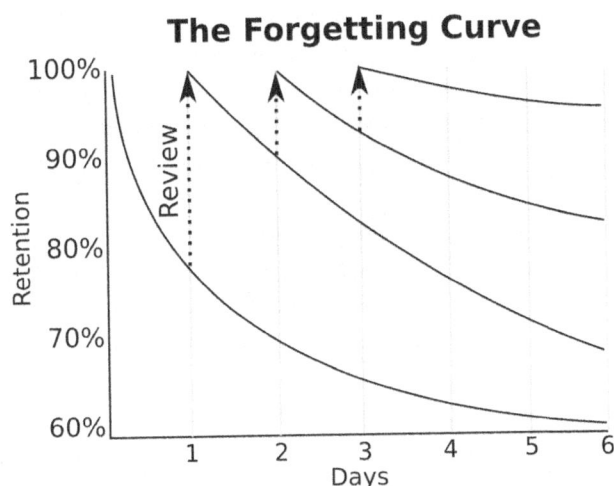

Figure 1.1: Ebbinghaus Forgetting Curves

Those curves represent how much newly learned information was retained in memory after the passage of time, measured in days. The curve that starts on day zero shows that after seven days, almost all the newly learned material has vanished. The dotted line arrows and their curves show what happened when the learned material was reviewed after one, two, and three days. Notice that each time the information is reviewed, the curve flattens, so that less information is lost over the same amount of time.

Those forgetting curves have been tested over and over since they were discovered by Bernard Ebbinghaus in 1885; for most information, they're quite accurate. Of course, there's the occasional vivid event that stays with us forever, but for information without a strong emotional component – such as the material on the Extra exam – those curves will tend to hold true.

By the way, those curves have nothing to do with how smart you are, nor with how great you think your memory is or isn't. As we understand more about the brain, it's becoming clear that those forgetting curves aren't a design flaw, they're essential to keeping our memories organized and useful.

The way to beat the forgetting curve is daily review, at least for the first few days after you learn new material. That's why those regular study sessions and practice exams are important to your success. Without those reviews, almost everything you learn on Monday will be almost gone by the next Monday.

For more tips on studying for ham exams, go to the Teaching Videos section of fasttrack-ham.com. There are a number of videos there you may find useful as you work through this material.

Take your time and savor the journey, but don't drag out this process forever; it just makes it more difficult. You don't need a year to get your Extra.

A word about practice exams, particularly for younger folks. I've worked with way too many people who, at some point in life, probably in their school career, got the notion that exams tell you how smart or talented you are.

TESTS DON'T TELL YOU THAT.

Tests just tell you the current state of your ability to pass that particular exam. That's it. If you take a practice exam on a topic and bomb out, or if you're scratching your head and are really challenged by something on the exam, it doesn't mean you don't have what it takes to get your Extra ticket. It doesn't mean you don't have the talent. (Talent only makes a real difference after you learn a skill, and are at the top of your game. Not before.) It just means you need to engage more intensely with that topic, or maybe find a different way of looking at it. I've tried to make things as clear as I could in this program, but maybe I missed on some topic and didn't get it explained in a way that clicked with you. That's on me, not on you. Don't abandon hope. Google is your friend, as is Wikipedia, YouTube, and the rest of the web. If you're struggling with a topic, use those resources. I have complete confidence that any of us who passed our General exam is fully capable of passing the Extra.

This is a QR code:

fasttrackham.com

If you use the QR Code Reader function built into most smartphones and tablets, the QR codes will take you directly to a useful web page. That one points toward fasttrackham.com.[1]

One last thought – don't get frozen by perfectionism. The object of the game is to get your Extra ticket. The passing score for the Extra exam is 74%. According to no less an authority than the Federal Communications Commission of The United States of America, you only need to be at 74% mastery of the material in this program to qualify 100% for your Extra Class license. Let's get going!

---

[1]On Android phones, the QR Code Reader is on the page that pulls down from the top and includes you WiFi connection, WiFi Calling, and other functions. QR Code Readers are also available as apps from the appropriate app store for your device.

Chapter 2

Operating Standards

If it's your job to eat a frog, it's best to do it first thing in the morning. And if it's your job to eat two frogs, it's best to eat the biggest one first.

- Mark Twain

In the spirit of eating the frog, we'll start with the Rules and Regulation questions, so we can look forward to the rest of the journey.

Staying in the Bands

E1A02 (D) [97.301, 97.305] When using a transceiver that displays the carrier frequency of phone signals, which of the following displayed frequencies represents the lowest frequency at which a properly adjusted LSB emission will be totally within the band?
    A. The exact lower band edge
    B. 300 Hz above the lower band edge
    C. 1 kHz above the lower band edge
    D. **3 kHz above the lower band edge**

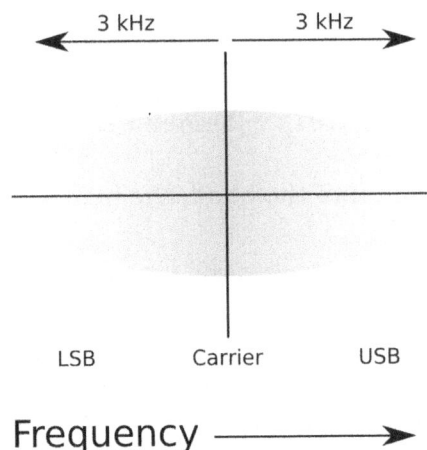

Figure 2.1: SSB Bandwidth

The sidebands are created at frequencies equal to the carrier frequency plus and minus the modulating frequency. Since our SSB phone signal typically includes frequencies up to about 3 kHz, lower sideband signals stretch down the dial about 3 kHz. Your center frequency needs to be not lower than **3 kHz above the lower band edge.** (See figure 2.1.)

Of course, upper sideband signals are just the opposite of those lower sideband signals – they stretch up the band 3 kHz from the center frequency, so your carrier needs to be set no higher than 3 kHz below the upper band edge.

E1A01 (A) [97.305, 97.307(b)] Which of the following carrier frequencies is illegal for LSB AFSK emissions on the 17 meter band RTTY and data segment of 18.068 to 18.110 MHz?

    A. **18.068 MHz**

    B. 18.100 MHz

    C. 18.107 MHz

    D. 18.110 MHz

Figure 2.2 shows the whole 17-meter band, which stretches from 18.068 MHz to 18.168 MHz.

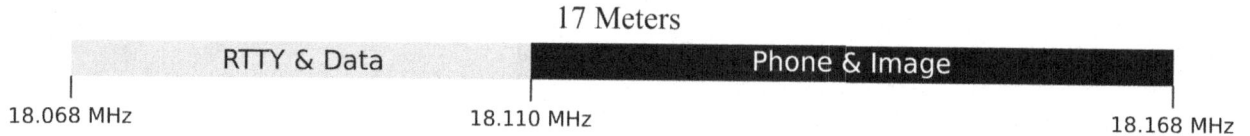

Figure 2.2: 17-meter Band

An LSB (lower sideband) signal with a carrier frequency of 18.068 will stretch down the dial beyond the lower boundary of the 17-meter band. LSB signals on the remaining frequencies listed will, of course, stay within the band.

E1A03 (C) [97.305, 97.307(b)] What is the maximum legal carrier frequency on the 20 meter band for transmitting USB AFSK digital signals having a 1 kHz bandwidth?

    A. 14.070 MHz

    B. 14.100 MHz

    C. **14.149 MHz**

    D. 14.349 MHz

For Extra Class operators, the 20-meter RTTY & data band goes from 14.000 MHz to 14.150 MHz, with the segments shown in Figure 2.3.

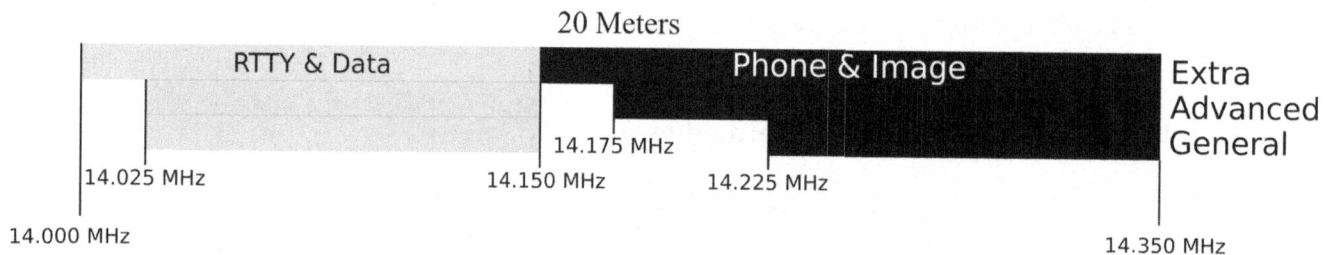

Figure 2.3: 20-Meter Band

To fit a 1 kHz wide USB signal into the band, the highest frequency we can use is 1 kHz below 14.150, MHz; **14.149 MHz**.

E1A04 (C) [97.301, 97.305] With your transceiver displaying the carrier frequency of phone signals, you hear a DX station calling CQ on 3.601 MHz LSB. Is it legal to return the call using lower sideband on the same frequency?

    A. Yes, because the DX station initiated the contact

    B. Yes, because the displayed frequency is within the 75 meter phone band segment

    C. **No, the sideband will extend beyond the edge of the phone band segment**

    D. No, U.S. stations are not permitted to use phone emissions below 3.610 MHz

Now we're in the 80-meter band.

You can see in Figure 2.4 that the low edge of the phone segment of the band is 3.600 MHz. If we respond to the DX call on 3.601 MHz, our LSB signal **will extend beyond the edge of the Phone band segment** and down to 3.598 MHz in the RTTY & Data band segment.

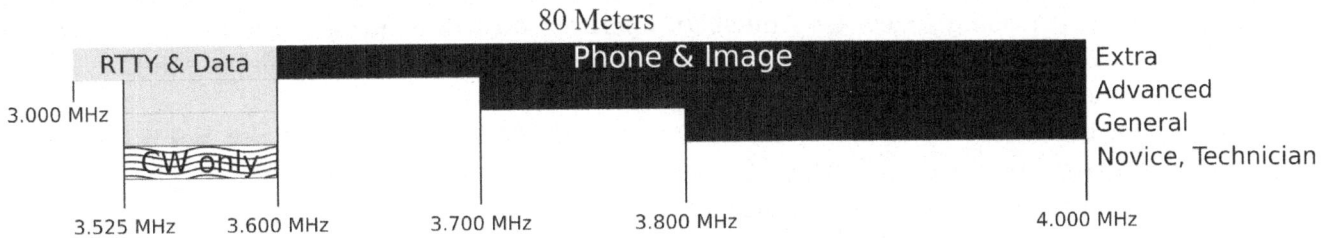

Figure 2.4: 80-Meter Band

The Low Bands

E1A07 (C) [97.313(k)] What is the maximum power permitted on the 2200 meter band?

    A. 50 watts PEP

    B. 100 watts PEP

    C. **1 watt EIRP (Equivalent isotropic radiated power)**

    D. 5 watts EIRP (Equivalent isotropic radiated power)

In 2017 the FCC authorized amateur radio operators with licenses of General class and above to transmit on the 2200-meter, 135.7 kHz to 137.8 kHz band. Operations there are limited in power to **1 watt EIRP**. You may recall that the 60-meter band has an "ERP" (Effective Radiated Power) limitation. That 60-meter limit is "relative to a dipole." For the 2200-meter band the limit is "relative to an isotropic antenna."

That may sound like a tiny amount of power, but amateurs experimenting on this band report that it is, in fact, quite challenging to create the equipment to radiate even this much effective radiated power.

E1A14 (D) [97.313(l)] Except in some parts of Alaska, what is the maximum power permitted on the 630 meter band?

    A. 50 watts PEP

    B. 100 watts PEP

    C. 1 watt EIRP

    D. **5 watts EIRP**

When it opened up the 2200-meter band, the FCC also authorized use of the 630-meter band. That one has a power limit of **5 watts EIRP**. (In Alaska it is 1 watt EIRP.)

E1C12 (D) [97.305(c)] On what portion of the 630 meter band are phone emissions permitted?

    A. None

    B. Only the top 3 kHz

    C. Only the bottom 3 kHz

    D. **The entire band**

The "map" of the 630-meter band is one solid bar from 472.0 kHz to 479.0 kHz. RTTY, Data, Phone and Image transmissions are allowed across the entire band. All 7 kHz of it!

E1C13 (C) [97.303(g)] What notifications must be given before transmitting on the 630 meter or 2200 meter bands?

    A. A special endorsement must be requested from the FCC

    B. An environmental impact statement must be filed with the Department of the Interior

    C. **Operators must inform the Utilities Technology Council of their call sign and coordinates of the station**

    D. Operators must inform the FAA of their intent to operate, giving their call sign and distance to the nearest runway

Transmissions on both of those low bands could, at least in theory, impact data transmission on power lines, known as PLC (Power Line Carrier.) Because of that, the FCC requires that licensees wanting to operate on those bands notify the **Utilities Technology Council of their call sign and the coordinates of the station.**

E1C14 (B) [97.303(g)] How long must an operator wait after filing a notification with the Utilities Technology Council before operating on the 2200 meter or 630 meter band?

    A. Operators must not operate until approval is received

    B. **Operators may operate after 30 days, providing they have not been told that their station is within 1 km of PLC systems using those frequencies**

    C. Operators may not operate until a test signal has been transmitted in coordination with the local power company

    D. Operations may commence immediately, and may continue unless interference is reported by the UTC

Once notice is provided to the UTC, there's a 30-day waiting period, during which time the UTC may notify the operator that they are within a kilometer of a PLC system. If that notice is given, no transmitting can be done on those bands from that location. Obviously, filing that notice is something you'd want to do well before setting up a station.

The 60-Meter Band

E1A05 (C) [97.313] What is the maximum power output permitted on the 60 meter band?

    A. 50 watts PEP effective radiated power relative to an isotropic radiator

    B. 50 watts PEP effective radiated power relative to a dipole

    C. **100 watts PEP effective radiated power relative to the gain of a half-wave dipole**

    D. 100 watts PEP effective radiated power relative to an isotropic radiator

You'll recall from your General exam studies that 30 and 60 meters are the HF bands with additional restrictions. One of these restrictions is power. On 60 meters the limit is **100 watts PEP effective radiated power relative to the gain of a half-wave dipole**.

Effective radiated power, or ERP, is the power coming out of the part of the antenna with the most gain, taking into account transmitter output power, feed line losses, reflected power losses, and the gain of the antenna **relative to a half-wave dipole**.

E1A06 (B) [97.15] Where must the carrier frequency of a CW signal be set to comply with FCC rules for 60 meter operation?

    A. At the lowest frequency of the channel

    B. **At the center frequency of the channel**

    C. At the highest frequency of the channel

    D. On any frequency where the signal's sidebands are within the channel

The 60-meter band is the band that is split into channels – for CW, we have five specific frequencies we can use in that band. The lowest is 5332 kHz, and that's defined as **the center frequency of the channel.** (See figure 2.5.)

CW on 60 Meters

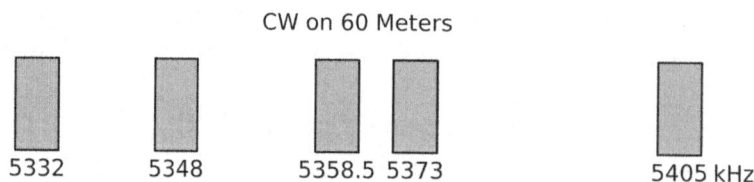

5332   5348   5358.5 5373   5405 kHz

Figure 2.5: CW on the 60-Meter Band

E1C01 (D) [97.303] What is the maximum bandwidth for a data emission on 60 meters?

    A. 60 Hz

    B. 170 Hz

    C. 1.5 kHz

    D. **2.8 kHz**

The maximum bandwidth for a data emission on 60 meters is **2.8 kHz** because the maximum bandwidth of *any* emission on 60 meters is **2.8 kHz**.

Speaking of bandwidth ...

Bandwidth

E1C07 (D) [97.3(a)(8)] At what level below a signal's mean power level is its bandwidth determined according to FCC rules?

    A. 3 dB

    B. 6 dB

    C. 23 dB

    D. **26 dB**

We've talked about bandwidth since early on in the Technician course. What, precisely, does the FCC actually mean when they talk about bandwidth? What are we supposed to measure? How far down into the noise do we need to go before whatever is there no longer counts as a signal?

It turns out the FCC has defined this for us very precisely, right there in Part 97, where it says, bandwidth is "The width of a frequency band outside of which the mean power of the transmitted signal is attenuated at least 26 dB below the mean power of the transmitted signal within the band."

Message Forwarding Systems

E1A08 (B) [97.219] If a station in a message forwarding system inadvertently forwards a message that is in violation of FCC rules, who is primarily accountable for the rules violation?

    A. The control operator of the packet bulletin board station

    B. **The control operator of the originating station**

    C. The control operators of all the stations in the system

    D. The control operators of all the stations in the system not authenticating the source from which they accept communications

Some years back, this rule was different. At that time, it was a little – let's say, flexible. It was the originating station, for sure, but it might just maybe be any other station that forwarded that message, even if the message was forwarded automatically with no knowledge on the part of the operator. They realized that was unworkable and rewrote the rule, so now it's **the control operator of the originating station** who holds primary responsibility.

E1A09 (A) [97.219] What action or actions should you take if your digital message forwarding station inadvertently forwards a communication that violates FCC rules?

    A. **Discontinue forwarding the communication as soon as you become aware of it**

    B. Notify the originating station that the communication does not comply with FCC rules

    C. Notify the nearest FCC Field Engineer's office

    D. All these choices are correct

Just because the control operator of the originating station is on the hook doesn't mean you shouldn't take action if you inadvertently forward an illegal message. The action you should take is to **discontinue forwarding the communication as soon as you become aware of it**. Yes, there's a bit of "closing the barn door after the horses have eloped to Hawaii" here, but that's what you're supposed to do.

Operating on Aircraft and Boats

E1A10 (A) [97.11] If an amateur station is installed aboard a ship or aircraft, what condition must be met before the station is operated?

    A. **Its operation must be approved by the master of the ship or the pilot in command of the aircraft**

    B. The amateur station operator must agree not to transmit when the main radio of the ship or aircraft is in use

    C. The amateur station must have a power supply that is completely independent of the main ship or aircraft power supply

    D. The amateur operator must have an FCC Marine or Aircraft endorsement on his or her amateur license

If you're installing an amateur station on your own boat, I don't imagine this will present a problem, but what if you're taking an Alaskan cruise and want to chat with a few Canadian hams on the way up the Inner Passage? You can't just bust out your handy-talky – or string up a random wire HF antenna up by the swimming pool! – and commence communicating. You need permission from the **master of the ship or** (if the vessel in question is an airplane) **the pilot in command of the aircraft** before the station is operated.

    Your chances of being allowed to operate your radio on board a commercial flight are about zero, though we've never had any problem carrying radios on board in our carry-ons. We hear mixed reports about various cruise lines. Some welcome operations on board, some strictly forbid it.

E1A11 (B) [97.5] Which of the following describes authorization or licensing required when operating an amateur station aboard a U.S.-registered vessel in international waters?

    A. Any amateur license with an FCC Marine or Aircraft endorsement

    B. **Any FCC-issued amateur license**

    C. Only General class or higher amateur licenses

    D. An unrestricted Radiotelephone Operator Permit

Assuming you have the approval of the master of the vessel, you're free to operate within your license privileges when you're aboard a U.S.-registered vessel in international waters. **Any FCC issued amateur license** holder has that privilege. Just note that key phrase, "U.S. registered." The rules change if that freighter you're riding to Curacao was registered in Panama.

E1A13 (B) [97.597.5] Who must be in physical control of the station apparatus of an amateur station aboard any vessel or craft that is documented or registered in the United States?

    A. Only a person with an FCC Marine Radio License grant

    B.**Any person holding an FCC issued amateur license or who is authorized for alien reciprocal operation**

    C. Only a person named in an amateur station license grant

    D. Any person named in an amateur station license grant or a person holding an unrestricted Radiotelephone Operator Permit

So long as the vessel is documented or registered in the United States, **any person holding an FCC issued amateur license or who is authorized for alien reciprocal operation** may operate the physical controls of the station apparatus of an amateur station aboard that vessel or craft.

    Having "alien reciprocal" privileges means the US has an agreement with the country that issued the alien's license that says, "Your licenses are okay to use here, so long as our hams get to use theirs over in your country." That means your license is valid in Canada, Australia, Japan, and Iceland, among a bunch of others. When aliens operate with reciprocal privileges here, they're to add a suffix to their station ID that indicates their approximate geographical place of operation. So, for instance, if an Icelander came to Washington State, they'd add /W7 to their call sign, to indicate they were operating in the 7 region. We're required to do something similar when, for instance, we're operating in Canada. If you're going international, be sure to check to see if you have reciprocal privileges and what the regulations are where you're going.

Key Concepts in This Chapter

- The lowest frequency at which a properly adjusted LSB emission will be totally within the band is 3 kHz above the lower band edge.

- 18.068 MHz is an illegal frequency for LSB AFSK emissions on the 17-meter band.

- The maximum legal carrier frequency on the 20-meter band for transmitting USB AFSK digital signals having a 1 kHz bandwidth is 14.149 MHz.

- If you hear a DX station calling CQ on 3.601 MHz, LSB, it is not legal to return the call on the same frequency.

- The maximum power permitted on the 2200-meter band is 1 watt EIRP.

- Except for some parts of Alaska, the maximum power permitted on the 630-meter band is 5 watts EIRP.

- On the 630-meter band, phone emissions are permitted across the entire band.

- Before transmitting on the 2200-meter or 630-meter bands, operators must inform the Utilities Telecom Council, the UTC, of the call sign and coordinates of the station. After filing notice with the UTC, operators may operate after 30 days, providing they have not been told that their station is within 1 kilometer of PLC systems using those frequencies.

- The maximum power output permitted on the 60-meter band is 100 watts PEP effective radiated power relative to the gain of a half-wave dipole.

- To comply with FCC rules for 60-meter operation, the carrier frequency of a CW signal must be set at the center frequency of the channel.

- The maximum bandwidth for a data emission on 60 meters is 2.8 kHz.

- According to FCC rules, the bandwidth of a signal is determined at 26 dB below its mean power level.

- If a station in a message forwarding system inadvertently forwards a message that is in violation of FCC rules, the control operator of the originating station is primarily responsible.

- If your digital message forwarding station inadvertently forwards a communication that violates FCC rules, you should discontinue forwarding the communication as soon as you become aware of it.

- If an amateur station is installed aboard a ship or aircraft, its operation must be approved by the master of the ship or the pilot in command of the aircraft.

- When operating an amateur station aboard a U.S.-registered vessel in international waters, the required licensing is any FCC-issued amateur license.

- Aboard any vessel or craft that is documented or registered in the United States, the person in physical control of the station apparatus of an amateur station can be any person holding an FCC issued amateur license or who is authorized for alien reciprocal operation.

Authentic Alien Ham Radio Operator

If you're committed to active learning, go to http://fasttrackham.com, make yourself an account (it's free) and take the Practice Exam for Chapter 2. The site works on desktop computers, tablets, and mobile phones. Don't "wait until you feel like you know it" to start taking practice exams – *use* the practice exams to learn it. Remember to treat the exam as an open-book exam. Practice being right. Don't guess – look it up! For you convenience, here's a QR code (readable by most Android and Apple smart phones) that will take you to the account creation page. Create your new account, log in, and get started. Once you are logged in, the next QR code will take you directly to the first practice exam.

Chapter 3

Miscellaneous Rules

Spread Spectrum Rules

E1F01 (B) [97.305] On what frequencies are spread spectrum transmissions permitted?
    A. Only on amateur frequencies above 50 MHz
    B. Only on amateur frequencies above 222 MHz
    C. Only on amateur frequencies above 420 MHz
    D. Only on amateur frequencies above 144 MHz

Spread spectrum has tremendous potential for us, but the regulations might be a part of the reason this technology hasn't caught on like wildfire.

We can only use spread spectrum technology **on amateur frequencies above 222 MHz**, so for most of us, that would limit it to the 1.25 cm or 70 cm bands – radios that work on higher frequencies than that are specialty items with the notable exception of the repurposed WiFi routers used to build amateur radio mesh networks. Those use spread spectrum.

We're also limited to 10 watts and are a bit limited in the types of emissions we can use.

We'll go more deeply into just what spread spectrum is and how it is created in Chapter 49.

E1F09 (D) [97.311] Which of the following conditions apply when transmitting spread spectrum emissions?
    A. A station transmitting SS emission must not cause harmful interference to other stations employing other authorized emissions
    B. The transmitting station must be in an area regulated by the FCC or in a country that permits SS emissions
    C. The transmission must not be used to obscure the meaning of any communication
    D. **All these choices are correct**

On a few bands, we are officially "secondary users" where we must take special care not to interfere with primary users and must accept interference from those primary users. With spread spectrum, we're, in essence, secondary users no matter what frequency we are using. We **must not cause harmful interference with other stations employing other authorized emissions**.

The transmitting station must be in an area regulated by the FCC or in a country that permits SS emissions.

We also can't use a spread spectrum system **to obscure the meaning of any communication**.

Canadian Ham Privileges

E1F02 (C) [97.107] What privileges are authorized in the U.S. to persons holding an amateur service license granted by the government of Canada?
    A. None, they must obtain a U.S. license

B. All privileges of the Amateur Extra Class license

C. **The operating terms and conditions of the Canadian amateur service license, not to exceed U.S. Amateur Extra Class privileges**

D. Full privileges, up to and including those of the Amateur Extra Class license, on the 80, 40, 20, 15, and 10-meter bands

We have a friendly ham radio relationship with our neighbors up north. We can operate with our licenses when we're in Canada and Canadians can use their licenses when they visit here. No permits, no writing letters to the embassy, just take your radio and a copy of your license.

They have a different licensing system than we do – only two levels, Basic and Advanced – so the licenses don't translate directly to US equivalents, but when they're here, they get whatever privileges they have in Canada, **not to exceed U.S. Extra Class privileges**.

External RF Power Amplifier Rules

E1F03 (A) [97.315] Under what circumstances may a dealer sell an external RF power amplifier capable of operation below 144 MHz if it has not been granted FCC certification?

A. **It was purchased in used condition from an amateur operator and is sold to another amateur operator for use at that operator's station**

B. The equipment dealer assembled it from a kit

C. It was imported from a manufacturer in a country that does not require certification of RF power amplifiers

D. It was imported from a manufacturer in another country and was certificated by that country's government

The FCC takes a rather dim view of putting high power RF amplifiers capable of operating below 144 MHz in the hands of unlicensed people. Why under 144 MHz? Let's see, what's down there under the VHF bands? There's the 10-meter band ....oh, and the 11 meter Citizens Band! That vile den of iniquity!

As every trucker knows, there are plenty of citizens out there blasting the Citizens Band with illegal high power RF amps. When I was driving a truck, I regularly met drivers who at least claimed to be running 800 watts on their CB – and that's somewhat believable, because semi-truck alternators are pretty beefy. 800 watts on a service that's supposed to be limited to 4 watts! In 2006, the FCC decided they'd had enough of that malarkey and came up with regulation 97.315, part of which is the answer to this question.

Under normal circumstances, any RF power amplifier made or sold must have FCC certification – and they specifically designed the requirements for certification to prevent those amplifiers from being capable of being driven by off-the-shelf CB radios. The exception is if the amplifier **was purchased in used condition from an amateur operator and is sold to another amateur operator for use at that operator's station**.

Has it solved the problem? Naaaahhh. But at least the knuckleheads have to work a little harder to break the law.

E1F11 (D) [97.317] Which of the following best describes one of the standards that must be met by an external RF power amplifier if it is to qualify for a grant of FCC certification?

A. It must produce full legal output when driven by not more than 5 watts of mean RF input power

B. It must be capable of external RF switching between its input and output networks

C. It must exhibit a gain of 0 dB or less over its full output range

D. **It must satisfy the FCC's spurious emission standards** when operated at the **lesser of 1500 watts or its full output power**

There are some interesting wrong answers for this one, including the one that says they must "exhibit a gain of 0 dB or less over its full output range." That's the same gain you'd get if you used a concrete block for an amplifier. Hmmm ... come to think of it, I'm pretty sure I owned one that would qualify. But no, the amplifier **must satisfy the FCC's spurious emission standards when operated at the lesser of 1500 watts or its full output power**.

If you go to purchase an RF amplifier, how do you know if it is FCC certified? It's supposed to have a clearly visible label saying so that includes an FCC ID number.

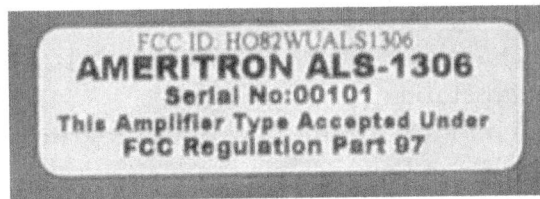

Figure 3.1: FCC Certification

E1C10 (A) [97.307] What is the permitted mean power of any spurious emission relative to the mean power of the fundamental emission from a station transmitter or external RF amplifier installed after January 1, 2003 and transmitting on a frequency below 30 MHz?

    A.**At least 43 dB below**
    B. At least 53 dB below
    C. At least 63 dB below
    D. At least 73 dB below

Okay, is there a different standard for spurious emissions based on something about January 1, 2003? Indeed, there is. Back in 2001, the Commission decided to simplify and slightly tighten the standards, and they gave folks a couple of years warning so they'd have time to get into compliance. Pre-2003 installations are grandfathered in. (Note, that's *installations:* that glorious boat anchor of a 1956 vintage power amp you just bought at the swap meet still has to meet 2003 specs for spurious emissions.)

The **43 dB limit** is for HF power amps and transmitters. Different limits apply to VHF and UHF amplifiers. The "mean" power is another way to say the average power, so the average power of any spurious emission must be **43 dB below** the average power of the "fundamental emission" – in other words, your carrier. Put another way, that spurious emission can only be 0.005% as powerful as your average power.

Geographic Restrictions

E1F04 (A) [97.3] Which of the following geographic descriptions approximately describes "Line A"?

    A. **A line roughly parallel to and south of the border between the U.S.and Canada**
    B. A line roughly parallel to and west of the U.S. Atlantic coastline
    C. A line roughly parallel to and north of the border between the U.S. and Mexico
    D. A line roughly parallel to and east of the U.S. Pacific coastline

Line A? If you're one of us "almost to the 49th parallel" folks, you need to be aware of this one.

Think of Line A as the UHF radio US/Canadian border. It's **a line roughly parallel to and south of the U.S. – Canadian border.** If you find yourself less than 100 miles from the Canadian border, and you're thinking of using your UHF radio, you need to look up the precise location of Line A where you are. North of Line A, we're not to transmit on the frequencies from 420 to 430 MHz, because that's spectrum used by non-hams up in Canada.

E1F05 (D) [97.303] Amateur stations may not transmit in which of the following frequency segments if they are located in the contiguous 48 states and north of Line A?

    A. 440 MHz - 450 MHz
    B. 53 MHz - 54 MHz
    C. 222 MHz - 223 MHz
    D. **420 MHz - 430 MHz**

And that explains why you won't find any UHF repeaters between **420 MHz and 430 MHz** up here near the Canadian border.

Special Temporary Authority

E1F06 (A) [1.931] Under what circumstances might the FCC issue a Special Temporary Authority (STA) to an amateur station?
A. **To provide for experimental amateur communications**
B. To allow regular operation on Land Mobile channels
C. To provide additional spectrum for personal use
D. To provide temporary operation while awaiting normal licensing

An STA, Special Temporary Authority, is permission from the FCC to transmit in some particular way that would normally be illegal. Amateur radio has a long history of experimentation and innovation, but not much of that would have happened without the FCC issuing STA's.

STA's for amateur radio are not common occurrences – the latest grant I could find was in 2012 – but the FCC faces a veritable blizzard of applications from other users, because "file for an STA" is what one does when the FCC rules say one can't do what one wants to do! One researcher found that about three-quarters of the applications weren't for experiments at all but for such things as temporary installations for television coverage of a special event. In our case, however, the purpose of STA's is **to provide for experimental amateur communications**.

Prohibited and Limited Messages

E1F07 (D) [97.113] When may an amateur station send a message to a business?
A. When the total money involved does not exceed $25
B. When the control operator is employed by the FCC or another government agency
C. When transmitting international third-party communications
D. **When neither the amateur nor his or her employer has a pecuniary interest in the communications**

This rule actually came into play in my own life, back in my truck driving days. We had several hams at our office, and we talked my dispatcher into getting his license. He got it, and promptly fired up his radio to chat with a couple of us. That was fine – neither he nor I had **a pecuniary interest in the communications**, meaning neither of us was talking about anything that would make either of us any money. It was fine, that is, until he said something like, "Hey, Michael, are you going to be able to take that load to Canada for me tomorrow?" Uhhhhhh .....sorry boss, we can't have that conversation on the ham radio.

E1F08 (A) [97.113] Which of the following types of amateur station communications are prohibited?
A. **Communications transmitted for hire or material compensation, except as otherwise provided in the rules**
B. Communications that have political content, except as allowed by the Fairness Doctrine
C. Communications that have religious content
D. Communications in a language other than English

Aside from station identification, which must be in English, we can communicate in other languages perfectly legally. I'm not sure how wise it is, but we can also discuss politics and religion if we choose to. But it is illegal to use your ham radio for **communications transmitted for hire or material compensation, except as otherwise provided in the rules**.

As you'll recall from your Technician exam, the rules don't provide much in the way of "otherwise" on this point.

Auxiliary Station Operators

E1F10 (B) [97.201] Who may be the control operator of an auxiliary station?
 A. Any licensed amateur operator
 B. Only Technician, General, Advanced or Amateur Extra Class operators
 C. Only General, Advanced or Amateur Extra Class operators
 D. Only Amateur Extra Class operators

The answer to this one is **only Technician, General, Advanced or Amateur Extra Class operators** may be control operators for auxiliary stations. Well, gee, isn't that the same as that answer that says, "*Any* licensed amateur operator?" No it isn't – there are still about 7,000 Novice licenses out there and novices may not be control operators of auxiliary stations.

Key Concepts in This Chapter

- Spread spectrum transmissions are only permitted on amateur frequencies above 222 MHz.

- A station transmitting SS emissions must not cause harmful interference to other stations employing other authorized emissions.

- Amateur spread spectrum stations must be in an area regulated by the FCC or in a country that permits SS emissions.

- Spread spectrum transmissions must not be used to obscure the meaning of any communication.

- When Canadian amateur operators visit the U.S. they have privileges that match the operating terms and conditions of the Canadian amateur service license, not to exceed U.S. Amateur Extra Class privileges.

- A dealer may sell an external RF power amplifier capable of operation below 144 MHz even if it has not been granted FCC certification if it was purchased in used condition from an amateur operator and is sold to another amateur operator for use at that operator's station.

- One of the standards an external RF power amplifier must meet if it is to qualify for a grant of FCC certification is that it must satisfy the FCC's spurious emission standards when operated at the lesser of 1500 watts or its full output power.

- The permitted mean power of any spurious emission relative to the mean power of the fundamental emission from a station transmitter or external RF amplifier installed after January 1, 2003 and transmitting on a frequency below 30 MHz is 43 dB below.

- Line A is a line roughly parallel to and south of the border between the U.S. and Canada. If you are north of Line A you may not transmit on frequencies between 420 and 430 MHz.

- The FCC might issue a Special Temporary Authority to an amateur station to provide for experimental amateur communications.

- An amateur station may send a message to a business when neither the amateur nor his or her employer has a pecuniary interest in the communications.

- Communications transmitted for hire or material compensation, except as otherwise provided in the rules, are prohibited.

- Only Technician, General, Advanced or Amateur Extra Class operators may be control operators of auxiliary stations.

Chapter 4

Special Station Restrictions

Spurious Emissions

E1B01 (D) [97.3] Which of the following constitutes a spurious emission?

    A. An amateur station transmission made without the proper call sign identification

    B. A signal transmitted to prevent its detection by any station other than the intended recipient

    C. Any transmitted signal that unintentionally interferes with another licensed radio station

    D. **An emission outside the signal's necessary bandwidth that can be reduced or eliminated without affecting the information transmitted**

At the Extra Class level, they ask us to get quite specific about this definition. For us, a spurious emission isn't just a transmission that's using excessive bandwidth – though that's evil enough. A spurious emission is **an emission outside its necessary bandwidth that can be reduced or eliminated without affecting the information transmitted.**

This doesn't mean you have *carte blanche* to use any amount of bandwidth you want, since using less would definitely "affect the information transmitted" – those regulations still stand. They just want you to know precisely what constitutes a *spurious* emission.

A spurious emission might be created by a poor transmitter that generates lots of harmonics, hooked to a multi-band antenna. Now your 40-meter phone signal at, say, 7.126 MHz is showing up as a harmonic at 14.252 MHz – that's a spurious emission. Your signal is **outside its necessary bandwidth** and that 14.252 MHz spurious emission could definitely be **reduced or eliminated without affecting the information transmitted**.

Protected Areas & Bands

E1B03 (A) [97.13] Within what distance must an amateur station protect an FCC monitoring facility from harmful interference?

    A. **1 mile**

    B. 3 miles

    C. 10 miles

    D. 30 miles

Way back when, the FCC maintained monitoring facilities all over the country, but they're mostly gone now. There was one near me, up by the Canadian border, near the lovely little community of Ferndale, WA. It's a field now, though on Google Earth you can still see the base of the tower. It's at 48° 57' 20.4" N 122° 33' 17.6" W if you love radio trivia and want to check it out.

There's only one FCC monitoring facility left, and it's near Laurel, Maryland, at 39° 09' 54.4" N, 76° 49' 15.9" W. If you're within 1 mile of the intersection of I-95 and the Patuxent

19

Freeway, you're probably within a mile of that facility, so do whatever you need to do (the rules don't say what that is) to avoid harmfully interfering with them.

They haven't stopped monitoring the rest of the country, by the way. They just do it with mobile units now.

E1B04 (C) [97.13, 1.1305-1.1319] What must be done before placing an amateur station within an officially designated wilderness area or wildlife preserve, or an area listed in the National Register of Historic Places?

    A. A proposal must be submitted to the National Park Service

    B. A letter of intent must be filed with the National Audubon Society

    C. **An Environmental Assessment must be submitted to the FCC**

    D. A form FSD-15 must be submitted to the Department of the Interior

This sounds rather daunting, but hams do, from time to time, place amateur stations in officially designated wilderness areas, wildlife preserves and areas listed in the National Register of Historical Places. For instance, August has become the semi-official month of setting up and operating from lighthouses all over the world, including the US. I can't be 100% certain, but I'm pretty sure a lot of those lighthouses are on the National Register of Historical Places, so I presume the hams operating there knew that to operate there, **an Environmental Assessment must be submitted to the FCC.**

This is *not* an Environmental Impact Study. It's really a questionnaire, and in my simple language, it asks, "What are you planning to do? Have you thought this through in terms of the environment? Have you considered alternative sites? Are you going to mess stuff up? Is anybody upset about you putting up this infernal contraption?" The Commission will review your Assessment and either issue what's called a "FONSI" – I'm not kidding – which stands for Finding of No Significant Impact, which means you're free to proceed, or they'll find that an Environmental Impact Study is warranted, in which case the FCC prepares the Environmental Impact Study.

E1B05 (C) [97.3] What is the National Radio Quiet Zone?

    A. An area in Puerto Rico surrounding the Arecibo Radio Telescope

    B. An area in New Mexico surrounding the White Sands Test Area

    C. **An area surrounding the National Radio Astronomy Observatory**

    D. An area in Florida surrounding Cape Canaveral

In eastern West Virginia, near the Virginia border, there's a big radio astronomy observatory. It's near the town of Green Bank. Imagine a big square, about 110 miles on a side, around that spot, and that's roughly the location of the National Radio Quiet Zone. The purpose of the zone is to create radio quiet so the radio telescope can hear what it needs to hear.

So far as the FCC rules are concerned, all we're prohibited from doing inside the zone is building a beacon station. However, there are state laws that are a good deal more restrictive, and common courtesy says respect what those folks out at the observatory are trying to get done. They don't seem to be terribly concerned about mobile stations, but if you're passing through there with your mobile, consider at least dropping to minimum power. By the way, if you travel the East Coast a lot, it's not unlikely you will pass through the zone – good sized chunks of the I-81, I-64 and I-79 freeways pass through it.

Hams have operated in the zone for Field Day and similar events, and done it with the permission of the radio astronomers who, let's face it, probably count a lot of hams in their number. What the astronomy center really seems to want is simply that people coordinate with them so interference is avoided.

E1B06 (A) [97.15] Which of the following additional rules apply if you are installing an amateur station antenna at a site at or near a public use airport?

    A. **You may have to notify the Federal Aviation Administration and register it with the FCC as required by Part 17 of the FCC rules**

    B. You must submit engineering drawings to the FAA

C. You must file an Environmental Impact Statement with the EPA before construction begins

D. You must obtain a construction permit from the airport zoning authority

The general rule for our tower heights is "no more than 200 feet above the ground under the antenna." Above that height, we must notify the FCC and the FAA of our intentions. (We may also collide with some local zoning ordinances, and other problems, but those are beyond the scope of this question.)

Near airports, that 200 foot guideline changes, and a shorter antenna may trigger the requirement that **you notify the Federal Aviation Administration and register your antenna with the FCC as required by Part 17 of FCC rules.**

The trigger for the limitation is the runway length of the airport. Different runway lengths lead to different requirements, but here's an example: If the runway you are near is longer than 1 kilometer (3280 feet) and the airport is within 6.1 km (3.79 miles) of your proposed installation, your antenna may be no higher than 1 meter (3.28 feet) above the airport elevation for every 100 meters (328 feet) from the nearest runway. So let's say you are 2 miles from an airport with a 3300 foot runway. Your antenna cannot be taller than 105.57 feet before you need to start doing paperwork.

E1B08 (D) [97.121] What limitations may the FCC place on an amateur station if its signal causes interference to domestic broadcast reception, assuming that the receivers involved are of good engineering design?

   A. The amateur station must cease operation

   B. The amateur station must cease operation on all frequencies below 30 MHz

   C. The amateur station must cease operation on all frequencies above 30 MHz

   D. **The amateur station must avoid transmitting during certain hours on frequencies that cause the interference**

If your station is causing interference to your neighbor's television, for instance, the FCC has the right to order you not to transmit during the hours from 8:00 PM to 10:30PM, local time, and on Sunday, additionally, from 10:30 AM until 1:00 PM local time. They also reserve to themselves the right to direct you to take "such steps as may be necessary to minimize interference to stations operating in other services," not necessarily just domestic broadcast stations. One of those steps might be to order that **the amateur station must avoid transmitting during certain hours on frequencies that cause the interference.**

E1B12 (A) [97.303(b)] What must the control operator of a repeater operating in the 70 cm band do if a radiolocation system experiences interference from that repeater?

   A. **Cease operation or make changes to the repeater to mitigate the interference**

   B. File an FAA NOTAM (Notice to Airmen) with the repeater system's ERP, call sign, and six-character grid locator

   C. Reduce the repeater antenna HAAT (Height Above Average Terrain)

   D. All these choices are correct

We are secondary users on the 70 cm band, so if a radiolocation system experiences interference from one of our 70 cm repeaters, we must **cease operation or make changes to the repeater to mitigate the interference.**

There are primary users on that band; the band is used by the military and other Federal agencies for a number of radar applications, multi-function position-location communications systems, and test range telecommand and flight termination systems.

Local Antenna & Tower Regulations

E1B11 (B) [97.15] What does PRB-1 require of regulations affecting amateur radio?

A. No limitations may be placed on antenna size or placement

B. Reasonable accommodations of amateur radio must be made

C. Amateur radio operations must be permitted in any private residence

D. Use of wireless devices in a vehicle is exempt from regulation

"PRB" stands for "Private Radio Bureau", which is the part of the FCC that oversees most of amateur radio. PRB-1 is a ruling the FCC made back in 1985 that requires state and local antenna regulations to "accommodate reasonably amateur communications, and to represent the minimum practicable regulation to accomplish the local authority's legitimate purpose."

This has no relationship whatsoever to any homeowner's association agreement into which you might have entered; you probably signed away your PRB-1 rights when you signed up for the HOA.

E1B07 (C) [97.15] To what type of regulations does PRB-1 apply?

A. Homeowners associations

B. FAA tower height limits

C. State and local zoning

D. Use of wireless devices in vehicles

PRB-1 applies only to **state and local** "entities." If you have a situation to which you think PRB-1 applies, I'd urge you to visit the ARRL's web page on PRB-1.

http://www.arrl.org/prb-1

RACES Operation

E1B09 (C) [97.407] Which amateur stations may be operated under RACES rules?

A. Only those club stations licensed to Amateur Extra Class operators

B. Any FCC-licensed amateur station except a Technician Class

C. Any FCC-licensed amateur station certified by the responsible civil defense organization for the area served

D. Any FCC-licensed amateur station participating in the Military Auxiliary Radio System (MARS)

RACES is the Radio Amateur Civil Emergency Service. That's the amateur radio emergency service that is administered locally by your local emergency preparedness people, whomever they may be, so the station must be **certified by the responsible civil defense organization for the area served.**

E1B10 (A) [97.407] What frequencies are authorized to an amateur station operating under RACES rules?

A. **All amateur service frequencies authorized to the control operator**

B. Specific segments in the amateur service MF, HF, VHF and UHF bands

C. Specific local government channels

D. Military Auxiliary Radio System (MARS) channels

No enhanced frequency privileges are granted just because a control operator is operating under RACES rules. Good thing you're going to be an Amateur Extra – you'll be extra useful in times of need! In your case, **all amateur service frequencies authorized to the control operator** will be all the amateur service frequencies.

Key Concepts in This Chapter

- A spurious emission is an emission outside the signal's necessary bandwidth that can be reduced or eliminated without affecting the information transmitted.

- Amateur stations within one mile of an FCC monitoring facility must protect that facility from harmful interference.

- Before placing an amateur station within an officially designated wilderness area or wildlife preserve, or in an area listed in the National Register of Historic Places, an Environmental Assessment must be submitted to the FCC.

- The National Radio Quiet Zone is an area surrounding the National Radio Astronomy Observatory.

- If you are installing an amateur station antenna at a site at or near a public use airport you may have to notify the Federal Aviation Administration and register it with the FCC as required by Part 17 of the FCC rules.

- If your station is interfering with domestic broadcast reception, the FCC might require that you avoid transmitting during certain hours on frequencies that cause the interference.

- If a radiolocation system experiences interference from one of our 70 cm repeaters, we must cease operation or make changes to the repeater to mitigate the interference.

- The FCC ruling known as PRB-1 requires reasonable accommodations of amateur radio by state and local entities.

- PRB-1 applies to state and local zoning.

- Any FCC-licensed amateur station certified by the responsible civil defense organization for the area served may be operated under RACES rules.

- An amateur station operating under RACES rules may be operated on all amateur service frequencies authorized to the control operator.

Are you keeping up with the Practice Exams? Research shows that memories are strengthened by recalling them regularly. If you just have a few minutes, get out your phone and try a Chapter Quiz. No time to look up answers? No problem – skip the questions you don't know.

Chapter 5

Automatic and Remote Control & Operating in Foreign Countries

Automatic & Remote Control

E1C03 (B) [97.3, 97.109] How do the control operator responsibilities of a station under automatic control differ from one under local control?

    A. Under local control there is no control operator

    B. Under automatic control the control operator is not required to be present at the control point

    C. Under automatic control there is no control operator

    D. Under local control a control operator is not required to be present at a control point

The only way your responsibilities as a licensee change when you are control operator for an automatically controlled station is that you no longer are **required to be present at the control point.**

    Our club's HamWAN link that connects our repeater to the internet is under automatic control, but our trustee is still the control operator.

E2C01 (D) What indicator is required to be used by U.S.-licensed operators when operating a station via remote control and the remote transmitter is located in the U.S.?

    A. / followed by the USPS two letter abbreviation for the state in which the remote station is located

    B. /R# where # is the district of the remote station

    C. / followed by the ARRL Section of the remote station

    D. **No additional indicator is required**

There are a few situations where we need to add a "slash" identifier to our call signs. Those slash identifiers are called "self-assigned indicators." When we operate from inside Canada, for instance, we add "slash" then the appropriate call sign suffix for the Canadian region where we are operating, so if I was operating in British Columbia, I'd be AF7KB/VA7. (I'm also required to at least once during the conversation identify my location, such as "Langley, BC.")

    We also need to add a slash identifier when we operate on our new license upgrade's bands during that time between when we pass an upgrade exam and when our new status is listed in the FCC database. So when you run out the door after your Extra exam and fire up the rig to talk on 14.152 MHz in the Extra portion of the 20-meter band, you'll add "slash AE" to your call sign.

    However, the situation described in this question is not a time when a slash identifier is required. Operating a station by remote control when the station is located in the US is simply operating a station. **No additional indicator is required.**

E1C08 (B) [97.213] What is the maximum permissible duration of a remotely controlled station's transmissions if its control link malfunctions?

    A. 30 seconds

    B. **3 minutes**

    C. 5 minutes

    D. 10 minutes

When you are building your remote controlled station, you must include a fail-safe mechanism such that if the control link malfunctions, the station shuts itself down within **3 minutes**.

E1C05 (A) [97.221(c)(1),[97.115(c)] When may an automatically controlled station originate third party communications?

    A. **Never**

    B. Only when transmitting RTTY or data emissions

    C. When agreed upon by the sending or receiving station

    D. When approved by the National Telecommunication and Information Administration

Honestly, I'm at a bit of a loss as to how an automatically controlled station *could* originate third party communications, but in any case, it is **never** allowed. Given all the rules about third-party communications, it's easy to see the FCC would want a licensed human being making the decisions about what third-party messages to transmit.

Foreign Operations

E1C04 (A) What is meant by IARP?

    A. **An international amateur radio permit that allows U.S. amateurs to operate in certain countries of the Americas**

    B. The internal amateur radio practices policy of the FCC

    C. An indication of increased antenna reflected power

    D. A forecast of intermittent aurora radio propagation

Operating from another country is possible, potentially great fun, and sometimes complicated.

There are three potential paths to being allowed to fire up your radio on foreign soil. There are potential questions about all of them on the exam. The path you take will depend on the country you plan to visit.

1. A *reciprocal permit*. If we have a reciprocal agreement with a country, you can operate in that country, but you don't necessarily have automatic permission to do so. The procedure varies by country – and it is a long list of countries that spans the globe but doesn't cover most of Europe and South America. Often, you need a letter of permission from their embassy or some similar official permission before you may operate.

2. A *CEPT License*. CEPT is The European Conference of Postal and Telecommunications Administrations, and as you might guess, CEPT covers most of Europe. CEPT provides reciprocal privileges to US General, Advanced and Extra class operators – Technicians are out of luck. To operate in countries covered by the CEPT agreement, you just need to carry some paperwork, including your amateur license, proof of US citizenship, and a copy – in English, German, and French – of the current FCC Notice regarding CEPT. Fear not, you can download that FCC Notice from the ARRL or the FCC.

3. The one mentioned in this question, an *IARP*, which is an *International Amateur Radio Permit*. An IARP **allows U.S. amateurs to operate in certain countries of the Americas.** At the moment – these things are subject to change – the countries that accept IARP's are Argentina, Bolivia, Brazil, Canada, Chile, El Salvador, Guatemala, Haiti, Mexico, Panama, Paraguay, Peru, Trinidad and Tobago, the U.S., Uruguay, and Venezuela.

You apply for an IARP through the ARRL. There are two classes of IARP. Class 1 requires an FCC amateur license of any class plus "a knowledge of the International Morse Code." It carries all amateur frequency and operating privileges and is equivalent of our Amateur Extra class license. No standard for what sort of "knowledge of the International Morse Code" is required, but authorities advise you to be ready to demonstrate competence if asked. (In fact, the application doesn't ask about code proficiency at all.)

A Class 2 IARP requires no code knowledge and carries privileges equivalent to the US Technician class license.

Finally, Canada is in a special class. Though they accept IARP's, as a US amateur you do not need one to operate in Canada – we have automatic reciprocal privileges in Canada.

E1C11 (A) [97.5] Which of the following operating arrangements allows an FCC-licensed U.S. citizen to operate in many European countries, and alien amateurs from many European countries to operate in the U.S.?

    A. **CEPT agreement**
    B. IARP agreement
    C. ITU reciprocal license
    D. All these choices are correct

Amateur operations by US amateurs in Europe are covered by the **CEPT agreement.**

When you're taking the exam, if you are completely baffled by a question and one of the choices is "all these choices are correct", don't go for it. Out of 44 possible "all these choices are correct" on the exam, only 18 times is that actually the correct answer. You'll slightly improve your odds by avoiding that answer *if you don't know the correct answer.* (But you're a *Fast Track* student, so you'll know it. Right?)

E1C06 (C) Which of the following is required in order to operate in accordance with CEPT rules in foreign countries where permitted?

    A. You must identify in the official language of the country in which you are operating
    B. The U.S. embassy must approve of your operation
    C. **You must bring a copy of FCC Public Notice DA 16-1048**
    D. You must append "/CEPT" to your call sign

There may be a new notice now with a different number by the time you take your trip, but just remember you must carry an **FCC Public Notice** when you operate under CEPT, and you'll get this question right.

E1C02 (C) [97.117] Which of the following types of communications may be transmitted to amateur stations in foreign countries?

    A. Business-related messages for non-profit organizations
    B. Messages intended for users of the maritime satellite service
    C. **Communications incidental to the purpose of the amateur service and remarks of a personal nature**
    D. All of these choices are correct

This question is in every question pool from Technician on, and it didn't disappoint us by disappearing at the Extra level! The key to getting this one is still that phrase, **remarks of a personal nature.**

Key Concepts in This Chapter

- The only way the control responsibilities of a station under automatic control differ from one under local control is that under automatic control the control operator is not required to be present at the control point.

- No additional indicator is required when a U.S. licensed operator is operating a station in the United States via remote control.

- If a remotely controlled station's control link malfunctions, the station must stop transmitting within three minutes.
- Automatically controlled stations may never originate third party communications.
- An IARP is an international amateur radio permit that allows U.S. amateurs to operate in certain countries of the Americas.
- The CEPT agreement allows an FCC licensed U.S. citizen to operate in many European countries.
- If you operate under CEPT rules in Europe, you must bring along a copy of an appropriate FCC public notice, listed in the test as DA 16-1048.
- The types of communications that may be transmitted to amateur stations in foreign countries are limited to communications incidental to the purpose of the amateur service and remarks of a personal nature.

Once you complete the Practice Exam for this chapter, consider challenging yourself with Progress Check #1 at fasttrackham.com. If you find any groups where your knowledge needs improvement, you can focus on those groups with the Group Drills.

Chapter 6

Volunteer Examiner Program

Each level of ham exam asks some questions about the Volunteer Examiner program, but at the Extra level, the questions get more detailed. That makes sense, since the vast majority of VE's are Extra class license holders so the test committee pretty obviously wants Extras well grounded on the topic.

E1E03 (C) [97.521] What is a Volunteer Examiner Coordinator?

    A. A person who has volunteered to administer amateur operator license examinations

    B. A person who has volunteered to prepare amateur operator license examinations

    C. **An organization that has entered into an agreement with the FCC to coordinate, prepare, and administer amateur operator license examinations**

    D. The person who has entered into an agreement with the FCC to be the VE session manager

Repeat after me: VEC's are organizations. VE's are the examiners. Specifically, a Volunteer Examiner Coordinator is **an organization that has entered into an agreement with the FCC to coordinate amateur operator license examinations.** There are a number of them around the country, including the ARRL, W5YI, and The Anchorage VEC.

E1E02 (C) [97.523] Who does Part 97 task with maintaining the pools of questions for all U.S. amateur license examinations?

    A. The VEs

    B. The FCC

    C. **The VECs**

    D. The ARRL

I often hear hams refer to "the official FCC questions." There's no such thing. The FCC got itself out of the amateur radio testing business over 35 years ago.

    The question pools are created and maintained by the VEC's; specifically, they are created by committees of amateur radio operators, and curated and distributed by the National Conference of VEC's.

E1E04 (D) [97.509, 97.525] Which of the following best describes the Volunteer Examiner accreditation process?

    A. Each General, Advanced and Amateur Extra Class operator is automatically accredited as a VE when the license is granted

    B. The amateur operator applying must pass a VE examination administered by the FCC Enforcement Bureau

    C. The prospective VE obtains accreditation from the FCC

    D. **The procedure by which a VEC confirms that the VE applicant meets FCC requirements to serve as an examiner**

You're not automatically accredited as a VE when you upgrade to Extra. The FCC is not in the business of accrediting VE's, but VEC's are – that's one of the main things they do.

Want to be a VE? Great! You can get accredited by the ARRL by downloading an open-book exam, filling it out, and mailing it in with your application. Other VEC's may have different procedures, and you'll want to be accredited by the VEC affiliated with the testing organization you plan to serve.

E1E05 (B) [97.503] What is the minimum passing score on all amateur operator license examinations?

    A. Minimum passing score of 70%

    B. **Minimum passing score of 74%**

    C. Minimum passing score of 80%

    D. Minimum passing score of 77%

It's not actually possible to score exactly **74%** on the Technician or General exams – you'll either get 71.4% or lower, or 74.3% or higher, but **74%** *is* possible on the Amateur Extra exam. You need to get 37 questions correct on the 50 question exam to get that **74%**.

E1E06 (C) [97.509] Who is responsible for the proper conduct and necessary supervision during an amateur operator license examination session?

    A. The VEC coordinating the session

    B. The FCC

    C. **Each administering VE**

    D. The VE session manager

The FCC rules on this are quite specific: If you are an **administering VE** for a ham exam, *you are the one* responsible for the proper conduct and necessary supervision of that exam. The VEC *may* have a "VE session manager", and that person *may* do certain things, but in the end, the responsibility for proper conduct and necessary supervision falls to **each administering VE**.

E1E07 (B) [97.509] What should a VE do if a candidate fails to comply with the examiner's instructions during an amateur operator license examination?

    A. Warn the candidate that continued failure to comply will result in termination of the examination

    B. **Immediately terminate the candidate's examination**

    C. Allow the candidate to complete the examination, but invalidate the results

    D. Immediately terminate everyone's examination and close the session

If someone taking a ham exam "fails to comply with the examiner's instructions," the orders are a VE should **immediately terminate the candidate's examination**.

Ouch! Seems a bit harsh, doesn't it? However, it does set a standard that spares the VE's a lot of potential worry and second-guessing. My guess is this is left over from the days when the FCC administered the tests, and they probably found it was a standard that worked.

E1E08 (C) [97.509] To which of the following examinees may a VE not administer an examination?

    A. Employees of the VE

    B. Friends of the VE

    C. **Relatives of the VE as listed in the FCC rules**

    D. All these choices are correct

That list of **relatives of the VE in the FCC rules** includes; his or her spouse, children, grandchildren, stepchildren, parents, grandparents, stepparents, brothers, sisters, stepbrothers, stepsisters, aunts, uncles, nieces, nephews, and in-laws.

E1E09 (A) [97.509] What may be the penalty for a VE who fraudulently administers or certifies an examination?

    A. **Revocation of the VE's amateur station license grant and the suspension of the VE's amateur operator license grant**

    B. A fine of up to $1000 per occurrence

C. A sentence of up to one year in prison

D. All these choices are correct

They're not going to put you in prison, nor even empty your wallet, but fraudulently administering or certifying a ham exam could lead to the **revocation of the VE's amateur station license grant and the suspension of the VE's amateur operator license grant**. In simple English, they yank your ham ticket and most likely never give it back.

E1E10 (C) [97.509] What must the administering VEs do after the administration of a successful examination for an amateur operator license?

A. They must collect and send the documents to the NCVEC for grading

B. They must collect and submit the documents to the coordinating VEC for grading

C. **They must submit the application document to the coordinating VEC according to the coordinating VEC instructions**

D. They must collect and send the documents to the FCC according to instructions

Ever wonder what happened to your paperwork after you took your Technician and/or General exam? The administering VE's had to **submit the application document to the coordinating VEC** according to the coordinating VEC instructions. Notice it goes to the VEC, such as the ARRL or W5YI – not to the FCC, and not to the NCVEC, the National Conference of Volunteer Examiner Coordinators; the organization of VEC organizations.

E1E11 (B) [97.509] What must the VE team do if an examinee scores a passing grade on all examination elements needed for an upgrade or new license?

A. Photocopy all examination documents and forward them to the FCC for processing

B. **Three VEs must certify that the examinee is qualified for the license grant and that they have complied with the administering VE requirements**

C. Issue the examinee the new or upgrade license

D. All these choices are correct

When you passed that Technician and/or General exam, three VE's signed your "Form 605 (Figure 6.1)" – that application you filled out, swearing, "I CERTIFY THAT I HAVE COMPLIED WITH THE ADMINISTERING VE REQUIREMENTS IN PART 97 OF THE COMMISSION'S RULES AND WITH THE INSTRUCTIONS PROVIDED BY THE COORDINATING VEC AND THE FCC." (You know it's serious, it's in ALL CAPS!) When an applicant passes an exam, **three VE's must certify that the examinee is qualified for the license grant and that they have complied with the administering VE requirements.**

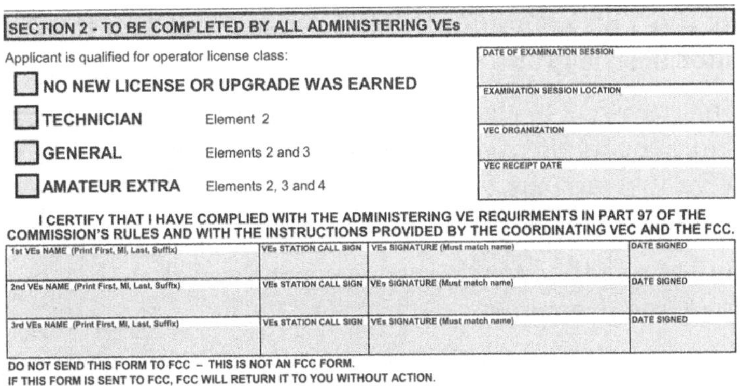
SECTION 2 - TO BE COMPLETED BY ALL ADMINISTERING VEs

Applicant is qualified for operator license class:

- ☐ NO NEW LICENSE OR UPGRADE WAS EARNED
- ☐ TECHNICIAN — Element 2
- ☐ GENERAL — Elements 2 and 3
- ☐ AMATEUR EXTRA — Elements 2, 3 and 4

DATE OF EXAMINATION SESSION

EXAMINATION SESSION LOCATION

VEC ORGANIZATION

VEC RECEIPT DATE

I CERTIFY THAT I HAVE COMPLIED WITH THE ADMINISTERING VE REQUIRMENTS IN PART 97 OF THE COMMISSION'S RULES AND WITH THE INSTRUCTIONS PROVIDED BY THE COORDINATING VEC AND THE FCC.

1st VEs NAME (Print First, MI, Last, Suffix)	VEs STATION CALL SIGN	VEs SIGNATURE (Must match name)	DATE SIGNED
2nd VEs NAME (Print First, MI, Last, Suffix)	VEs STATION CALL SIGN	VEs SIGNATURE (Must match name)	DATE SIGNED
3rd VEs NAME (Print First, MI, Last, Suffix)	VEs STATION CALL SIGN	VEs SIGNATURE (Must match name)	DATE SIGNED

DO NOT SEND THIS FORM TO FCC – THIS IS NOT AN FCC FORM.
IF THIS FORM IS SENT TO FCC, FCC WILL RETURN IT TO YOU WITHOUT ACTION.

Figure 6.1: Form 605

E1E12 (A) [97.509] What must the VE team do with the application form if the examinee does not pass the exam?

A. **Return the application document to the examinee**

B. Maintain the application form with the VEC's records

C. Send the application form to the FCC and inform the FCC of the grade

D. Destroy the application form

Good news – if you fail a ham exam, no record is kept of that event. If an applicant does not pass an exam, the VE's are required to **return the application document to the examinee**.

E1E01 (A) [97.527] For which types of out-of-pocket expenses do the Part 97 rules state that VEs and VECs may be reimbursed?

    A. **Preparing, processing, administering and coordinating an examination for an amateur radio operator license**
    B. Teaching an amateur operator license examination preparation course
    C. No expenses are authorized for reimbursement
    D. Providing amateur operator license examination preparation training materials

The correct answer, **preparing, processing, administering and coordinating an examination for an amateur radio license** is a direct quote of the entire rule regarding VE and VEC reimbursement of expenses. VE's and VEC's can be reimbursed for out-of-pocket expenses, such as the rental for a room, making copies of the exams, copying the completed paperwork, sending the paperwork to the VEC, etc. They can't turn it into a for-profit business, though.

Key Concepts in This Chapter

- A Volunteer Examiner Coordinator is an organization that has entered into an agreement with the FCC to coordinate amateur operator license examinations.

- Part 97 tasks the VEC's with maintaining the pools of questions for all U.S. amateur license examinations.

- The Volunteer Examiner Accreditation process is the procedure by which a VEC confirms that the VE applicant meets FCC requirements to serve as an examiner.

- The minimum passing score on amateur operator license examinations is 74%.

- Each administering VE is responsible for the proper conduct and necessary supervision during an amateur operator license examination session.

- If a candidate fails to comply with the examiner's instructions during an amateur operator license examination, the VE should immediately terminate the candidate's examination.

- A VE may not administer an examination to relatives of the VE as listed in the FCC rules.

- A VE who fraudulently administers or certifies an examination is subject to a penalty of revocation of the VE's amateur station license grant and the suspension of the VE's amateur operator license grant.

- After the administration of a successful examination for an amateur operator license, the VE's must submit the application document to the coordinating VEC according to the coordinating VEC instructions.

- When you score your passing grade on your Extra Class license exam, three VEs must certify that you are qualified for the license grant and that they have complied with the administering VE requirements.

- If a candidate does not pass the exam, the VE team is to return the application document to the examinee.

- VE's and VEC's may be reimbursed for costs of preparing, processing, administering and coordinating an examination for an amateur radio license.

Chapter 7

Amateur Satellites – Hams in Space, Episode III

On the General exam, amateur satellites were barely mentioned, but the Extra question pool contains about two dozen questions on the topic, and you'll have at least one question on your exam about satellites. Satellites are also called "space stations" on the exam.

Telemetry & Telecommand

E1D01 (A) [97.3 E1D01 (A) [97.3] What is the definition of telemetry?

A. **One-way transmission of measurements at a distance from the measuring instrument**

B. Two-way transmissions in excess of 1000 feet

C. Two-way transmissions of data

D. One-way transmission that initiates, modifies, or terminates the functions of a device at a distance

We use two different terms for communications with a satellite that aren't the actual ham to ham transmissions. The stuff going from us to the satellite, to tell it what to do, is the *telecommands*.

The stuff this question asks about, the communications from the satellite to the satellite controller is *telemetry*, which is **one way transmission of measurements at a distance from the measuring instrument**. Indeed, the roots of the word *telemetry*, "tele" and "meter", mean "measurement at a distance."

E1D03 (B) [97.3] What is a space telecommand station?

A. An amateur station located on the surface of the earth for communication with other earth stations by means of earth satellites

B. **An amateur station that transmits communications to initiate, modify or terminate functions of a space station**

C. An amateur station located in a satellite or a balloon more than 50 kilometers above the surface of the Earth

D. An amateur station that receives telemetry from a satellite or balloon more than 50 kilometers above the surface of the Earth

Here's that opposite number to telemetry, telecommand. Think of a telecommand station as Mission Control for the satellite. The telecommand station **transmits communications to initiate, modify or terminate functions of a space station.**

You might think a ham radio telecommand station would look like Figure 7.1, but those amateur satellites are only up about 100 miles. You could telecommand a ham satellite from your back yard with a minimal amount of antenna and, say, 10 watts – *if* you had the command codes.

Figure 7.1: Radio Telescope

E1D02 (A) [97.211(b)] Which of the following may transmit special codes intended to obscure the meaning of messages?

    A. **Telecommand signals from a space telecommand station**
    B. Data containing personal information
    C. Auxiliary relay links carrying repeater audio
    D. Binary control characters

Nobody, including the FCC, wants some unauthorized someone to be able to control a satellite, so the rules say it is just fine for **telecommand signals from a space telecommand station** to use special codes to obscure the meaning of messages.

E1D12 (A) [97.207(e), 97.203(g)] Which of the following amateur stations may transmit one-way communications?

    A. **A space station, beacon station, or telecommand station**
    B. A local repeater or linked repeater station
    C. A message forwarding station or automatically controlled digital station
    D. All these choices are correct

As amateurs, we're generally prohibited from transmitting one-way communications; in other words, broadcasting. Some exceptions to the rule are **space stations, beacon stations, and telecommand stations.**

E1D05 (D) [97.213(d)] What must be posted at the station location of a station being operated by telecommand on or within 50 km of the earth's surface?

    A. A photocopy of the station license
    B. A label with the name, address, and telephone number of the station licensee
    C. A label with the name, address, and telephone number of the control operator
    D. **All these choices are correct**

When you launch that APRS-equipped weather balloon, be sure you include all the paperwork.

E1D06 (A) [97.215(c)] What is the maximum permitted transmitter output power when operating a model craft by telecommand?

    A. **1 watt**
    B. 2 watts
    C. 5 watts
    D. 100 watts

From what I understand from enthusiasts, modern RC radios are so capable, there's no problem for this use of amateur radio to solve, but for the exam, remember that when operating a model craft by telecommand, the power limit is **1 watt.**

E1D07 (A) [97.207] Which HF amateur bands have frequencies authorized for space stations?

    A. **Only the 40, 20, 17, 15, 12 and 10-meter bands**
    B. Only the 40, 20, 17, 15 and 10-meter bands

C. Only the 40, 30, 20, 15, 12 and 10-meter bands

D. All HF bands

The right answer is the only one that includes **17 meters** *and* **12 meters**, and that's probably as good a way as any to keep this one in your head.

Just because the **40, 20, 17, 15, 12, and 10-meter** bands are *authorized* for space stations doesn't mean you'll find any satellite action there. There's a little bit of 15-meter and a bit more 10-meter going on, but it's mostly VHF/UHF.

E1D08 (D) [97.207] Which VHF amateur bands have frequencies authorized for space stations?

    A. 6 meters and 2 meters

    B. 6 meters, 2 meters, and 1.25 meters

    C. 2 meters and 1.25 meters

    D. **2 meters**

6 meters isn't authorized for space stations, nor is the 220 MHz, 1.25 meter band. Of the choices given, only **2 meters** is available for space stations.

E1D09 (B) [97.207] Which UHF amateur service bands have frequencies authorized for space stations?

    A. 70 cm only

    B. **70 cm and 13 cm**

    C. 70 cm and 33 cm

    D. 33 cm and 13 cm

There are UHF frequencies available for space stations in the **70 cm and 13 cm** bands. 70 cm is, of course, our familiar 420 MHz band. 13 cm is way up there in the 2 GHz range.

E1D10 (B) [97.211] Which amateur stations are eligible to be telecommand stations of space stations (subject to the privileges of the class of operator license held by the control operator of the station)?

    A. Any amateur station designated by NASA

    B. **Any amateur station so designated by the space station licensee**

    C. Any amateur station so designated by the ITU

    D. All these choices are correct

The **space station licensee** gets to choose the telecommand station. Of course, the operation of that station is subject to the privileges of the class of operator license held by the control operator.

E1D11 (D) [97.209] Which amateur stations are eligible to operate as earth stations?

    A. Any amateur station whose licensee has filed a pre-space notification with the FCC's International Bureau

    B. Only those of General, Advanced or Amateur Extra Class operators

    C. Only those of Amateur Extra Class operators

    D. **Any amateur station, subject to the privileges of the class of operator license held by the control operator**

An earth station is simply any amateur station communicating via a space station. No special permit is needed – you point your antenna at the right part of the sky at the right time, listen for the channel to clear, push the button and go for it! **Any amateur station, subject to the privileges of the class of operator license held by the control operator**, is eligible to operate as an earth station.

E1D04 (A) [97.119(a)] Which of the following is required in the identification transmissions from a balloon-borne telemetry station?

    A. **Call sign**

    B. The output power of the balloon transmitter

    C. The station's six-character Maidenhead grid locator

    D. All these choices are correct

You're not required to send any of those things in the wrong answers while you're down here on the ground, but you are required to identify your station with your call sign. Even 50 kilometers off the surface, it's the same.

Key Concepts in This Chapter

- Telemetry is one-way transmission of measurements at a distance from the measuring instrument.

- A space telecommand station is an amateur station that transmits communications to initiate, modify or terminate functions of a space station.

- Space stations and telecommand stations may transmit one-way communications. So may a beacon station.

- A photocopy of the station license, a label with the name, address, and telephone number of the station licensee, and a label with the name, address, and telephone number of the control operator must be posted at the station location of a station being operated by telecommand on or within 50 km of the earth's surface.

- The only amateur service HF bands with frequencies authorized for space stations are the 40, 20, 17, 15, 12 and 10-meter bands. For the exam, remember 17 and 12 meters.

- In the VHF band, only the 2-meter band has frequencies available for space stations.

- In the UHF band, the 70 cm and 13 cm bands have frequencies available for space stations.

- Any amateur station so designated by the space station licensee is eligible to be the telecommand station of a space station.

- Any amateur station, subject to the privileges of the class of operator license held by the control operator is eligible to operate as an earth station.

- The identification transmissions from a balloon-borne telemetry station must include the station's call sign.

Chapter 8

More Amateur Radio in Space

Here's some good news: A lot of these questions on satellites are repeats from your Technician exam. Just in case some of that information has slipped your mind, we'll go over them anyway....

E2A01 (C)] What is the direction of an ascending pass for an amateur satellite?

    A. From west to east

    B. From east to west

    C. **From south to north**

    D. From north to south

When we talk about "ascending" and "descending" passes, we're imagining the satellite circling the globe on, at least roughly, a pole to pole orbit, with the South pole being at the "bottom" and the North pole at the "top." Thus, an ascending pass for an amateur satellite means the satellite is appearing to the south and is headed north. An ascending path simply means a path **from south to north.** (See figure 8.1)

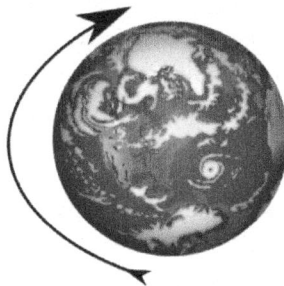

Figure 8.1: Ascending Pass

E2A04 (B) What is meant by the term mode as applied to an amateur radio satellite?

    A. Whether the satellite is in a low earth or geostationary orbit

    B. **The satellite's uplink and downlink frequency bands**

    C. The satellite's orientation with respect to the earth

    D. Whether the satellite is in a polar or equatorial orbit

With space stations, our terminology changes just a bit. We're used to using mode to indicate things like CW or SSB. Amateur radio satellite "modes" are what we usually call "bands." They're indicated by single letters, with the uplink mode listed first, then the downlink mode, and those mode indicators tell you the frequency range they're using, such as VHF/UHF.

E2A05 (D) What do the letters in a satellite's mode designator specify?

    A. Power limits for uplink and downlink transmissions

    B. The location of the ground control station

    C. The polarization of uplink and downlink signals

    D. **The uplink and downlink frequency ranges**

37

The first letter of the mode indicates the satellite's receive band, the second its transmit band. If a satellite lists its modes as V/U, that means it is receiving on VHF, transmitting on UHF.

Maybe a way to remember this is that the satellite has to "hear" you before it can transmit your signal, so the receive mode comes first.

E2A09 (A) What do the terms "L band" and "S band" specify regarding satellite communications?

    A. **The 23 centimeter and 13 centimeter bands**

    B. The 2-meter and 70 centimeter bands

    C. FM and Digital Store-and-Forward systems

    D. Which sideband to use

What we call the **23 centimeter band** is what the science world calls the L-Band. For them, it's from 1.0 to 2.0 GHz, but for amateur satellites, we're only concerned with 1.26 GHz to 1.27 GHz.

The S-Band is wayyyyyyyy up there, from 2.0 to 4.0 GHz. We call our part of the S-band the **13 centimeter band**.

E2A07 (D) Which of the following types of signals can be relayed through a linear transponder?

    A. FM and CW

    B. SSB and SSTV

    C. PSK and packet

    D. **All these choices are correct**

When we say a transponder is "linear", we're saying the output waveform matches the input waveform – what goes in is what comes out. Since this is a linear transponder, all of the modes listed in the answers, including FM, CW, SSB, SSTV, PSK and Packet will work through that transponder.

Despite the fact that FM will, technically, work, FM and other bandwidth hog modes are unwelcome on a linear transponder equipped satellite. For FM there are FM satellites that operate more like repeaters.

E2A08 (B) Why should effective radiated power to a satellite that uses a linear transponder be limited?

    A. To prevent creating errors in the satellite telemetry

    B. **To avoid reducing the downlink power to all other users**

    C. To prevent the satellite from emitting out-of-band signals

    D. To avoid interfering with terrestrial QSOs

Let's start this by saying making electricity in space is both difficult and expensive – and the ham radio community isn't funded like the big players who can hang vast acreage of space-grade solar panels on their gigantic satellites. Ours are small, relatively inexpensive satellites with just enough solar panels to make them work. Power is always an issue. You can see AMSAT's former president, Joe Spier, K6WAO, holding a full scale model of an amateur satellite in Figure 8.2.

Remember, that transponder is linear – whatever comes in is what it's going to try to send out. If your way-too-powerful signal overloads the transponder, it is going to start throttling back the power to save some watts – and that's going to affect all the other users on the satellite at the moment. We limit our power when working satellites **to avoid reducing the downlink power to all other users**.

Working satellites just about requires the ability to operate in full-duplex mode, so you can hear your own signal coming back from the satellite. Experienced satellite operators say to

Figure 8.2: Full Scale Model of Typical Amateur Radio Satellite

AO-7 Modes and Frequencies			
Uplink		**Downlink**	
Mode	**Frequency**	**Mode**	**Frequency**
USB	145.850 MHz	USB	29.400 MHz
USB	145.855 MHZ	USB	29.405 MHz
...and so forth, in .005 MHz steps to ...			
USB	145.950 MHz	USB	29.500 MHz

Table 8.1: Non-inverting Transponder Frequencies

adjust your power down until you just start to fade out, then bump it up just a bit and you'll be set up correctly.

E2A02 (D) Which of the following occurs when a satellite is using an inverting linear transponder?

    A. Doppler shift is reduced because the uplink and downlink shifts are in opposite directions

    B. Signal position in the band is reversed

    C. Upper sideband on the uplink becomes lower sideband on the downlink, and vice versa

    D. **All these choices are correct**

Let's start here: A linear transponder is not a repeater. It seems somewhat like one, but it isn't. A repeater 'hears" on a single frequency and "talks" on another single frequency. Functionally, a repeater is a receiver and a transmitter. The receiver turns the received RF into audio and passes along the audio to the transmitter, which turns the audio into RF.

That's not at all what a transponder does. A transponder is capable of receiving and transmitting across a band of frequencies simultaneously. That band of frequencies is called the satellite's passband.

For instance, Table 8.1 shows a few representative frequency pairs for satellite AO-7's non-inverting transponder.

You can see that a transmission received on 145.850 MHz will go back out on 29.400 MHz.

AO-7 Modes and Frequencies			
Uplink		**Downlink**	
Mode	Frequency	Mode	Frequency
LSB	432.125 MHz	USB	145.975 MHz
LSB	432.130 MHZ	USB	145.970 MHz
...and so forth, in .005 MHz steps to ...			
LSB	432.175 MHz	USB	145.925 MHz

Table 8.2: Inverting Transponder Frequencies

One received on a higher frequency will go back out on a higher frequency. Additionally, whatever mode is received will be what is sent back out. That's how a non-inverting linear transponder works.

Contrast that with the frequencies for AO-7's inverting transponder in Table 8.2.

The modes are switching from LSB on the uplink to USB on the downlink. As the uplink frequencies go up, the downlink frequencies are going down. That's what an inverting transponder does. **The signal positions in the passband are reversed**, and **lower sideband on the uplink becomes upper sideband on the downlink**.

One benefit of this is that Doppler shift is reduced because the uplink and downlink shifts are in opposite directions – the more the Doppler shift makes the frequency I'm transmitting seem higher at the satellite, the more the transponder lowers the frequency on which it is transmitting. It doesn't completely eliminate Doppler shift, but it makes it easier to manage.

E2A03 (D) How is the signal inverted by an inverting linear transponder?

A. The signal is detected and remodulated on the reverse sideband

B. The signal is passed through a non-linear filter

C. The signal is reduced to I and Q components and the Q component is filtered out

D. **The signal is passed through a mixer and the difference rather than the sum is transmitted**

An inverting linear transponder inverts a signal by **passing it through a mixer. The difference rather than the sum is transmitted.**

The RF received by a transponder is never demodulated and turned into audio. Instead, all the RF it "hears" across its passband goes into a mixer where, much like in the IF (intermediate frequency) section of a receiver, it is mixed with a signal from an oscillator, yielding sum and difference signals.

Figure 8.3 shows a simplified block diagram of a non-inverting linear transponder. For simplicity, I've made this one an up-converting transponder. Incoming signals go into a mixer and get mixed with a 300 MHz signal from a local oscillator, just like in the IF section of a receiver. This produces a sum signal of 445 MHz and a difference signal of 155 MHz. Those both go to a filter, where the 155 MHz signal gets removed and out the signal goes on 445.000 MHz.

If we wanted to make a non-inverting down-converter (and who doesn't!) we'd need two oscillators and two mixers. We'd take the sum from the first mixer and the difference from the second.

If that transponder in Figure 8.3 picks up a signal at 146 MHz, it will send it out at 446.000 MHz.

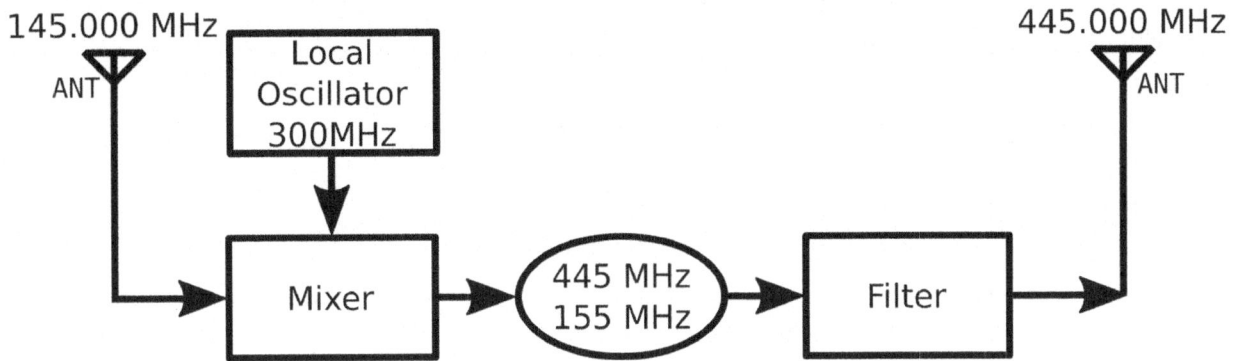

145.000 MHz ANT

Local Oscillator 300MHz

Mixer

445 MHz 155 MHz

Filter

445.000 MHz ANT

Figure 8.3: Non-inverting Linear Transponder

Figure 8.4 shows an inverting linear transponder. Looks familiar, doesn't it? All that has changed is the filter. Rather than filtering out the difference signal and passing the sum signal, now things are reversed. The sum signal of 445 MHz is filtered out and what's left is the difference signal at 155 MHz.

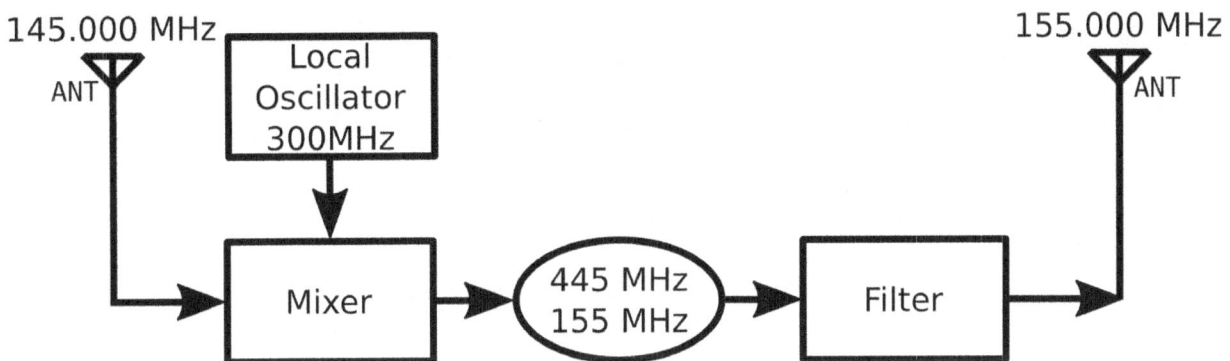

145.000 MHz ANT

Local Oscillator 300MHz

Mixer

445 MHz 155 MHz

Filter

155.000 MHz ANT

Figure 8.4: Inverting Linear Transponder

As you see, our 145 MHz input gives us a 155 MHz output. A 146 MHz input to this inverting transponder would give us a 154 MHz output. Because higher input frequencies produce lower output frequencies and vice versa, single-sideband signals are automatically converted from LSB to USB, or the reverse.

(Yes, 155 MHz is outside the ham bands; we'd fix that by adjusting the oscillator frequency.)

E2A11 (B) What type of antenna can be used to minimize the effects of spin modulation and Faraday rotation?

    A. A linearly polarized antenna

    B. **A circularly polarized antenna**

    C. An isotropic antenna

    D. A log-periodic dipole array

They've thrown some fancy science words into this one, but just remember if a transmitting antenna is spinning, your ideal receiving antenna is **a circularly polarized antenna**. "Spin modulation" is the result of the satellite spinning. Faraday rotation is the rotation of the signal's polarization caused by the magnetic fields present in the ionosphere. (Among Faraday's many discoveries was that a magnetic field could affect the polarization of light – later we'd discover that electromagnetic waves were just another form of light.)

Commercial FM broadcasters use circularly polarized transmitting antennas for a similar reason. Most car aerials are vertical, but some are horizontal, and most table radio antennas are horizontal. How to transmit so both can hear the station well? Circular polarization to the rescue!

E2A06 (A) What are Keplerian elements?

A. **Parameters that define the orbit of a satellite**

B. Phase reversing elements in a Yagi antenna

C. High-emission heater filaments used in magnetron tubes

D. Encrypting codes used for spread spectrum modulation

Johannes Kepler (Fig. 8.5) was the guy who worked out a set of equations to mathematically describe and predict orbital mechanics, so in his honor, we call the set of eight numbers that defines an orbiting body's orbit its **Keplerian elements.** If you really enjoy filling tablets of paper with penciled calculations, you could work out a satellite's location by hand using those elements, but it is a bit more practical to just plug them into one of the calculators you'll find on the web, or just use a satellite locator app on your smart phone.

Figure 8.5: Johannes Kepler

Kepler worked out those laws of planetary motion in about 1605. That was a rather remarkable achievement, considering Newton wouldn't come up with his Law of Universal Gravitation until about 80 years later.

E2A10 (B) What type of satellite appears to stay in one position in the sky?

A. HEO

B. **Geostationary**

C. Geomagnetic

D. LEO

Whenever you watch a cable television network, or listen to satellite radio, it's coming from a **geostationary** satellite. If you orbit a satellite at about 22,000 miles up, its orbital period is 24 hours, so it seems to hang stationary above a single spot on earth; it is *geostationary*.

Since the first edition of this program, amateurs have managed to launch and deploy one geostationary amateur radio satellite, the Es'hail-2, which carries two amateur radio transponders operating in the 2 GHz band for uplinks and the 10 GHz band for downlinks. The satellite's coverage area extends roughly from Brazil, across Africa, the Middle East, and Europe, and east to Thailand.

Despite numerous valiant attempts, we've yet to get a geostationary amateur satellite up that serves the US, but we're getting closer and one may be deployed by the time you read this.

E2A12 (C) What is the purpose of digital store-and-forward functions on an amateur radio satellite?

A. To upload operational software for the transponder

B. To delay download of telemetry between satellites

C. **To store digital messages in the satellite for later download by other stations**

D. To relay messages between satellites

Let's say I want to use a ham satellite to send a message to someone over in Belgium. I wait until the satellite passes overhead, and send up my message. But – we have a small problem. That satellite can't see my friend in Belgium. In fact, it turns out the satellite won't be visible to Belgium until tomorrow afternoon. We need a system that lets the satellite hang onto my message until it is in position to send it to Belgium! That's what digital store-and-forward does. **It stores digital messages in the satellite for later download by other stations.**

E2A13 (B) Which of the following techniques is normally used by low earth orbiting digital satellites to relay messages around the world?

    A. Digipeating

    B. **Store-and-forward**

    C. Multi-satellite relaying

    D. Node hopping

Obviously, this question is just a reverse of the previous question. Low earth orbiting satellites use **store-and-forward** to relay messages around the world.

Key Concepts in This Chapter

- The direction of an ascending pass for an amateur satellite is from south to north.

- As applied to an amateur radio satellite, "mode" refers to the satellite's uplink and downlink frequency bands. The mode is indicated by the letters of the satellite's mode designator.

- If a satellite is operating on the L and S bands, those are the 23 centimeter and 13 centimeter bands.

- If a satellite has a linear transponder, FM, CW, SSB, SSTV, PSK and Packet signals can all be relayed through it.

- Effective radiated power to a satellite which uses a linear transponder should be limited to avoid reducing the downlink power to all other users.

- When a satellite is using an inverted linear transponder, Doppler shift is reduced because the uplink and downlink shifts are in opposite directions. Also, the signal position in the band is reversed, and upper sideband on the uplink becomes lower sideband on the downlink and vice versa.

- The signal in an inverting linear transponder is inverted by passing it through a mixer. The difference rather than the sum is transmitted.

- A circularly polarized antenna can be used to minimize the effects of spin modulation as well as Faraday rotation.

- Keplerian elements are parameters that define the orbit of a satellite.

- Geostationary satellites appear to stay in one position in the sky.

- The purpose of digital store-and-forward functions on an amateur satellite is to store digital messages in the satellite for later download by other stations

Chapter 9

Television Practices

Despite my repeated assertions over the years that it was "just a fad," this television thing seems to be here to stay.

Figure 9.1: Radio with Pictures

Hams have been experimenting with and using television since at least as far back as 1938, long before there were commercial television stations. Broadly speaking, we have two amateur television systems; slow scan and fast scan, commonly known as SSTV and FSTV. Slow scan sends still pictures in the same bandwidth as an SSB phone signal, and operates mostly in the HF bands. Fast scan takes a lot of bandwidth – up to 6 MHz – and is for frequencies from the 70 cm band up. In the early days of amateur television, TV – and especially fast scan TV -- was a truly exotic thing for hams to be able to produce. The camera alone was huge and hair-raisingly expensive, and that was just the start of the equipment you had to purchase, build, and/or scrounge up. Today, we have television cameras, video recorders and even video editing software in our smart phones, making video almost comically easy to create. (Not necessarily good video, mind you, but video nonetheless.)

Aside from the fun of experimenting with the medium, both SSTV and FSTV can be great assets when we assist with communications for public events or when we have the opportunity to be the "eyes on the ground" in a disaster area.

One spectacular use for amateur FSTV that has evolved in the past few years is its use in First Person Viewpoint radio control model flying.

Fast Scan Amateur Television (FSTV)

E2B01 (A) How many times per second is a new frame transmitted in a fast-scan (NTSC) television system?
  A. **30**
  B. 60
  C. 90
  D. 120

The way this question is written, you might get the idea that NTSC is some sort of abbreviation or synonym for fast-scan television, but it isn't. It's the name of a television standard. In the pre-digital TV era, there were three analog television standards in use around the globe. The United States, Japan, Thailand, The Philippines and a few Pacific Island nations used NTSC, known in the biz as NITT-see, which stands for the industry group that created the standard, the National Television Standards Committee.

NTSC transmits **30 frames per second**.

E2B02 (C) How many horizontal lines make up a fast-scan (NTSC) television frame?

    A. 30

    B. 60

    C. **525**

    D. 1080

In the days of the now long-gone cathode ray tube television, the picture was created by a glowing dot that would trace a series of horizontal lines across the screen as the brightness of the dot varied. Because of the speed the dot was moving and the fact that the cathode ray tube's face kept glowing for a bit after the dot had moved on, the illusion of a solid picture was created.

On today's televisions, there's no flying dot, just glowing LCD's or LED's in most cases.

There are **525** horizontal lines in a NTSC fast-scan TV frame. 30 and 60 lines would be far too few to create much of a picture.

Figure 9.2: Michael Faraday at 30 Lines

Figure 9.3: 60 Lines

Figure 9.4: 525 Lines

1080 would be high-definition television, which would require more than double the bandwidth of the **525** line NTSC standard. (See figures 9.2, 9.3, 9.4)

E2B03 (D) How is an interlaced scanning pattern generated in a fast-scan (NTSC) television system?

    A. By scanning two fields simultaneously

    B. By scanning each field from bottom to top

    C. By scanning lines from left to right in one field and right to left in the next

    D. **By scanning odd numbered lines in one field and even numbered lines in the next**

NTSC fast-scan television alternates sending **odd numbered lines and even numbered lines**. Each set of lines is called a "field" and two fields make up a frame. So fast-scan TV is sending 60 fields per second for a frame rate of 30 frames per second.

E2B06 (A) What is vestigial sideband modulation?

    A. **Amplitude modulation in which one complete sideband and a portion of the other are transmitted**

    B. A type of modulation in which one sideband is inverted

    C. Narrow-band FM modulation achieved by filtering one sideband from the audio before frequency modulating the carrier

    D. Spread spectrum modulation achieved by applying FM modulation following single-sideband amplitude modulation

Video signals have some characteristics not usually shared by audio signals. First, they are bandwidth hogs – anything that will reduce bandwidth without destroying quality is in high demand. Second, they contain a lot of low frequency information that is the luminance, or brightness information.

The video signal is transmitted using simple amplitude modulation. As we know from our phone experience, using single sideband is a great way to reduce bandwidth, but it requires a rather expensive and finicky receiver on the other end – as we also know.

The NTSC system compromises between "real" SSB and straight amplitude modulation by using vestigial sideband modulation. That just means one sideband is a "partial" sideband.

To reduce bandwidth, and leave the receiver system fairly simple, the NTSC system filters one sideband to remove the high frequency information, while leaving the other sideband intact. **One complete sideband and a portion of the other are transmitted.** This cuts down the bandwidth of the signal by the amount the filtered sideband is reduced. The carrier is not suppressed, it is transmitted with the full and partial sidebands. In effect, then, the system operates as SSB for the low frequency parts of the signal, but full carrier AM for the less predominant high frequencies. Think of it as "hybrid single-sideband."

E2B05 (C) Which of the following describes the use of vestigial sideband in analog fast-scan TV transmissions?

    A. The vestigial sideband carries the audio information

    B. The vestigial sideband contains chroma information

    C. **Vestigial sideband reduces bandwidth while allowing for simple video detector circuitry**

    D. Vestigial sideband provides high frequency emphasis to sharpen the picture

Because the carrier is not suppressed, the video receiver stage of a television set needs no BFO, nor any product detector. It's basically a simple AM receiver. **Vestigial sideband reduces bandwidth while allowing for simple video detector circuitry.**

E2B07 (B) What is the name of the signal component that carries color information in NTSC video?

    A. Luminance

    B. **Chroma**

    C. Hue

    D. Spectral intensity

In NTSC video signals, the picture information consists of two signals. One contains the brightness information, and that's called the "luminance" signal. The other contains the color information, and that one is called the **chroma** signal.

The actual brightness of any given pixel in a color TV picture is the sum of all the colors. The blue and red components are from the **chroma** signal, and the green is the difference between the luminance and the **chroma**.

E2B08 (A) What technique allows commercial analog TV receivers to be used for fast-scan TV operations on the 70 cm band?

    A. **Transmitting on channels shared with cable TV**

    B. Using converted satellite TV dishes

    C. Transmitting on the abandoned TV channel 2

    D. Using USB and demodulating the signal with a computer sound card

Televisions sold in the U.S. are required to have tuners that will receive both analog and digital signals. It turns out that cable channels 58 through 61 use frequencies in the 70 cm ham band, so hams have been using those frequencies as well as a few cable channel frequencies in the 33 cm band for their analog TV operations.

Slow Scan Amateur Television (SSTV)

Another form of amateur television is slow-scan TV.

When amateur slow scan TV started, back in the 1950's, it took heroic efforts to accomplish. Hams had to invent and make most of the equipment, including printers to capture the image.

Today, we can use our personal computers hooked into our transceivers and with not too much effort send and receive full color SSTV images.

SSTV can be used on any band that permits phone transmissions, from the HF bands clear on up through UHF and beyond. It's easy on bandwidth, too – 3 kHz will do. It's not what you'd call high-speed, though. Figure on around two minutes or more to send or receive a picture.

The International Space Station crews regularly transmit SSTV pictures from orbit, so having the capability to at least receive SSTV can be great fun.

E1A12 (C) What special operating frequency restrictions are imposed on slow scan TV transmissions?

    A. None; they are allowed on all amateur frequencies

    B. They are restricted to 7.245 MHz, 14.245 MHz, 21.345 MHz, and 28.945 MHz

    C. **They are restricted to phone band segments**

    D. They are not permitted above 54 MHz

While you can legally transmit slow-scan TV on any phone band segment, the band plan restricts it to certain frequencies.

HF SSTV Calling Frequencies		
Band	**Frequency**	**Mode**
160 Meters	1.89 MHz	LSB
80 Meters	3.845 MHz (3.730 in Europe)	LSB
40 Meters	7.170 MHz (7.165 in Europe)	LSB
30 Meters	10.132 MHz	USB. Use narrow mode MP73N
20 Meters	14.230 MHz	USB
15 Meters	21.340 MHz	USB
12 Meters	24.975 MHz	USB
10 Meters	28.680 MHz	USB

Table 9.1: HF SSTV Calling Frequencies

E2B11 (B) What is the function of the Vertical Interval Signaling (VIS) code sent as part of an SSTV transmission?
    A. To lock the color burst oscillator in color SSTV images
    B. **To identify the SSTV mode being used**
    C. To provide vertical synchronization
    D. To identify the call sign of the station transmitting

There are around a half-dozen different families of transmission systems, or modes, for SSTV. Most of the families have several different variations of system within the family. For instance, the most popular mode in the US seems to be "S1", for Scottie 1, but the Scottie family also includes S2, S3, S4 and one called DX. (You can guess what that one's for.)

You could waste a lot of time trying to guess which mode a signal was using – and meanwhile, you'd miss the picture. That's why the Vertical Interval Signaling code is sent as part of an SSTV transmission – **to identify the SSTV mode being used**.

SSTV pictures begin with a calibration header. The next thing transmitted, immediately after the calibration header, is the Vertical Interval Signaling code. That code basically contains two digits that tell the receiver the system the sender is using.

E2B10 (A) What aspect of an analog slow-scan television signal encodes the brightness of the picture?
    A. **Tone frequency**
    B. Tone amplitude
    C. Sync amplitude
    D. Sync frequency

Each pixel of an SSTV picture is sent as a series of audio tones, with separate tones for each color. The **tone frequency** indicates the brightness.

E2B04 (A) How is color information sent in analog SSTV?
    A. **Color lines are sent sequentially**
    B. Color information is sent on a 2.8 kHz subcarrier
    C. Color is sent in a color burst at the end of each line
    D. Color is amplitude modulated on the frequency modulated intensity signal

SSTV transmits color pictures by sending three lines of information for each line in the picture; each separate line represents either the red, green, or blue components of the line. The three lines are combined into one to form a single scan line in the final picture.

E2B12 (A) What signals SSTV receiving software to begin a new picture line?

    A. **Specific tone frequencies**

    B. Elapsed time

    C. Specific tone amplitudes

    D. A two-tone signal

Specific tone frequencies are also used to tell the receiver to begin a new line of the picture. Everything in SSTV runs off audio tone frequencies – the brightness, the control signals, and the color. Because of this, SSTV can work on single-sideband just like phone, since so far as the radio is concerned, there's no difference between those tones and your voice. It's all audio.

E2B09 (D) What hardware, other than a receiver with SSB capability and a suitable computer, is needed to decode SSTV using Digital Radio Mondiale (DRM)?

    A. A special IF converter

    B. A special front end limiter

    C. A special notch filter to remove synchronization pulses

    D. **No other hardware is needed**

Digital Radio Mondiale, or DRM – not to be confused with DMR, Digital Mobile Radio – is a system for digital data transmission developed by the non-profit DRM Consortium. It's based on a simple premise: bandwidth is scarce and computer power is now abundant.

DRM uses data compression techniques to pack a lot of information into a little bandwidth – much the same way your computer can shrink a word processor file using WinZip or similar programs, or the way it can shrink a large .wav sound file down to a small .mp3 file. Amateurs are using DRM for slow-scan television, or SSTV.

The software to accomplish this is open source and free, and the only equipment required is your SSB receiver and a suitable computer. **No other hardware is needed**.

E1B02 (A) Which of the following is an acceptable bandwidth for Digital Radio Mondiale (DRM) based voice or SSTV digital transmissions made on the HF amateur bands?

    A. **3 kHz**

    B. 10 kHz

    C. 15 kHz

    D. 20 kHz

The acceptable bandwidth for a DRM based voice or SSTV digital transmission on the HF amateur bands is the same as for any SSB signal – **3 kHz.**

Key Concepts in This Chapter

- In a fast-scan (NTSC) television system, a new frame is transmitted 30 times per second. NTSC is the video standard used by North American Fast Scan Amateur TV stations.

- A fast-scan (NTSC) television frame is made up of 525 lines.

- An interlaced scanning pattern in an NTSC television system is generated by scanning odd numbered lines in one field and even numbered lines in the next.

- Vestigial sideband modulation is amplitude modulation in which one complete sideband and a portion of the other are transmitted. An advantage of using vestigial sideband for standard fast-scan TV transmissions is vestigial sideband reduces bandwidth while allowing for simple video detector circuitry.

- The name of the signal component that carries color information in NTSC video is "chroma."

- Transmitting on channels shared with cable TV allows commercial analog TV receivers to be used for fast-scan TV operations on the 70 cm band.

- Slow scan TV transmissions are restricted to phone band segments and their bandwidth can be no greater than that of a voice signal of the same modulation type.

- The function of Vertical Interval Signaling (VIS) code sent as part of an SSTV transmission is to identify the SSTV mode being used.

- The aspect of an analog slow-scan television signal that encodes the brightness of the picture is the tone frequency.

- To send color information in analog SSTV, color lines are sent sequentially.

- In SSTV, specific tone frequencies signal the SSTV receiving software to begin a new picture line.

- To receive SSTV Digital Radio Mondiale, or DRM, you just need a receiver with SSB capability and a suitable computer. No other hardware is needed.

- DRM based voice or SSTV digital transmissions on the HF bands are limited to the same bandwidth as any SSB signal – 3 kHz.

Once you have mastered the Practice Exam for this chapter, you can challenge yourself with Progress Check #2.

Chapter 10

Contesting & DX'ing

Contesting

E2C02 (A) Which of the following best describes the term "self-spotting" in regards to HF contest operating?

A. **The often prohibited practice of posting one's own call sign and frequency on a spotting network**

B. The acceptable practice of manually posting the call signs of stations on a spotting network

C. A manual technique for rapidly zero beating or tuning to a station's frequency before calling that station

D. An automatic method for rapidly zero beating or tuning to a station's frequency before calling that station

A spotting network is an aid for contesters. It can be run on the Internet or on appropriate digital mode ham networks. The idea is that participants in a contest list the call signs and QTH's of hams they have heard participating in the contest so that others can also work those hams. Getting your call sign added to a spotting network is, then, a boost for your own contest results, so adding your *own* call sign to the network is **generally prohibited** and, in any case, bad form.

E2C03 (A) From which of the following bands is amateur radio contesting generally excluded?

A. **30 meters**

B. 6 meters

C. 2 meters

D. 33 centimeters

The **30 meter** band is one of what are known as the WARC bands. WARC stands for the World Administrative Radio Conference. In 1979, the WARC recommended global allocations of what we now know as the **30 meter** band as well as the 17- and 12-meter bands for amateur radio. Included in the recommendation was the "suggestion" (heh) that due to the narrow bandwidth of these bands they not be used for contesting. The entire **30 meter** band, for instance, is only 50 kHz wide.

This isn't quite the law, but it does show up in some official recommendations. For example, the IARU Region 1 HF Manager's Handbook flatly states, "Contest activity shall not take place on the 10, 18 and 24 MHz bands."

Because of this semi-official prohibition, the WARC bands serve as useful HF frequencies for non-contesters to use during heavy contesting days.

E2C06 (C) During a VHF/UHF contest, in which band segment would you expect to find the highest level of SSB or CW activity?

A. At the top of each band, usually in a segment reserved for contests

B. In the middle of each band, usually on the national calling frequency

C. **In the weak signal segment of the band, with most of the activity near the calling frequency**

D. In the middle of the band, usually 25 kHz above the national calling frequency

Contesting on the VHF/UHF bands isn't about hitting the local repeater, it's about working simplex contacts – sending and receiving on the same frequency – at a distance. So VHF/UHF contesters avoid the local repeaters and their QSO's are usually conducted on frequencies **near the calling frequency.**

E2C07 (A) What is the Cabrillo format?

A. **A standard for submission of electronic contest logs**

B. A method of exchanging information during a contest QSO

C. The most common set of contest rules

D. The rules of order for meetings between contest sponsors

Imagine the complexity of scoring a contest. Say there are a mere 20 operators involved in a contest claiming an average of only five contacts in the contest. That's 200 verifications to perform – "A says they talked to B. Does B say they talked to A?" -- and 200 individual contact scores to calculate and enter to get to the final results.

If you are even a casual contester, you know those example numbers are laughably low. Now imagine the real world, where there might be thousands of hams across the globe participating. Scoring the contest without a computer could be a real nightmare, and even scoring it with a computer could be a nightmare if each entry had to be manually entered in a database.

That's where the Cabrillo format, **a standard for submission of electronic contest logs,** saves the day. It standardizes the reporting of contest contacts so they can go seamlessly – and electronically -- into a computer to generate quicker scoring.

DX'ing

E3A06 (B) What might help to restore contact when DX signals become too weak to copy across an entire HF band a few hours after sunset?

A. Switch to a higher frequency HF band

B. **Switch to a lower frequency HF band**

C. Wait 90 minutes or so for the signal degradation to pass

D. Wait 24 hours before attempting another communication on the band

This is just Ionospheric Propagation 101.

It's a pretty good bet that as we rotate under the ionosphere into approaching night, propagation conditions are going to change. The most likely change is that higher frequencies will shut down and lower frequencies will open up. So what might help restore a contact that has vanished is to **switch to a lower frequency HF band**.

E2C10 (D) Why might a DX station state that they are listening on another frequency?

A. Because the DX station may be transmitting on a frequency that is prohibited to some responding stations

B. To separate the calling stations from the DX station

C. To improve operating efficiency by reducing interference

D. **All these choices are correct**

There are several reasons we might transmit on one frequency and listen on another when we're hunting DX. We might be transmitting **on a frequency that is prohibited to some responding stations.** We might want to **separate the calling stations from the DX station**, so we don't create a pileup on top of the DX station. Or, we might do it **to improve operating efficiency by reducing interference. All the answers for this question are correct.**

E2C11 (A) How should you generally identify your station when attempting to contact a DX station during a contest or in a pileup?

A. **Send your full call sign once or twice**

B. Send only the last two letters of your call sign until you make contact

C. Send your full call sign and grid square

D. Send the call sign of the DX station three times, the words "this is", then your call sign three times

Every year in Washington State we have a QSO party and contest called The Salmon Run. The challenge is to work all the counties in Washington State. While hams from all over participate, it's a pretty big deal in our state's ham community, often involving club gatherings featuring burgers and adult beverages.

The western side of our state is very populated, but the eastern side less so. In fact, there are some counties on that side of the state that contain mostly wheat and rocks – and neither of those is very good at operating a ham radio. Those counties can be challenging to work, but typically a handful of generous souls will spend a day zipping from county to county with their HF mobile rigs, so folks can get those counties worked. (The mobile operators also rack up a ton of points, of course.)

When they fire up and announce their QTH is Douglas County, there's going to be a pileup. Everybody in earshot is going to jump on that frequency at once, creating momentary chaos.

So what's proper for you as the operator trying to add that Douglas County QSO to your score? "Give up" obviously isn't the answer. Neither is some long, drawn out monologue like "send the call sign of the DX station three times, the words 'this is', then your call sign three times." About the time you get to the second repetition of your call sign, everybody listening is going to be thinking, "Well, at least we know who this windbag is."

Just wait for a pause, **send your call sign once or twice**, and listen for a reply. The whole point in a contest atmosphere is to keep things moving along efficiently, so keep it brief.

Confirming Contacts

E2C08 (B) Which of the following contacts may be confirmed through the U.S. QSL bureau system?

A. Special event contacts between stations in the U.S.

B. **Contacts between a U.S. station and a non-U.S. station**

C. Repeater contacts between U.S. club members

D. Contacts using tactical call signs

The QSL bureau system is a system of call-area based bureaus who volunteer to help hams exchange QSL cards at relatively low cost, compared to paying international postage for each individual card. They are clearing houses for inbound and outbound foreign QSL cards.

It's all based on a pretty simple fact – it's cheaper to ship things in bulk, especially overseas. So the bureaus consolidate all the cards that come in for you until they have enough to make an economical shipment, then ship them off to you. A similar system gets your return QSL card to the ham in Argentina. As you can see, you give up speed – often *lots* of speed – for economy in this system.

E2C05 (B) What is the function of a DX QSL Manager?

A. To allocate frequencies for DXpeditions

B. **To handle the receiving and sending of confirmation cards for a DX station**

C. To run a net to allow many stations to contact a rare DX station

D. To relay calls to and from a DX station

For hams who are working a high volume of DX, such as a DXpedition, receiving and sending QSL cards from and to all their contacts can get to be downright burdensome. For those people, there are QSL Managers – folks who provide the service of tracking down the addresses of the contacts, keeping track of received QSL cards, and generally handling the **receiving and sending of confirmation cards for a DX station.**

Key Concepts in This Chapter

- Self-spotting is the often prohibited practice of posting one's own call sign and frequency on a spotting network.

- Amateur radio contesting is generally excluded from the 30 meter band.

- During a VHF/UHF contest, the band segment where you would expect to find the highest level of activity is in the weak signal segment of the band, with most of the activity near the calling frequency.

- The Cabrillo format is a standard for submission of electronic contest logs.

- If DX signals become too weak to copy across an entire HF band a few hours after sunset, it might help to switch to a lower frequency HF band.

- A DX station might state they are listening on another frequency because the DX station may be transmitting on a frequency that is prohibited to some responding stations, to separate the calling stations from the DX station or to improve operating efficiency by reducing interference.

- Generally, when attempting to contact a DX station during a contest or in a pileup, you should send your full call sign once or twice.

- Contacts between a U.S. station and a non-U.S. station may be confirmed through the U.S. QSL bureau system.

- The function of a DX QSL manager is to handle the receiving and sending of confirmation cards for a DX station.

Chapter 11

Operating Methods: VHF/UHF, Part 1

In this chapter and the next we will explore all sorts of propagation and modes for VHF/UHF.

Tropospheric Ducting

Tropospheric ducting is a VHF/UHF phenomenon. It happens as a result of temperature inversions in the troposphere, the lower atmosphere.

If you've ever watched the flight data display on an airline flight, you know the temperature up there at 35,000 feet gets down in the -60° (F) range. Normally, as we go up, the temperature goes down. Every now and then though, a layer of warm air gets sandwiched between a cold layer on the surface and one higher up. So, as we go up, first things get colder, then suddenly warm again, then colder all the way up. That layer of warm air has less density, and thus a different refractive index than the cold air above and below, so it forms a sort of pipeline for radio waves of the correct frequency.

Sometimes, VHF and UHF signals can travel for hundreds of miles in these pipelines, ricocheting from the top to bottom and back to the top, but never getting back down to earth (or out into space) until the end of the duct. That's tropospheric ducting.

E3A10 (B) Which type of atmospheric structure can create a path for microwave propagation?

    A. The jet stream

    B. **Temperature inversion**

    C. Wind shear

    D. Dust devil

The paths – we call them ducts – we use for microwave propagation via tropospheric ducting are created by **temperature inversions**. And, by the way, they don't just work for microwaves – I've experienced some spectacular tropospheric ducting of over 700 miles way down on the commercial FM band, around 90 MHz.

E3A05 (C) Tropospheric propagation of microwave signals often occurs in association with what phenomenon?

    A. Grayline

    B. Lightning discharges

    C. **Warm and cold fronts**

    D. Sprites and jets

Temperature inversions can occur for a number of reasons, but the most common is the collision of **warm fronts and cold fronts**. The less dense warm air floats up on top of the colder air and there's the temperature inversion, and there's the tropospheric propagation.

57

E3A07 (C) Atmospheric ducts capable of propagating microwave signals often form over what geographic feature?
    A. Mountain ranges
    B. Forests
    C. **Bodies of water**
    D. Urban areas

Famous ham radio teacher Gordon West held, and may still hold, the all-time distance record for tropospheric propagation DX – all the way from Southern California to Hawaii through a remarkably stable temperature inversion that had formed between the two locations.

Bodies of water are particularly conducive to temperature inversions because there's nothing there to create turbulence or updrafts that would tend to break up the inversion.

E3A11 (B) What is a typical range for tropospheric propagation of microwave signals?
    A. 10 miles to 50 miles
    B. **100 miles to 300 miles**
    C. 1200 miles
    D. 2500 miles

While the record distance for tropospheric propagation is about 2,600 miles, more typically it's good for **100 to 300 miles**. The high end is most likely to occur over bodies of water.

E3A04 (D) What do Hepburn maps predict?
    A. Sporadic E propagation
    B. Locations of auroral reflecting zones
    C. Likelihood of rain-scatter along cold or warm fronts
    D. **Probability of tropospheric propagation**

Oh, boy, here we go into some of the REALLY arcane stuff that's on the Extra exam.

You'll search in vain through meteorological literature and web sites for the term "Hepburn map." Indeed, you can search the whole Internet for the term and it barely exists. (Although you will learn a lot about Audrey and Katharine Hepburn, so there's that.)

Hepburn maps come from a web site run by one William Hepburn, a Canadian meteorologist and DX listening enthusiast, and are displayed here:

http://www.dxinfocentre.com/tropo.html

You point your browser at the site, and you can instantly see if conditions in your area are favorable for tropospheric ducting.

Auroral Propagation

E3A12 (C) What is the cause of auroral activity?
    A. The interaction in the F2 layer between the solar wind and the Van Allen belt
    B. An extreme low-pressure area in the polar regions
    C. **The interaction in the E layer of charged particles from the Sun with the earth's magnetic field**
    D. Meteor showers concentrated in the extreme northern and southern latitudes

If you place a magnet under some paper and sprinkle on some iron filings, you'll see the magnetic lines of force surrounding the magnet. Since the earth is a big magnet, it has lines of magnetic force surrounding it too, in a pattern that resembles the pattern you'll see with a bar magnet under your paper.

Those lines of force are called the magnetosphere, or just the earth's magnetic field. Most of the time, the magnetosphere guides most of the particles in the solar wind around the earth. The particles' most likely point of entry is the poles, because of the shape of that magnetic field. In the gust of solar wind that follows a coronal mass ejection, the magnetosphere gets distorted. More particles come whistling in. As those particles collide with air molecules, they

58

knock electrons loose, forming ions; charged molecules. When an electron recombines with an ion, a photon is released. Get enough of that going on and the sky lights up with aurora.

E3A13 (A) Which of these emission modes is best for auroral propagation?

    A. **CW**

    B. SSB

    C. FM

    D. RTTY

Of the choices given, if you take into consideration that auroral signals are often distorted and weak, it should seem right to you that **CW** would be the best choice among the answers given. The others would almost certainly be far too garbled and distorted by the interaction with the aurora.

Meteor Scatter

As the earth moves through space, it is constantly encountering chunks of matter. Most of these chunks are the size of a grain of dust, a few are larger. When these chunks enter the atmosphere, they get heated by friction and as they blast through the atmosphere, all that energy leaves a trail of ions that persists, sometimes for fractions of a second, sometimes longer.

As early as 1929, Hantaro Nagaoka, a physicist in Japan, noticed that short bursts of long distance propagation seemed to correlate with meteor activity. This was confirmed in 1944 when James Stanley Hey, researching a system to detect German V2 rockets headed for London, found that signals from 30 MHz to 50 MHz were reflected by meteor trails.

For hams, meteor scatter propagation opens up the possibility of long-distance contacts on VHF frequencies.

E3A08 (A) When a meteor strikes the earth's atmosphere, a cylindrical region of free electrons is formed at what layer of the ionosphere?

    A. **The E layer**

    B. The F1 layer

    C. The F2 layer

    D. The D layer

As you can see in Figure 11.1, when an incoming meteor reaches the atmosphere, it passes first through the F layers of the ionosphere, where air molecules are very few and far between. There's not enough air there to create friction to burn up the meteor. Then comes the E layer. What we call the E layer, meteorologists call the mesosphere – they regard it as a distinct layer, and air is a good deal more dense there, so that incoming meteor gets a lot hotter as it decelerates, and its enormous kinetic energy gets turned into heat energy.

That's where the magic that creates meteor propagation happens. The average meteor enters the atmosphere traveling close to 100,000 feet per *second* relative to the earth. By comparison, a .50 caliber sniper rifle bullet pokes along at a mere 3,000 feet per second. All that kinetic energy compresses the air, quickly heating it to very high temperatures that rip some electrons off the air molecules, and boils minerals off the meteor. That leaves a cylinder of ionized air and mineral gas molecules hanging in the E layer behind the meteor. That cylinder is what we use as a trampoline for our meteor propagation signals.

By the time the meteor makes it down to the D layer – well, in almost every case, it isn't a meteor any more, it's just a bit of ash that has given up all its kinetic energy and will just waft down to earth, joining the tons of other meteor debris that landed that day.

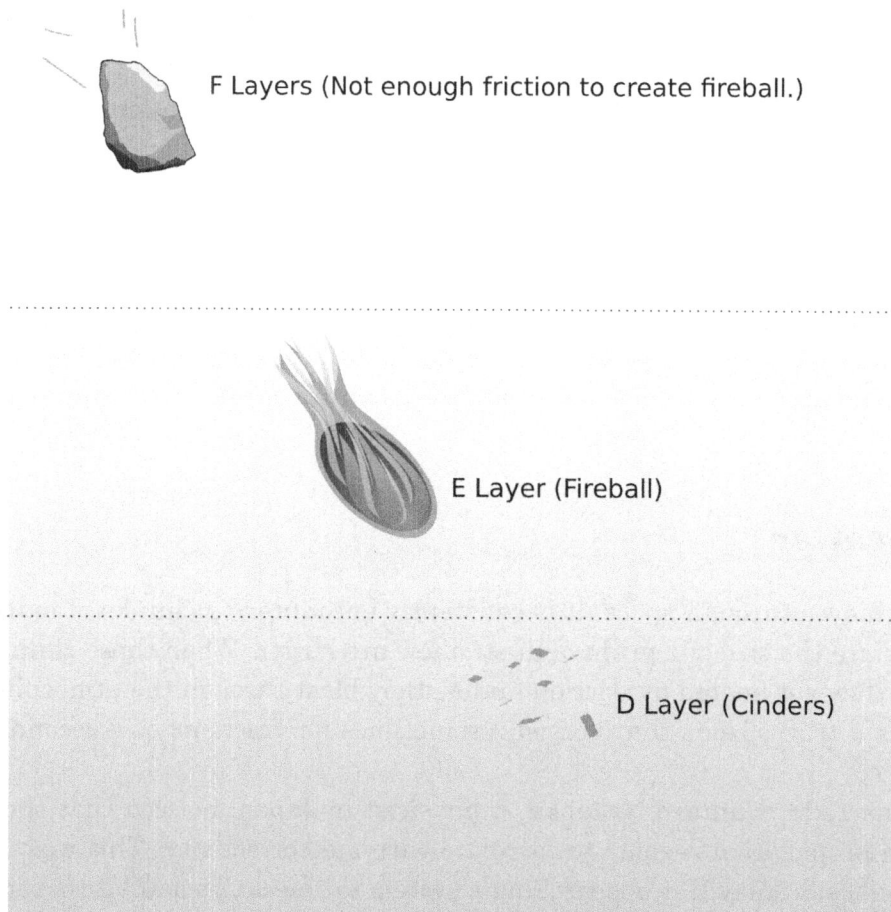

F Layers (Not enough friction to create fireball.)

E Layer (Fireball)

D Layer (Cinders)

Figure 11.1: Meteor Descent

E3A09 (C) Which of the following frequency ranges is most suited for meteor scatter communications?

A. 1.8 MHz - 1.9 MHz
B. 10 MHz - 14 MHz
C. **28 MHz - 148 MHz**
D. 220 MHz - 450 MHz

Almost every propagation challenge comes down to finding the Goldilocks frequency – the wavelength that is just the right frequency to bounce off the target we have in mind. Not so low that the target won't reflect it because the wave's too long, but not so high that it shoots right through the target. We want the frequency that's jusssssssssst right.

As we know, the denser the layer of ionosphere we want to use for skip, the higher the frequency we want to use. During the day, when the D layer is present, sometimes we can use 10 meters to work skip off that. At night, we're more likely to be using the F layer, and lower frequencies.

Think of that cylinder of ions trailing out behind the meteor as a miniature, very dense bit of ionosphere. Yes, it's up in the E layer, but it's going to act like a particularly dense piece of D layer.

In the case of meteor scatter, the Goldilocks frequency generally turns out to be in the very top of the HF range or the low to medium VHF range, from 10 meters up to 2 meters, or **28 MHz – 148 MHz**.

E2D01 (B) Which of the following digital modes is designed for meteor scatter communications?

A. WSPR
B. **MSK144**

C. Hellschreiber

D. APRS

The first practical users of meteor scatter for communications seem to have been the Canadians, who built a system that communicated from a base in Saskatchewan to Toronto by listening for a 90 MHz carrier pointed at the area of the sky where meteors were most likely. That carrier served the same purpose as our beacon stations. When the transmitting station "heard" the carrier, it would send a burst of high-speed data – managed, in those days, by playing back a magnetic tape recording of the data at high speed. It worked! They used the system for about eight years.

Notice the nature of meteor scatter propagation – it occurs in brief bursts. For that reason, working meteor scatter is a game of "get in and get out quickly." We need a system that will fire off quick blasts of data, getting as much message into as little time as possible, with high reliability. A mode known as **MSK144** – **M**inimum **S**hift **K**eying using **144** bit message frames – was designed to do precisely that. In fact, **MSK144** was designed specifically for working meteor scatter.

E2D02 (D) Which of the following is a good technique for making meteor scatter contacts?

    A. 15-second timed transmission sequences with stations alternating based on location

    B. Use of special digital modes

    C. Short transmissions with rapidly repeated call signs and signal reports

    D. **All these choices are correct**

All these answers address the fundamental nature of meteor scatter, which is that it is a short duration phenomenon. **15 second timed transmission sequences with stations alternating based on location** is a useful system. In fact, there is a web site devoted to coordinating such transmissions called pingjockey.net. (Meteor scatter operators refer to those bursts of propagation as pings.) **The use of high-speed CW or digital modes** is, in fact, just about the only way meteor scatter propagation communication *can* work, and **short transmissions with rapidly repeated call signs and signal reports** are the norm. For this question, all the answers are correct.

Key Concepts in This Chapter

- An atmospheric structure that can create a path for microwave propagation is a temperature inversion.

- Tropospheric propagation of microwave signals often occurs along a collision between warm and cold fronts.

- Atmospheric ducts capable of propagating microwave signals often form over bodies of water.

- The typical range for tropospheric propagation of microwave signals is 100 miles to 300 miles.

- Hepburn maps predict the probability of tropospheric propagation.

- The cause of auroral activity is the interaction in the E layer of charged particles from the Sun with the earth's magnetic field.

- Of the choices of CW, SSB, FM and RTTY, CW is the best for aurora propagation.

- When a meteor strikes the earth's atmosphere, a cylindrical region of free electrons is formed at the E layer of the ionosphere.

- The most suitable frequency range for working meteor scatter is 28 MHz to 148 MHz -- 10 meters through 2 meters.

- The digital mode especially designed for use for meteor scatter signals is MSK144.

- Good techniques for making meteor scatter contacts include 15 second timed transmission sequences with stations alternating based on location, use of high-speed CW or digital modes, and short transmissions with rapidly repeated call signs and signal reports.

Chapter 12

Operating Methods: VHF/UHF, Part 2

Earth-Moon-Earth Communications

E3A01 (D) What is the approximate maximum separation measured along the surface of the Earth between two stations communicating by EME?
   A. 500 miles, if the moon is at perigee
   B. 2000 miles, if the moon is at apogee
   C. 5000 miles, if the moon is at perigee
   D. **12,000 miles, if the moon is visible by both stations**

EME stands for earth-moon-earth – it's another way to say moon bounce.

The distance around Earth at the equator is just shy of 25,000 miles, and from the Moon you can see a whole hemisphere of Earth. (See figure 12.1.)

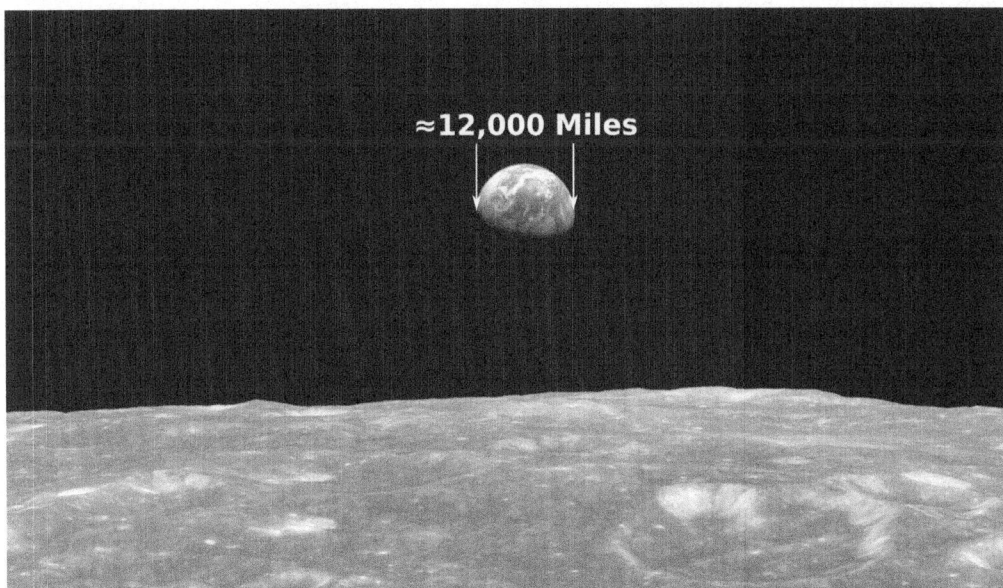

Figure 12.1: Earth Seen From the Moon

Anywhere on the half of the earth that is pointed at the moon can be contacted by moon bounce. Assuming the moon is visible from both stations, they can, at least in theory, communicate by moon bounce, so the maximum separation is about **12,000 miles**, halfway around the earth.

E3A02 (B) What characterizes libration fading of an EME signal?

    A. A slow change in the pitch of the CW signal

    B. **A fluttery irregular fading**

    C. A gradual loss of signal as the sun rises

    D. The returning echo is several hertz lower in frequency than the transmitted signal

We all know the moon goes around the earth about once a month, and that we always see the same side of the moon. (That's because the moon is what they call tidally locked to the earth.)

However, if you stuck a rod in the surface of the moon pointing exactly at the center of the earth, you'd find it wouldn't *quite* always point at the same spot. In fact, we don't just see 50% of the moon. Over the course of a lunar month, we really can see very close to 59% of the moon's surface. Why? The moon, like everything else in orbit, isn't in a perfectly circular orbit, it's in an elliptical one, so it's constantly speeding up or slowing down relative to the earth's rotation. It oscillates, just a little, relative to us. Like if you pointed your nose at a doorknob but slowwwwwwly moved your head just a little bit from side to side.

When we bounce signals off the moon, we're bouncing them off a slowly moving target, and most likely we're bouncing off a spot that's an uneven surface. As that uneven surface moves relative to our antenna, the signal might exhibit a **fluttery, irregular fading**. Remember, we're working with very weak signals here, so it doesn't take much to upset our electromagnetic applecart.

E3A03 (A) When scheduling EME contacts, which of these conditions will generally result in the least path loss?

    A. **When the moon is at perigee**

    B. When the moon is full

    C. When the moon is at apogee

    D. When the MUF is above 30 MHz

Apogee and perigee are the points in the moon's orbit when it is, respectively, farthest from and closest to the earth. Knowing that, it should be pretty obvious that the best time to bounce signals off the moon is **when the moon is at perigee** – its time of closest approach to earth. (See figure 12.2.)

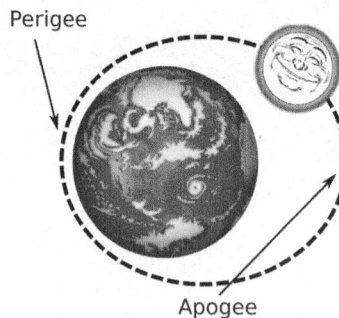

Figure 12.2: Perigee

On average, the moon is about 238,000 miles from earth. But at **perigee** it can come as close as 224,000 miles, shaving 14,000 miles off the outward bound signal's trip, and another 14,000 off the inbound trip.

Some people remember apogee and perigee by remembering that when something is at apogee it is "<u>a</u>far" or "<u>a</u>way."

E2D03 (D) Which of the following digital modes is especially useful for EME communications?

    A. MSK144

    B. PACTOR III

    C. Olivia

    D. **JT65**

The fundamental problem with earth-moon-earth propagation is distance; it's literally "from here to the moon and back!" The average distance to the moon is about 238,000 miles – that's a 476,000 mile round trip. Since signal strength diminishes with distance, and that's a whole lot of distance, the signal loss by the time the signal gets back here to earth will be on the order of 250 to 300 dB.

We need some way of communicating that will work even when the signal is just *almost* getting through, which should be impossible, but that's exactly what **JT65** does. It's a digital mode that throws the parameter of speed out the window in favor of slow but reliable performance in the worst of conditions. Amateur radio operator and, incidentally, Nobel Laureate in Physics, **J**oe **T**aylor created **JT65** specifically to explore EME – earth-moon-earth – propagation.

E2D05 (B) What is one advantage of using the JT65 mode?
　　A. Uses only a 65 Hz bandwidth
　　B. **The ability to decode signals which have a very low signal-to-noise ratio**
　　C. Easily copied by ear if necessary
　　D. Permits fast-scan TV transmissions over narrow bandwidth

Pretty much the whole point of JT65 is that it has **the ability to decode signals which have a very low signal to noise ratio**.

You need a computer to use JT65, but there are apps that let you use your smart phone as a JT65 terminal.

E2D09 (A) What type of modulation is used for JT65 contacts?
　　A. **Multi-tone AFSK**
　　B. PSK
　　C. RTTY
　　D. IEEE 802.11

JT65 uses **multi-tone AFSK** – Audio Frequency Shift Keying. Specifically, it uses a set of 44 different tones. One tone is a synchronization tone, the rest are used to send a limited character set that includes all the letters, numbers, and a bit of punctuation.

E2D06 (A) Which of the following describes a method of establishing EME contacts?
　　A. **Time synchronous transmissions alternately from each station**
　　B. Storing and forwarding digital messages
　　C. Judging optimum transmission times by monitoring beacons reflected from the moon
　　D. High-speed CW identification to avoid fading

Our usual way of doing business – "CQ CQ CQ ...." – is not practical when attempting earth-moon-earth contacts. It takes a bit of collaboration to make it work. The way EME contacts are established is, essentially, everybody agrees that "I'll transmit on such-and-such a frequency at such-and-such a time, for one minute and I'll be listening on such-and-such a frequency at another specific time for one minute while you send. Let's see if this thing works ...." That's a plain language way of saying we use **time synchronous transmissions alternately from each station.**

APRS

E2D04 (C) What technology is used to track, in real time, balloons carrying amateur radio transmitters?
　　A. Ultrasonics
　　B. Bandwidth compressed LORAN
　　C. **APRS**
　　D. Doppler shift of beacon signals

There are many things one could do with **APRS**, the Automatic Packet Reporting System, including track balloons carrying amateur radio transmitters. At the heart of how APRS works is a digital protocol and an informal network of digipeaters, all operating at the same 2-meter frequency. In North America, that frequency is 144.390. The data those digipeaters handle can also be fed to the Internet for global use.

Tracking people or objects is a very common use of **APRS.** Most APRS equipped radios contain a GPS, so location data can be regularly updated. That location data can be plotted on a map, via the internet. Here's a place to check it out if you've never seen it in action:

http://www.openaprs.net/

Some hams have set up weather stations that automatically report local conditions into APRS, and you'll see those pop up on that map, as well as any hams who have their APRS turned on. APRS can also handle messages and you can even use it to send e-mail.

Figure 12.3: APRS Radio Setup

Figure 12.3 shows a nice APRS setup, using a Kenwood receiver hooked to an AvMap G6 GPS that is displaying the APRS data mapped to the area. Some newer model radios, including the latest model of that Kenwood in the picture, include a GPS in the radio – no external unit needed.

E2D10 (C) How can an APRS station be used to help support a public service communications activity?

A. An APRS station with an emergency medical technician can automatically transmit medical data to the nearest hospital

B. APRS stations with General Personnel Scanners can automatically relay the participant numbers and time as they pass the check points

C. **An APRS station with a Global Positioning System unit can automatically transmit information to show a mobile station's position during the event**

D. All these choices are correct

Don't get sidetracked by the answer that says an EMT can automatically transmit medical data to the nearest hospital. While it's true there's actually no technical reason that couldn't be done, there are numerous practical, ethical, and legal reasons that that is just not going to happen.

Keep it simple. Think of APRS as ham radio plus GPS, and that will answer many of the questions on the exam about APRS.

E2D11 (D) Which of the following data are used by the APRS network to communicate your location?

    A. Polar coordinates

    B. Time and frequency

    C. Radio direction finding spectrum analysis

    D. **Latitude and longitude**

A moment's consideration should reveal that the only data type listed here that would be able to define a location is **latitude and longitude**, and, of course, that is what APRS uses.

E2D07 (C) What digital protocol is used by APRS?

    A. PACTOR

    B. 802.11

    C. **AX.25**

    D. AMTOR

This one we probably have to file under the "only need to know it for the test" category, because unless you are deeply steeped in the arcane details of network architecture and communications protocols, and plan to bang out some code to revolutionize the APRS system, you're wildly unlikely to use this knowledge. Just remember APRS uses **AX.25** for its packet transmission. (It's the only answer given that uses the same number of characters as "APRS.")

Put in plain terms, **AX.25** is the amateur radio version of X.25, a 1980's era digital packet protocol that is no longer widely used. It's still out there, though, and your computer probably supports it – good news for us!

There's a reason we use this protocol. APRS needs to accomplish some things that are a bit unusual for computer networks. It needs to communicate with multiple terminals at once. Sometimes it needs to go through a repeater, which requires a slightly different protocol. It needs to carry a lot of different types of data that will be translated differently at the receiver end, so the data must carry information about what protocol to use on that particular packet of information. It's a complex and, frankly, rather brilliant system, and **AX.25** serves the purpose well.

E2D08 (A) What type of packet frame is used to transmit APRS beacon data?

    A. **Unnumbered Information**

    B. Disconnect

    C. Acknowledgement

    D. Connect

The Automatic Packet Reporting System works by sending information in the form of what are called packets.

In a packet system, packets are, metaphorically speaking, carried inside frames. Frames contain all sorts of other data, including the address of the sender, the address of the intended receiver, the protocol used to encode the data, control characters to tell the receiver what to do with the data, etc. Think of the frame as the box the packet is traveling in, and the box has instructions printed on the outside about how to open it and what to do with what's inside.

In packet transmission technology, there are several possible types of frames. Different types of frames carry different types and amounts of data and require (or don't require) different responses from the receiver. The simple version is, there are numbered frames, supervisory frames, unnumbered frames – which don't carry information – and **unnumbered information** frames, which do carry information.

Unnumbered Information frames are simply part of the AX.25 protocol, and they're used to send the APRS beacon data. APRS beacon data is the data your radio sends out to announce – well, whatever you want, once you know how to program your beacon data. It can be your call sign, your location, your name, an announcement about an upcoming club meeting, or just about anything else.

What's significant about **unnumbered information** frames is that under AX.25 they do not require acknowledgement from the receiver. They're just sent out there and if you receive it great, and if not, another one will be along shortly. Because they don't require acknowledgement, they can be sent out to all the receivers listening at once without creating an unworkable blizzard of acknowledgements and repeat requests.

Broadband Mesh Networks

E2C09 (C) What type of equipment is commonly used to implement an amateur radio mesh network?

    A. A 2-meter VHF transceiver with a 1200 baud modem

    B. An optical cable connection between the USB ports of 2 separate computers

    C. **A wireless router running custom firmware**

    D. A 440 MHz transceiver with a 9600 baud modem

One of the first popular Wi-Fi routers was the LinkSys WRT54 series, and that's the weapon of choice for mesh networks.

Chances are good that you have owned or at least seen one of these ubiquitous blue boxes – if you're like me, you have an old one or two stuck in a drawer, "just in case." They're still making the WRT54 line, too, so even if you wanted to start your mesh network with brand new gear, you'd only be out about $40 for the router.

You take that **wireless router**, feed it some new, ham-written firmware so it's **running custom software**, and you have the heart of your own mesh network node.

E2C04 (B) Which of the following frequencies are sometimes used for amateur radio mesh networks?

    A. HF frequencies where digital communications are permitted

    B. **Frequencies shared with various unlicensed wireless data services**

    C. Cable TV channels 41 through 43

    D. The 60 meter band channel centered on 5373 kHz

Part of our 13 cm band overlaps "WiFi" channels 1 through 6, making it very simple to adapt off-the-shelf WiFi routers for amateur radio use – and we can run more power and better antennas!

E2C12 (C) What technique do individual nodes use to form a mesh network?

    A. Forward error correction and Viterbi codes

    B. Acting as store-and-forward digipeaters

    C. **Discovery and link establishment protocols**

    D. Custom code plugs for the local trunking systems

The techniques individual nodes use to form a mesh network are virtually identical to how your laptop forms a network with your home WifFi router; **discovery and link establishment protocols.**

Key Concepts in This Chapter

- The approximate maximum separation measured along the surface of the earth between two stations communicating by moon bounce is 12,000 miles, if the moon is visible by both stations.

- Libration fading of an EME signal is characterized by a fluttery irregular fading.

- When scheduling EME contacts, scheduling them for when the moon is at perigee will generally result in the least path loss.

- A digital mode that is especially useful for EME, earth-moon_earth communications is JT65. JT65 has the ability to decode signals which have a very low signal to noise ratio.

- JT65 uses multi-tone AFSK modulation.

- EME contacts use time synchronous transmissions alternately from each station, alternating transmissions at 1 minute intervals.

- The technology used to track, in real time, balloons carrying amateur radio transmitters is APRS. APRS stations with a GPS unit can automatically transmit latitude and longitude information to show a mobile station's position.

- The digital protocol used by APRS is AX.25. APRS uses Unnumbered Information packet frames to transmit APRS beacon data.

- Mesh networks commonly use a standard wireless router running custom software.

- Frequencies shared with various unlicensed wireless data services are sometimes used for amateur radio mesh networks.

- Individual nodes form a mesh network by using discovery and link establishment protocols.

70

Chapter 13

Operating Methods: Digital Modes for HF

E2E01 (B) Which of the following types of modulation is common for data emissions below 30 MHz?

A. DTMF tones modulating an FM signal

B. **FSK**

C. Pulse modulation

D. Spread spectrum

Of the possible answers, only one type of modulation is common for data emissions below 30 MHz, and that's **FSK**, or Frequency Shift Keying. We don't use FM on the HF bands, nor pulse modulation, and spread spectrum is totally illegal on the HF bands.

E2E04 (A) What is indicated when one of the ellipses in an FSK crossed-ellipse display suddenly disappears?

A. **Selective fading has occurred**

B. One of the signal filters is saturated

C. The receiver has drifted 5 kHz from the desired receive frequency

D. The mark and space signal have been inverted

FSK is "Frequency Shift Keying." One use of FSK is for sending RTTY signals.

Figure 13.1 shows a crossed ellipse display.

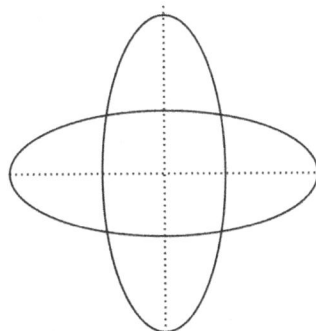

Figure 13.1: Crossed Ellipse Display

It's used to accurately tune an RTTY signal, and is often a feature of RTTY software.

You probably remember the "ones" and "zeroes" of an RTTY signal are called "Mark" and "Space", and one of the ellipses of the crossed ellipse display shows the strength of the Mark signal, while the other shows the strength of the Space signals.

To tune in the signal, you tune until both ellipses are at their maximum. This is the RTTY equivalent of fine-tuning an SSB phone signal until the received voice sounds right. However,

71

if **selective fading** is occurring, and one frequency is not being received as well as the other, one of the ellipses will remain smaller than the other or even disappear, no matter how long you twist on that tuning knob.

What's *selective fading*? It's an ionospheric propagation phenomenon that acts like the received signal was run through a very narrow notch filter. If conditions stack up just right – or wrong, depending on your perspective – one tone of the Mark and Space pair can be more affected than the other even though they are a mere 170 Hz apart.

E2E06 (C) What is the most common data rate used for HF packet?

  A. 48 baud
  B. 110 baud
  C. **300 baud**
  D. 1200 baud

300 baud isn't just "the most common" data rate used for HF packet, it's pretty much the *de facto* standard.

E2E13 (D) Which of these digital modes has the fastest data throughput under clear communication conditions?

  A. AMTOR
  B. 170 Hz shift, 45 baud RTTY
  C. PSK31
  D. **300 baud packet**

Of the modes listed, the fastest data throughput – assuming clear communication conditions – is 300 baud packet.

"Baud" is a measure of the symbol rate of a transmission system. Higher numbers are faster. We'll get more deeply into symbol rate in Chapter 48.

If you look at the wrong answers, you'll quickly eliminate the one that says "170 Hz shift, 45 baud RTTY." 300 baud is pretty obviously faster than 45 baud.

PSK31 is called PSK31 because it operates at 31 baud – lots slower than **300 baud**.

That leaves AMTOR. You may never have seen AMTOR in operation – it's not used much any more. It is robust, which means it is good at getting signals through in poor conditions, and it is quite error resistant – but slow. It is limited to 100 baud.

E2E08 (B) Which of the following HF digital modes can be used to transfer binary files?

  A. Hellschreiber
  B. **PACTOR**
  C. RTTY
  D. AMTOR

I suppose if one wanted to get ridiculously technical, any of the modes listed *could* transfer binary files. You *could* use Hellschreiber to send a fax of a printout of the file or use RTTY to send all the digits. You could transfer a binary file by CW if you had a lot of time on your hands and were willing to send an awfully long string of 1's and 0's very accurately.

However, the only mode listed that is actually designed to *practically* transfer binary files is **PACTOR**.

E2E05 (A) Which of these digital modes does not support keyboard-to-keyboard operation?

  A. **PACTOR**
  B. RTTY
  C. PSK31
  D. MFSK

Think of **PACTOR** as "e-mail over ham radio." When you use PACTOR, it's almost exactly like using e-mail. You compose a message, address it to an e-mail address, and hit Send. Incoming messages come in as complete messages, and you open and read them. The messages can even come to your desktop or mobile phone's e-mail viewer. The messages just happen to have completed at least the initial part of their journey on the ham bands.

The two modes listed in the wrong answers both function more like internet chat rooms. "MFSK" is not a digital mode at all but a type of modulation.

E2E09 (D) Which of the following HF digital modes uses variable-length coding for bandwidth efficiency?

A. RTTY

B. PACTOR

C. MT63

D. **PSK31**

PSK31 is a popular mode for keyboard-to-keyboard DX'ing on the HF bands. Its extraordinarily narrow bandwidth – 31 Hz – lets operators pack all their power into a narrow band, enhancing DX. The 31 Baud speed is a close match to average typing speed.

There's no particular technical reason why the other modes listed couldn't use variable-length coding for bandwidth efficiency, but that coding, known as Varicode, is a central feature of **PSK31** and not of the others. Varicode uses shorter symbol strings to represent the most-often-used characters.

E2E10 (C) Which of these digital modes has the narrowest bandwidth?

A. MFSK16

B. 170 Hz shift, 45 baud RTTY

C. **PSK31**

D. 300-baud packet

Its hard to imagine a radio signal fitting into a narrower bandwidth than 31 Hz, and that's what PSK31 manages. The "16" in MFSK16 does not refer to its bandwidth but to the fact that it uses 16 tones.

E2E11 (A) What is the difference between direct FSK and audio FSK?

A. **Direct FSK applies the data signal to the transmitter VFO, while AFSK transmits tones via phone**

B. Direct FSK occupies less bandwidth

C. Direct FSK can transmit faster baud rates

D. Only direct FSK can be decoded by computer

Direct FSK applies the data signal to the transmitter VFO, causing the carrier to be, in essence, frequency modulated. The difference between FSK and what we usually call Frequency Modulation is that FSK is toggling the carrier between two (or more) discrete frequencies, to signify 1's or 0's.

You won't find a direct FSK button on your HF transceiver. About the only amateur radio systems that use direct FSK are digital voice systems such as D-Star or System Fusion.

E2E12 (A) How do ALE stations establish contact?

A. **ALE constantly scans a list of frequencies, activating the radio when the designated call sign is received**

B. ALE radios monitor an internet site for the frequency they are being paged on

C. ALE radios send a constant tone code to establish a frequency for future use

D. ALE radios activate when they hear their signal echoed by back scatter

Anyone who has ever killed an evening sending CQ CQ CQ into the darkness with no reply but static knows that HF can be, at times, a less than robust means of communication.

ALE, Automatic Link Establishment stations turn HF into a considerably more reliable medium.

ALE is a system that runs from special software on a computer hooked into an HF radio. The radio must be one that can be controlled by the computer; in other words, the computer needs to be able to operate the VFO.

The computer causes the radio to **constantly scan a list of specific frequencies**, listening for calls from other ALE stations. The calls will be in the form of digital transmissions. The computer **activates the radio when the designated call sign** (the operator's) **is received**. (It can also alert the operator to CQ calls.)

It's possible to call specific stations by call sign, or just call CQ and see who is around. It's also possible to only be alerted by the computer when a particular call sign shows up.

Once contact is established with this system, the choice of modes is up to the operator. ALE can be operated as a keyboard-to-keyboard system, or the operators might choose to switch over to CW, SSB, or any other HF mode they choose.

E2E03 (C) How is the timing of FT4 contacts organized?
    A. By exchanging ACK/NAK packets
    B. Stations take turns on alternate days
    C. **Alternating transmissions at 7.5 second intervals**
    D. It depends on the lunar phase

FT4 is the follow-up to the wildly popular FT8 mode. FT8 was made specifically for skywave communication in dismal ionospheric conditions, of which we have had no shortage lately.

FT8 was, and is, difficult to use for contests because its rather terse messages are pre-formatted, of necessity, to, basically, "[Your call sign] [My call sign] [Signal report] [73]. Perfect for making lots of contacts in bad conditions, but different contests require different exchanges. FT4 addresses those issues, and also operates twice as quickly as FT8, because in contesting more contacts in less time is what it's all about.

FT8 operates on 15-second cycles; I transmit for 15 seconds, then I listen for 15 seconds while you transmit. FT4 cuts those cycles in half to 7.5 seconds. (The actual transmission lasts 4.48 seconds, but there's some slack in there for inaccurate clocks.)

E2E07 (D) Which of the following is a possible reason that attempts to initiate contact with a digital station on a clear frequency are unsuccessful?
    A. Your transmit frequency is incorrect
    B. The protocol version you are using is not supported by the digital station
    C. Another station you are unable to hear is using the frequency
    D. **All these choices are correct**

All the possible answers for this one are correct. Of course you won't be able to initiate contact with *any* station if your transmit frequency is incorrect. If the digital station is using a different protocol version than you are – you're out of luck. And it's possible that the digital station you're trying to contact can hear a station that is out of range for you to hear. **All these choices are correct**.

Key Concepts in This Chapter

- A common type of modulation common for data emissions below 30 MHz is FSK.
- If one of the ellipses in an FSK crossed-ellipse display suddenly disappears, it means selective fading has occurred.
- The most common data rate used for HF packet is 300 baud.
- Out of AMTOR, 170 Hz shift, 45 baud RTTY, PSK31, and 300 baud packet, 300 baud packet has the fastest data throughput.
- An HF digital mode that can be used to transfer binary files is PACTOR.
- Of RTTY, PACTOR, PSK31, and MFSK, PACTOR is the one that does not support keyboard-to-keyboard operation.
- An HF digital mode that uses variable-length coding for bandwidth efficiency is PSK31.
- Of MFSK16, RTTY, PSK31, and 300 baud packet, PSK31 has the narrowest bandwidth.
- The difference between direct FSK and audio FSK is direct FSK applies the data signal to the transmitter VFO.

- ALE stations establish contact by constantly scanning a list of frequencies, activating the radio when the designated call sign is received.

- The timing of FT4 contacts is organized around alternating transmissions at 7.5 second intervals.

- There are a number of reasons that attempts to initiate contact with a digital station on a clear frequency might be unsuccessful. Your transmit frequency might be incorrect, the protocol version you are using is not supported by the digital station, or another station you are unable to hear may be using the frequency.

When you have mastered the Practice Exam for this chapter, consider completing Progress Check #3 at fasttrackham.com. If you have been practicing regularly you should do quite well on it.

Chapter 14

Advanced Propagation

Electromagnetic Wave Characteristics

E3A14 (B) What is meant by circularly polarized electromagnetic waves?
    A. Waves with an electric field bent into a circular shape
    B. **Waves with a rotating electric field**
    C. Waves that circle the earth
    D. Waves produced by a loop antenna

Electromagnetic waves are always polarized. Some are vertically polarized, some horizontal, some are somewhere in between, and some are constantly rotating their polarization. Polarization is mostly a function of the transmitting antenna.

Even though polarization is a three-dimensional phenomenon, it's much easier – and just as useful – to think of polarization in two dimensions. Just imagine a piece of paper with a sine wave drawn on it. Hold the paper so it is vertical, with the sine wave running from left to right. That's a vertically polarized wave.

Now rotate the paper away from yourself so it is parallel to the ground. That's a horizontally polarized wave. Finally, spin the paper around and around on the horizontal axis – that's a circularly polarized wave, a wave with a rotating electric field. (The magnetic field is spinning right along with the electric field, but we don't concern ourselves much with that.) We get optimum reception when the polarization of the receiving antenna matches the polarization of the received wave.

There are a few ways we might create **waves with a rotating electric field**.

The easiest to imagine is simply by physically spinning the antenna. That's what happens with a telecommunications satellite antenna as the satellite spins.

More practically, down here on earth, we can construct an antenna in such a way that the signal sort of chases itself around and around the antenna, creating circular polarization.

One that's easy to construct is called a turnstile antenna, so-called simply because it looks a lot like a turnstile. (See figure 14.1.)

In its simplest form there's a radiating element that consists of two dipoles, usually on a vertical mast, mounted at a 90 degree angle to each other and fed 90° out of phase with each other, with a reflecting element mounted ¼ wavelength below the driven element that looks just like the driven element. It radiates horizontally polarized waves parallel to the antenna elements, and circularly polarized waves in the direction of the antenna mast. When constructed for UHF wavelengths, the turnstile is fairly light and easy to move around – perfect for working satellites.

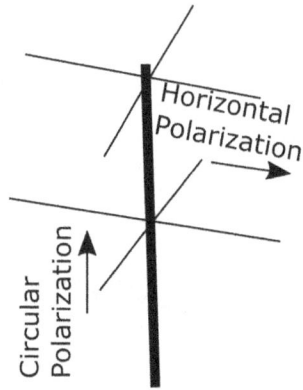

Figure 14.1: Turnstile Antenna

Radio Horizon

E3C06 (A) By how much does the VHF/UHF radio horizon distance exceed the geometric horizon?

A. **By approximately 15 percent of the distance**
B. By approximately twice the distance
C. By approximately 50 percent of the distance
D. By approximately four times the distance

When you first see the Sun at sunrise, you're seeing a refracted image of the Sun. (See figure 14.2, page 78.) The Sun isn't "really" where you see it. You're seeing an image that has been refracted by the atmosphere, bent around the curvature of the earth. If it wasn't for refraction, the Sun would still be below the horizon for a few minutes.

Figure 14.2: Atmospheric Refraction Creates False Sunrise

Since radio waves are just another form of light, they get refracted by the atmosphere, too, meaning our VHF/UHF radios can "see" just a little bit over the horizon, to the tune of **approximately 15 percent of the distance** to the horizon.

E3C14 (D) Why does the radio-path horizon distance exceed the geometric horizon?

A. E-region skip
B. D-region skip
C. Due to the Doppler effect
D. **Downward bending due to density variations in the atmosphere**

This question probably should specify that it applies more to VHF/UHF waves than to HF waves, since the shorter wavelengths of VHF/UHF are more easily refracted by the atmosphere,

78

but the answer, **downward bending due to density variations in the atmosphere** is still valid.

Ground-Wave Propagation

E3C12 (C) How does the maximum range of ground-wave propagation change when the signal frequency is increased?

    A. It stays the same

    B. It increases

    C. **It decreases**

    D. It peaks at roughly 14 MHz

The higher the frequency, the more it tends to travel in a straight line, so when the signal frequency is increased, the maximum distance of ground-wave propagation **decreases**.

What is this "ground-wave?" As your signal travels from your transmitting antenna to the receiving antenna, there are several paths it could take. It could go direct – if the receiving antenna is in sight of the transmitting antenna. If it is a VHF/UHF wave it might bounce off a few hills and buildings, then get to the receiver, and with a little luck it might even get to a receiver slightly over the horizon. An HF wave might bounce off the ionosphere as a sky wave. Or, special case, it might go directly but travel along close to the ground, following the curvature of Earth. If it travels along close to the ground, it will excite other waves in the ground as it passes, and those waves will have the effect of slightly slowing the bottom of the wave, making it tilt ever-so-slightly down. In that way, it can travel right over the horizon and keep going for quite a while. That's a ground wave, and it's useful for medium distance communications at relatively low frequencies, especially when skywave propagation is not available due to conditions.

Ground-wave propagation will change as the terrain changes. Even a good, soaking rain between you and the receiving station can alter the propagation.

Our ham frequencies above the 160-meter band are usually too high for long-distance ground-wave propagation except in extraordinary conditions. (The 80-meter band is often good for medium-distance ground-wave propagation, given an antenna system designed to create a strong ground-wave.)

E3C13 (A) What type of polarization is best for ground-wave propagation?

    A. **Vertical**

    B. Horizontal

    C. Circular

    D. Elliptical

With ground-wave propagation we want something to happen that is usually unwanted. We want our signal to interact with the ground. The downside of this is it suffers losses – it's losing a little energy as it interacts with the ground. The upside is we might be able to shoot a signal over the horizon when that's otherwise impossible.

If you picture a simple model of a horizontally polarized wave – a sine wave drawn on a piece of paper held parallel to the ground – you'll see that wave is never going to run into the ground – it's just going back and forth across the surface. So the polarization that is best for ground-wave propagation is **vertical** polarization. That's why every commercial AM radio transmitting antenna you have ever seen was a vertical antenna.

Transequatorial Propagation

E3B01 (A) What is transequatorial propagation?

A. **Propagation between two mid-latitude points at approximately the same distance north and south of the magnetic equator**

B. Propagation between points located on the magnetic equator

C. Propagation between a point on the equator and its antipodal point

D. Propagation between points at the same latitude

Transequatorial propagation is a special form of propagation that occurs **between two mid-latitude points at approximately the same distance north and south of the magnetic equator.** (See figure 14.3.) It is most likely to affect frequencies in the VHF range.

Figure 14.3: Transequatorial Propagation

The magnetic equator is not quite the same as the usual equator with which we are all familiar, because the magnetic north and south poles are not located at the same spots as the geographic north and south poles. The magnetic equator is offset from the geographic equator.

Transequatorial propagation is still not fully understood, but we do know it most commonly occurs in the late afternoon and evening, and works best in times of high sunspot activity.

Late afternoon/early evening transequatorial propagation – known as TEP – is most often useful on the 6 meter band. Late evening TEP tends to work best on 2 meters and can even open up on the 70-cm band. It also is mostly limited to stations that are equal distances north and south of that magnetic equator, and that are within about 20° of each other relative to a line drawn between the north and south magnetic poles.

E3B02 (C) What is the approximate maximum range for signals using transequatorial propagation?

A. 1000 miles

B. 2500 miles

C. **5000 miles**

D. 7500 miles

The range of afternoon TEP is from 3000 to 4000 miles, evening TEP is from 2000 to **5000 miles**.

E3B03 (C) What is the best time of day for transequatorial propagation?

A. Morning

B. Noon

C. **Afternoon or early evening**

D. Late at night

It rather depends on how one defines "early" evening, but the best time of day for transequatorial propagation is certainly not morning, nor noon, nor late at night, and those are the incorrect answers given. TEP can, under very good conditions, continue until around 11:00 PM, but that's unusual. TEP works best in the **afternoon or early evening**.

Extraordinary & Ordinary Waves

E3B04 (B) What is meant by the terms "extraordinary" and "ordinary" waves?

   A. Extraordinary waves describe rare long-skip propagation compared to ordinary waves, which travel shorter distances

   B. Independent waves created in the ionosphere that are elliptically polarized

   C. Long-path and short-path waves

   D. Refracted rays and reflected waves

The textbook answer to this one is, extraordinary waves are **independent waves created in the ionosphere that are elliptically polarized.** Let's see if we can make some sense out of that answer.

To get our heads around the concepts of extraordinary and ordinary waves, it's probably easiest to take a brief detour into optics.

Some substances are what is known as "birefractive." That means they refract light by different amounts depending on the polarization of the light. Calcite crystals are birefractive. They bend vertically polarized light rays by a different amount than they do horizontally polarized light rays.

When a randomly polarized beam of light passes through a birefractive substance, what comes out the other side is an elliptically polarized beam – simply put, it's stronger on, say, the horizontal axis than the vertical axis even though it retains some rotating polarization. Optics scientists call that altered beam an "extraordinary" beam.

Now let's relate that to radio and the ionosphere. Of course, our radio signals leave our antenna with a particular polarization. Let's say you're using a half-wave horizontal dipole. Your signal leaves the house with horizontal polarization and heads off to the ionosphere. However, that horizontally polarized signal consists of two circularly polarized components. One is right-hand circularly polarized, the other is left-hand circularly polarized. In the realm of terrestrial propagation, this makes no difference to us, because both components of the wave arrive at the horizontal receiving antenna at the same time. Once our signal is passing through the ionosphere on its way to its destination, things change. It turns out the ionosphere is often much like that chunk of calcite – it's birefractive, and radio waves travel through it at slightly different rates depending on whether their circular polarization is left- or right-handed. That means the clockwise spinning component of the wave arrives slightly out of phase with the counter-clockwise spinning component, creating fading.

Since the components are refracted differently, they also end up heading in different directions, and may find paths to a receiver that they wouldn't otherwise; we get some fading, but we also might get some reception that we otherwise wouldn't have had.

E3B07 (C) What happens to linearly polarized radio waves that split into ordinary and extraordinary waves in the ionosphere?

   A. They are bent toward the magnetic poles

   B. They become depolarized

   C. **They become elliptically polarized**

   D. They become phase-locked

There's nothing in particular for us to do about this – we can't adjust our transmitter or antenna to increase or decrease extraordinary waves. They're just a fact of physics, part of the science of propagation.

Long-Path Propagation

E3B05 (C) Which amateur bands typically support long-path propagation?

    A. Only 160 meters to 40 meters

    B. Only 30 meters to 10 meters

    C. **160 meters to 10 meters**

    D. 6 meters to 2 meters

Long-path propagation is propagation via a path 180° from what would be the shortest path to the receiver. For instance, from North America, the short path to India is roughly due North. But sometimes conditions are terrible over the North pole, so we can try pointing our directional antenna due South and try the long-path.

Making a long-path contact is going to involve a lot of "hops." In other words, the signal is typically going to go up to the ionosphere, come back down to Earth, bounce back up, and do that several times to get to the destination. The wavelengths that work best for that are in the range from **160 to 10 meters**.

E3B06 (B) Which of the following amateur bands most frequently provides long-path propagation?

    A. 80 meters

    B. **20 meters**

    C. 10 meters

    D. 6 meters

The *range* of best frequencies for long-path propagation is 160 to 10 meters, but the frequencies that work *most frequently* are in the **20-meter** band.

Sporadic-E Propagation

E3B09 (A) At what time of year is sporadic E propagation most likely to occur?

    A. **Around the solstices, especially the summer solstice**

    B. Around the solstices, especially the winter solstice

    C. Around the equinoxes, especially the spring equinox

    D. Around the equinoxes, especially the fall equinox

Sporadic E propagation, that hit or miss, unpredictable form of propagation that occurs for reasons not at all understood, is most likely to occur **around the solstices, especially the summer solstice**.

In the northern hemisphere, the summer solstice is the longest day of the year – around June 20th – and the winter solstice is the shortest day of the year – around December 21st.

E3B11 (D) At what time of day can sporadic-E propagation occur?

    A. Only around sunset

    B. Only around sunset and sunrise

    C. Only in hours of darkness

    D. **Any time**

Sporadic-E propagation is liable to occur at **any time**. That's why we call it sporadic!

We're probably a long way from fully understanding and being able to predict sporadic-E propagation. There are many theories, ranging from something to do with thunderstorms to ionospheric wind shear to meteor activity. There isn't even universal agreement that sporadic-E is a single phenomenon – there may be several types of sporadic-E.

Chordal Hop Propagation

E3B12 (B) What is the primary characteristic of chordal hop propagation?

A. Propagation away from the great circle bearing between stations

B. Successive ionospheric refractions without an intermediate reflection from the ground

C. Propagation across the geomagnetic equator

D. Signals reflected back toward the transmitting station

In much the same way a temperature inversion can create a tropospheric duct for VHF/UHF signals, sometimes layers of the ionosphere can create similar ducts for HF signals. When that occurs, it can cause what's called chordal hop propagation. **Successive ionospheric reflections** pass the signal along **without an intermediate reflection from the ground**. (See Figure 14.4, page 83.)

Figure 14.4: Normal Sky Wave (left) and Chordal Hop Propagation (right)

"Chord" in this case refers to the geometry term, not the musical one. A chord of a circle is a line that intersects any two points on the circle. On a chordal hop, we get a path like the one shown on the right in Figure 14.4.

E3B10 (A) Why is chordal hop propagation desirable?

A. The signal experiences less loss compared to multi-hop using Earth as a reflector

B. The MUF for chordal hop propagation is much lower than for normal skip propagation

C. Atmospheric noise is lower in the direction of chordal hop propagation

D. Signals travel faster along ionospheric chords

There's nothing magic happening on a chordal path. The desirability of the chordal path is simply a matter of distance. The path a chordal hop signal takes is much, much shorter than the path of a normal skip signal. A chordal hop bypasses all that going back down to earth and going back up to the ionosphere distance, and the inverse square law tells us that every time we cut the distance our signal travels in half it arrives four times more powerful. **The signal experiences less loss along the path compared to normal skip propagation.**

Propagation Software

E3C01 (B) What does the radio communication term "ray tracing" describe?
  A. The process in which an electronic display presents a pattern
  B. **Modeling a radio wave's path through the ionosphere**
  C. Determining the radiation pattern from an array of antennas
  D. Evaluating high voltage sources for x-rays

First, let me say this is really advanced stuff – chances are very good that you're not going to be using ray tracing software to study the ionosphere if you're not at least working on a PhD, or educated in this stuff to that level – a level that is way, WAY out beyond the scope of the Extra license. Obviously, they want you to be aware of it and maybe have a little appreciation of the science, and knowing a little about the "why's" behind ray tracing is useful for us as hams.

If you're familiar with computer graphics programs, you are probably aware of "ray tracing" as a certain way of generating images. This isn't really that. In radio propagation, ray tracing refers to **modeling a radio wave's path through the ionosphere**. You can see an image of some output from a ray tracing program in Figure 14.5.

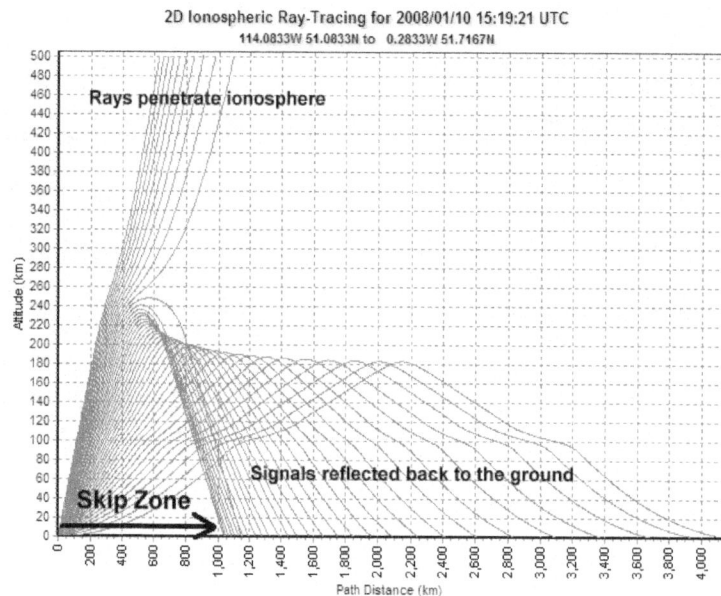

Figure 14.5: Ray Tracing Example

At the General level, we usually teach ionospheric propagation as what I'd call a billiard ball proposition. Radio wave goes up, bounces off the ionosphere like a billiard ball, radio wave goes back to earth. Repeat as necessary to reach the receiver. Simple enough, and not at all what's really going on.

First, radio waves don't usually reflect off the ionosphere, they get refracted. They get bent, then bent again and again until they're headed down out of the ionosphere and back toward earth. (Or not ...) We can imagine that radio wave taking a curved path through the ionosphere.

We can create a crude two-dimensional model of this using a fairly simple equation, but that's a simple two-dimensional model of a complex, three-dimensional process. What if we really want to model what's going on up there so that we can get a deeper understanding of how the ionosphere works, or predict DX possibilities? That gets into some heavy-duty math – integral calculus equations, and lots of them, all interacting with each other. That's why we use ray tracing programs, such as the Italian Geophysics and Volcanology Institute's IONORT software.

E3C11 (C) What does VOACAP software model?

    A. AC voltage and impedance

    B. VHF radio propagation

    C. **HF propagation**

    D. AC current and impedance

As you probably already noticed when you studied for the General exam, ionospheric propagation is a dizzyingly complex and occasionally mysterious topic. Even the parts we fully understand require lots of variables plugged into complex equations to create not-always-reliable predictions of when conditions will be favorable. If we were to really attempt to work it all out with pencil and paper we'd probably never have time to actually switch on the radio and have a conversation.

Happily, people have developed software applications for modeling **HF propagation**.. One of the best known is VOACAP, for Voice of America Coverage Analysis Program. If you're familiar with the Voice of America broadcasting program – now, by the way, almost defunct – you'll understand why the US government would have a keen practical interest in the topic of HF propagation.

The current version of the original VOACAP software itself isn't very current, and has challenges running on current operating systems, but it's available as an online resource, and you can still download the original software if you search around on the web for it. There are other more modern propagation modeling programs out there, but VOACAP's basic engine is at the core of many of them.

VOACAP software is not ray tracing software; it's HF propagation prediction software.

Key Concepts in This Chapter

- Circularly polarized electromagnetic waves are waves with a rotating electric field.

- The VHF/UHF radio horizon distances exceed the geometric horizon by approximately 15 percent of the distance. This is because of downward bending due to density variations in the atmosphere.

- As signal frequency increases, the maximum distance of ground-wave propagation decreases.

- The best type of polarization for ground-wave propagation is vertical polarization.

- Transequatorial propagation is propagation between two mid-latitude points at approximately the same distance north and south of the magnetic equator.

- The approximate maximum range for signals using transequatorial propagation is 5000 miles.

- The best time of day for transequatorial propagation is afternoon or early evening.

- Extraordinary waves are independent waves created in the ionosphere that are elliptically polarized. When linearly polarized radio waves split into ordinary and extraordinary waves in the ionosphere they become elliptically polarized.

- The amateur bands that typically support long-path propagation are those from 160 meters to 10 meters. That band that most frequently provides long-path propagation is the 20-meter band.

- The most likely time of year for Sporadic E propagation is around the solstices, especially the summer solstice. Sporadic E propagation is liable to happen at any time of day.

- The primary characteristic of chordal hop propagation is successive ionospheric reflections without an intermediate reflection from the ground.

- Chordal hop propagation is desirable because the signal experiences less loss along the path compared to normal skip propagation.

- The term ray tracing describes modeling a radio wave's path through the ionosphere.

- VOACAP software models HF propagation.

Chapter 15

Space Weather

E3C10 (B) What does the 304A solar parameter measure?

    A. The ratio of x-ray flux to radio flux, correlated to sunspot number

    B. **UV emissions at 304 angstroms, correlated to solar flux index**

    C. The solar wind velocity at 304 degrees from the solar equator, correlated to solar activity

    D. The solar emission at 304 GHz, correlated to x-ray flare levels

The 304A solar parameter is the measure of **UV** (ultraviolet) **emissions at 304 angstroms, correlated to the solar flux index**. Let's translate that.

The solar flux index was standardized back in the 1950's – it's a measure of the sun's radiation at the frequency of 2800 MHz, a wavelength of 10.7 cm. Think of it as a broad view of the Sun's activity. Very useful; it is one of the main inputs for propagation forecasting software. We have daily historical data going back some 60 years for comparison purposes, so the solar flux index isn't going away. However, it was created before we had satellites. We don't get much of the Sun's UV radiation down here on the surface, and the amount varies wildly with atmospheric conditions, so it is nearly impossible to get an accurate reading on the amount of UV coming from the Sun. But stick a satellite up above the atmosphere, and we can get a super-accurate reading of the activity in the UV (and x-ray) parts of the spectrum, and that turns out to be a more accurate and immediately useful measure than the solar flux index.

The "A" in 304 A stands for angstrom units, equal to one 10,000,000[th] of a millimeter. 304 angstroms is a wavelength of ultraviolet light. If you remember that the "A" stands for angstrom, you'll get the correct answer to this one. It's the only answer that mentions angstroms.

It's relevant to ionospheric propagation because about half of the ionization of the F layer is created by ultraviolet rays, so a higher 304A number means more ionization, which is usually good news for propagation.

E3C02 (A) What is indicated by a rising A or K index?

    A. **Increasing disruption of the geomagnetic field**

    B. Decreasing disruption of the geomagnetic field

    C. Higher levels of solar UV radiation

    D. An increase in the critical frequency

The A and K indexes – or "indices" if you want to be really proper – are measurements of the stability of the earth's geomagnetic field. A rising A or K index indicates **increasing disruption of the geomagnetic field**.

The K-index is measured at various magnetic observatories, so different places report different values. The K-index is, basically, a three-hour snapshot of the state of the magnetic field, so the K-index is a measure of the short-term stability of the earth's magnetic field. All the daily K-index values are combined mathematically to create the daily A-index, so it is a measure of the long-term stability of the magnetic field.

Why do we care about the stability of the field? Simple. Magnetic lines of force don't do much until they are in motion – then, they start transferring energy, creating electrical charges, which affect our radio signals.

K-index values from 0 to 1 indicate a quiet day in the magnetic field. Values of 1 to 5 indicate a minor or moderate solar storm. That's often good news for HF propagation. Values over 5 indicate major storms that severely hamper or black out HF altogether. A 9 would be a genuinely scary event that would wipe out all sorts of electronic devices, not to mention playing havoc with the electrical transmission system.

E3C03 (B) Which of the following signal paths is most likely to experience high levels of absorption when the A index or K index is elevated?
   A. Transequatorial
   B. **Polar**
   C. Sporadic-E
   D. NVIS

Think back to that bar magnet under a piece of paper with the iron filings sprinkled to show the magnetic lines of force. (Figure 15.1)

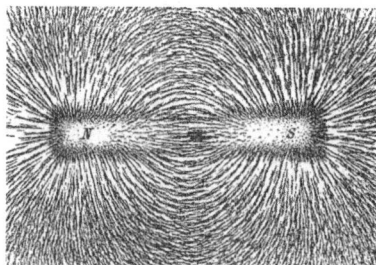

Figure 15.1: Magnetic Lines of Force

The lines of force converge near the poles of the magnet. So do the magnetic lines of force around the earth, and that affects the behavior of the ionosphere at the poles. As you'll recall from last chapter, in a solar storm those lines of force allow particles from the Sun to interact with the ionosphere, creating auroras. The effect of heavy aurora activity is to make transpolar propagation – sending signals over the **polar paths** – more problematic than other paths because the ionosphere is often more absorptive where the auroras occur. That's especially true when the A or K index is elevated, because that's when more aurora activity happens.

E3C04 (C) What does the value of Bz (B sub Z) represent?
   A. Geomagnetic field stability
   B. Critical frequency for vertical transmissions
   C. **Direction and strength of the interplanetary magnetic field**
   D. Duration of long-delayed echoes

The interplanetary magnetic field is mostly the Sun's magnetic field. It gets measured and quantified in several ways. Among those ways are the B indexes – B_X, B_Y, and B_Z. B_X measures the orientation and strength of the field on the X axis of an imaginary three-dimensional grid, with the X axis defined by the plane of the ecliptic – the plane traced by the earth's orbit. B_Y tells us about the Y axis, and B_Z – the one with the most effect on our own magnetosphere's events – shows the **direction and strength of the interplanetary magnetic field** on the Z axis. In other words, B_Z tells us the direction of the North/South axis of the Sun's magnetic field relative to our own field and how strong it is.

The interplanetary magnetic field is normally aligned the same as ours – pointing northward. Unlike earth's field, though, the Sun's can change orientation, especially when there is a lot of sunspot activity.

E3C05 (A) What orientation of Bz (B sub z) increases the likelihood that incoming particles from the sun will cause disturbed conditions?
   A. **Southward**
   B. Northward

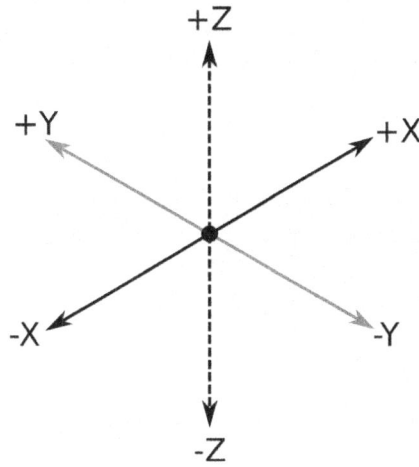

Figure 15.2: 3d Axes

C. Eastward

D. Westward

If you hold a pair of bar magnets so the north poles and south poles are across from each other, they repel, right? But if you reverse them, they attract. If you had your piece of paper and jar of iron filings, you'd see the combined lines of magnetic force of those magnets shift dramatically as you reversed the magnets. The same thing happens with the Sun's magnetic field's interaction with the earth's.

We say the earth's magnetic field points northward, so when the sun's magnetic field points **southward**, the magnets are attracting. Solar wind particles can more easily enter the upper atmosphere, and interact with the ionosphere to affect propagation, whether for good or ill.

E3C07 (D) Which of the following descriptors indicates the greatest solar flare intensity?

A. Class A

B. Class B

C. Class M

D. **Class X**

A solar flare is a stupendous release of energy caused by the sudden collapse of a magnetic line of force surrounding the sun.

Solar flares vary in their magnitude, and, of course, solar astronomers have a scale for those magnitudes.

Imagine you're melting some cheese on the stove for some tasty and nutritious nachos. One little bubble comes up to the surface. It barely breaks the surface, doesn't even go "bloop." That's about like a Class A solar flare.

Another bubble comes up, and it does bloop, rather audibly. It's 10 times more powerful than that A class bubble. That's a Class B.

The wrong answers don't include it, but there's a Class C, too, and it's 10 times more powerful than a B. We're not even close to the top of the scale.

The next stop on the scale is not D, but Class M. A Class M is one serious nacho cheese bubble. It bloops, it even splatters some cheese around the inside of the pot. It might even plop some out on the top of the stove. Class M's are another 10 times more powerful than C's.

You keep heating the cheese. Then you get a mega-bloop. A **Class X**. This thing shoots nacho cheese clear up to the ceiling! Another 10 times more powerful than M's, X's are a big solar deal. Now we're in the territory of a good-sized Coronal Mass Ejection, when a big chunk of "Sun stuff" gets hurled into space, traveling, sometimes, millions of miles per hour. (A NASA web site distinguishes solar flares from CME's which are related but different. They compare the Sun to a cannon; the solar flare is the muzzle flash, the CME is the cannonball.)

After the first level of X, known as X1, the sub-categories indicate how many times more powerful than an X1 the flare is. So an X2 is twice as powerful as an X1, an X3 is three times as powerful, etc. The biggest one that has been measured with this system occurred in 2003, and it overloaded the satellite-based sensors when it blew past X28. Some evidence suggests it eventually peaked at X45.

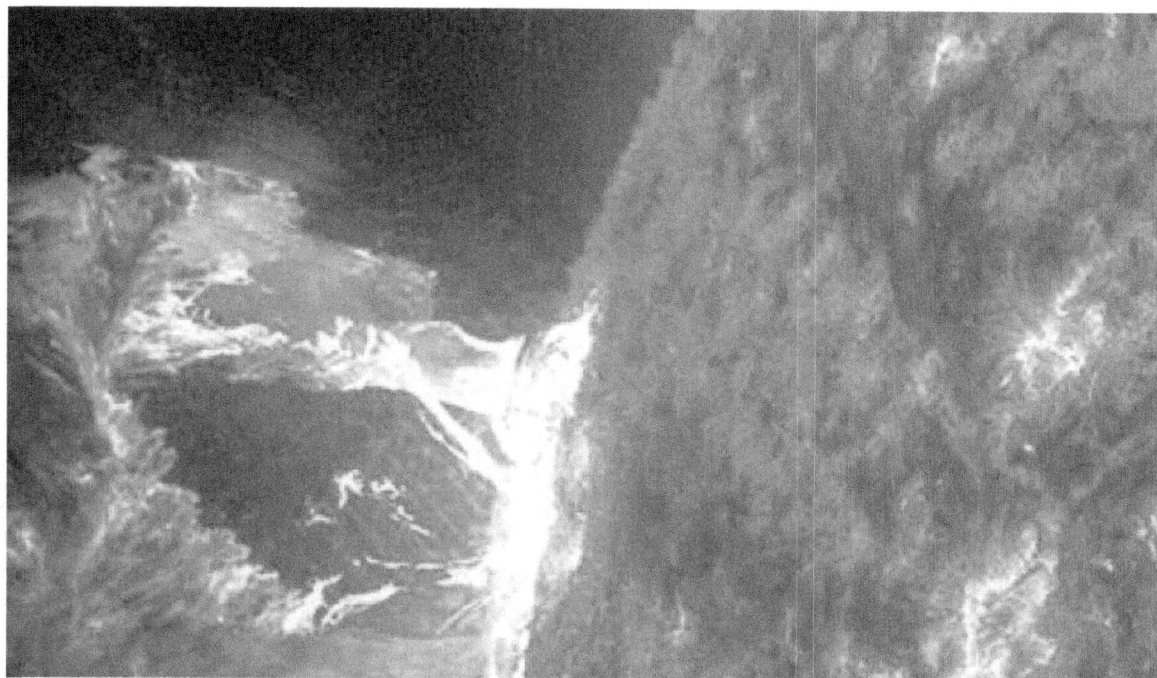

Figure 15.3: M-Class Solar Flare

Enjoy those nachos!

E3C09 (B) How does the intensity of an X3 flare compare to that of an X2 flare?
    A. 10 percent greater
    B. **50 percent greater**
    C. Twice as great
    D. Four times as great

An X3 flare is three times as powerful as an X1, or 50 percent more powerful than an X2.
    That solar flare in 2003 that went beyond X28 was roughly 28 times as powerful as an X1.
    We're very lucky that thing wasn't pointed at us when it went off.

E3C15 (B) What might be indicated by a sudden rise in radio background noise across a large portion of the HF spectrum?
    A. A temperature inversion has occurred
    B. **A solar flare has occurred**
    C. Increased transequatorial propagation is likely
    D. Long-path propagation is likely

"What's that noise in my headphones? It doesn't sound like lightning – is my neighbor running his power drill again? Doesn't sound quite like that, either – just, all of a sudden, I'm hearing tons of static across a bunch of bands... weird."
    What you might be hearing is the effects of a big solar flare that just erupted on the Sun! A sudden rise in radio background noise might indicate **a solar flare has occurred**.
    Man, it's always something, isn't it?

E3C08 (A) What does the space weather term "G5" mean?
    A. **An extreme geomagnetic storm**
    B. Very low solar activity
    C. Moderate solar wind

D. Waning sunspot numbers

The G scale of space weather was developed to be a less obscure way of communicating space weather – and the potential effects – to the general public, without the use of K indexes, A indexes, and the less well-known measures with which we, as hams, are more familiar.

G5 indicates **an extreme geomagnetic storm**. The scale doesn't go any higher. A geomagnetic storm is a temporary disturbance of the Earth's magnetic field, typically associated with solar flares and coronal mass ejections.

A G5 geomagnetic storm corresponds to a K index of 9.

Key Concepts in This Chapter

- The 304A solar parameter measures UV emissions at 304 angstroms, correlated to solar flux index.

- A rising A or K index indicates increasing disruption of the geomagnetic field.

- Polar paths are the signal paths most likely to experience high levels of absorption when the A or K index is elevated.

- The value Bz (B sub Z) represents the direction and strength of the interplanetary magnetic field. If Bz is oriented southward, it increases the likelihood that incoming particles from the Sun will cause disturbed conditions.

- The greatest solar flare activity is designated as Class X.

- The intensity of an X3 solar flare is 50% greater than that of an X2 flare.

- A sudden rise in radio background noise might indicate a solar flare has occurred.

- The space weather term G5 means an extreme geomagnetic storm.

Chapter 16

Test Equipment

Oscilloscopes

E4A01 (A) Which of the following limits the highest frequency signal that can be accurately displayed on a digital oscilloscope?
   A. **Sampling rate of the analog-to-digital converter**
   B. Amount of memory
   C. Q of the circuit
   D. All these choices are correct

The **sampling rate of the analog-to-digital converter** is how often per second the digital oscilloscope measures the waveform. Let's use a non-oscilloscope example. When I record audio in my studio, I use a sample rate that's the same as CD's use: 44.1 kHz. So 44,100 times per second, the computer takes a look at the audio that's present, measures it, and records that as a number. That's the analog-to-digital conversion. When the audio is played back, 44,100 times per second the computer reads a number, then raises or lowers the voltage in the sound generating circuit to match that number and the sound gets generated. That's digital-to-analog conversion.

A digital oscilloscope cannot accurately represent a frequency beyond its sampling rate. In fact, it cannot represent a frequency higher than one-half its sample rate. So if you want to look at a 6-meter band signal, you would need an oscilloscope with a sampling rate up around at least 100 MHz. Happily, these once exotic instruments have become much more affordable. The digital 'scope pictured in Figure 16.1 is a 200 MHz machine and costs about $300 from Chinese suppliers.

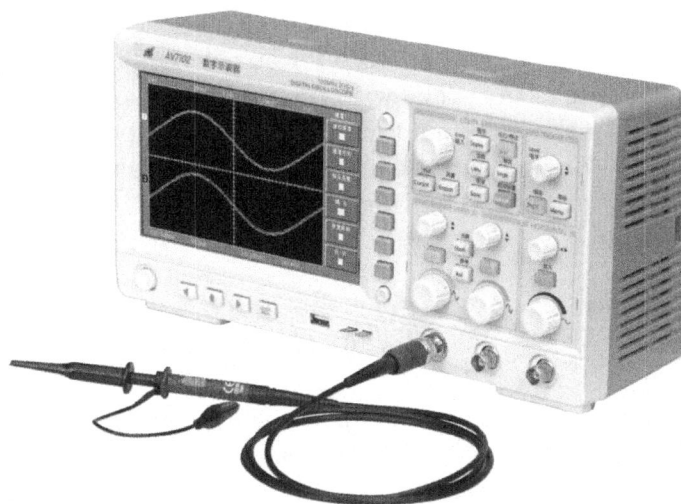

Figure 16.1: Digital Oscilloscope

E4A06 (A) What is the effect of aliasing on a digital oscilloscope caused by setting the time base too slow?

A. **A false, jittery low-frequency version of the signal is displayed**

B. All signals will have a DC offset

C. Calibration of the vertical scale is no longer valid

D. Excessive blanking occurs, which prevents display of the signal.

Bad Things happen if a digital system tries to sample a signal that is a higher frequency than the sampling rate which, in a digital oscilloscope, is set by setting the "time base." The bad thing that happens is "aliasing." We'll go into more depth on what, precisely, aliasing is in Chapter 44.

When aliasing occurs, **false signals are displayed**.

Quality instruments incorporate sophisticated anti-aliasing algorithms, but if you're seeing something on your digital 'scope that just doesn't make sense, it's worth asking yourself if you're just seeing some aliasing. Try reducing the input level – overloading digital circuits is another cause of aliasing.

E4A09 (A) Which of the following is good practice when using an oscilloscope probe?

A. **Keep the signal ground connection of the probe as short as possible**

B. Never use a high-impedance probe to measure a low-impedance circuit

C. Never use a DC-coupled probe to measure an AC circuit

D. All these choices are correct

There are at least two good reasons to **keep the signal ground connection of the probe as short as possible** when using an oscilloscope.

First, that ground wire can act as an antenna, introducing unwanted noise into your measurement.

Second, that ground wire acts as an inductor, especially at the radio frequencies we're usually measuring with our oscilloscopes. Effectively, you're adding an unknown inductance into the circuit you're trying to measure, so it makes sense to make that inductance as low a value as possible by making the ground connection as short as possible.

E4A04 (A) How is the compensation of an oscilloscope probe typically adjusted?

A. **A square wave is displayed and the probe is adjusted until the horizontal portions of the displayed wave are as nearly flat as possible**

B. A high frequency sine wave is displayed and the probe is adjusted for maximum amplitude

C. A frequency standard is displayed and the probe is adjusted until the deflection time is accurate

D. A DC voltage standard is displayed and the probe is adjusted until the displayed voltage is accurate

When we want to see an accurate picture of a waveform on an oscilloscope, we must "compensate the probe." Compensation is just another word for our old friend "impedance matching." We want to match the impedance of the probe with the impedance of the oscilloscope.

To do that, we need a square wave generator, which is so important to oscilloscope use that many 'scopes include one. We hook up the probe to the square wave generator, and adjust a variable capacitor in the probe until **the horizontal portions of the displayed wave are as nearly flat as possible.**

Spectrum Analyzers

E4A02 (B) Which of the following parameters does a spectrum analyzer display on the vertical and horizontal axes?

A. RF amplitude and time

B. **RF amplitude and frequency**

C. SWR and frequency

D. SWR and time

Before it is switched on, a spectrum analyzer looks a lot like an oscilloscope, but it performs a totally different function.

An oscilloscope graphs time across the X axis and amplitude on the Y axis. To use terminology that may show up on the exam, an oscilloscope represents a signal in the *time domain*. You can measure frequency with it, but only one frequency at a time. (You measure the frequency using a calibrated scale on the screen – or just jigger with the horizontal sweep speed until you're seeing one cycle and then read what the knob points at if you're not all that into precision. Not that I've ever done such a thing.)

A spectrum analyzer graphs *frequency* across the X axis and amplitude on the Y axis. It represents the signal in the *frequency domain*. What you see is a graphic representation of the spectrum of the frequencies being measured. There are spectrum analyzers that operate in the audio frequency range, but in this question they assume we're talking about an RF spectrum analyzer, so the correct answer is **RF amplitude and frequency**.

E4A03 (B) Which of the following test instruments is used to display spurious signals and/or intermodulation distortion products generated by an SSB transmitter?

    A. A wattmeter

    B. **A spectrum analyzer**

    C. A logic analyzer

    D. A time-domain reflectometer

Of the instruments listed, only a **spectrum analyzer** will display spurious signals and/or intermodulation distortion products in an SSB transmitter – or any other type of transmitter, for that matter.

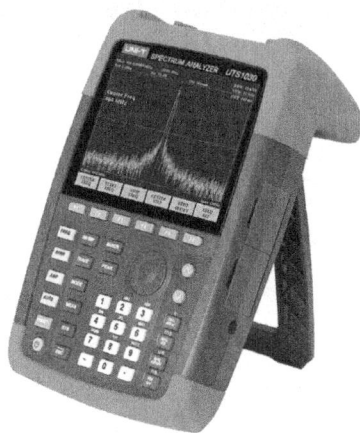

Figure 16.2: Handheld Portable Spectrum Analyzer

It's pretty simple to use, too, once you know the ins and outs of getting your transmitter's signal into the spectrum analyzer.

You hook it up, switch on the transmitter, and hit the push-to-talk button. Instantly, you'll have a display of the spectrum of the output of your transmitter. Getting one nice solid bar right on the carrier frequency? Nice! Go ahead and modulate the carrier using a two-tone generator that creates two non-harmonically related audio tones, keeping an eye on the spectrum analyzer. Wow, you can see the sideband! Oh ... you're also getting some other bars that show your transmitter is radiating strongly on multiple frequencies? There's yer problem, right there – those are spurious emissions, probably the result of intermodulation distortion. Something in your transmitter is non-linear – what goes in is not equaling what comes out.

Now you start tracing back through the stages of the transmitter with the spectrum analyzer, which takes some adjusting of how you're hooking it up.

In ancient times, before spectrum analyzers, we still could measure intermodulation distortion and spurious emissions, but it involved a lot of filters, calibration and a good oscilloscope. Spectrum analyzers take about an hour's worth of work and turn it into a few minutes worth of work.

Spectrum analyzers are typically a bit more expensive than oscilloscopes. For a quality stand-alone unit, with its own display, figure $500 to $1000, depending mostly on the frequency range. For a unit that plugs into your laptop and uses it for the display, though, you can find them for under $300.

E4B10 (B) Which of the following methods measures intermodulation distortion in an SSB transmitter?

A. Modulate the transmitter using two RF signals having non-harmonically related frequencies and observe the RF output with a spectrum analyzer

B. **Modulate the transmitter using two AF signals having non-harmonically related frequencies and observe the RF output with a spectrum analyzer**

C. Modulate the transmitter using two AF signals having harmonically related frequencies and observe the RF output with a peak reading wattmeter

D. Modulate the transmitter using two RF signals having harmonically related frequencies and observe the RF output with a logic analyzer

We'll cover intermodulation distortion in much more depth beginning on page 123. For now, intermodulation distortion is a form of distortion caused by the mixing of two or more frequencies, resulting in unwanted frequencies being added to the original signal.

Two of the answers to this one look very much alike, but one is correct while the other is dead wrong, so read them carefully when you're taking the exam.

To measure intermodulation distortion in an SSB transmitter, we **modulate the transmitter with two non-harmonically related *audio* frequencies and observe the RF output with a spectrum analyzer.**

Antenna Analyzers

E4A08 (D)

Which of the following measures SWR?

A. A spectrum analyzer

B. A Q meter

C. An ohmmeter

D. **An antenna analyzer**

Of all the test instruments hams have available to them, it looks to me like the second most popular, right behind the multimeter, is the **antenna analyzer**. By this point in your ham career you have almost certainly at least seen someone operate one of these handy devices. One is pictured in Figure 16.3.

Antenna analyzers solve tons of antenna questions instantly. You plug in your antenna, set the analyzer to the frequency you choose, and you get an instant readout of the SWR, the impedance, and whether the reactance is balanced or more capacitive or inductive; that tells you if the antenna is too long or too short. By tuning up and down you can determine the bandwidth of your antenna. Some will even double as a frequency meter, and with some you can even read the length of your coax. They are huge time savers and let you fine tune your antenna's performance for the absolute maximum.

E4A07 (B) Which of the following is an advantage of using an antenna analyzer compared to an SWR bridge to measure antenna SWR?

A. Antenna analyzers automatically tune your antenna for resonance

B. **Antenna analyzers do not need an external RF source**

C. Antenna analyzers display a time-varying representation of the modulation envelope

D. All these choices are correct

Antenna analyzers are free-standing, portable units. **They do not need an external RF source**. No other instruments needed – if you can hook the antenna to it (you might need an adapter) you can measure it. The analyzer contains its own very low power transmitter and analyzes the antenna by analyzing the behavior of that signal. Analyzers are versatile, but two things they will *not* do are listed as incorrect answers. They won't automatically tune your antenna for resonance, and they can't show you anything about the modulation envelope because they're not connected to your transmitter.

Figure 16.3: MFJ 269-C Antenna Analyzer

E4A11 (D) How should an antenna analyzer be connected when measuring antenna resonance and feed point impedance?

A. Loosely couple the analyzer near the antenna base

B. Connect the analyzer via a high-impedance transformer to the antenna

C. Loosely couple the antenna and a dummy load to the analyzer

D. **Connect the antenna feed line directly to the analyzer's connector**

This same question appeared in your General exam question pool, but this time they've concocted some fancier sounding wrong answers. The correct answer is still **connect the antenna feed line directly to the analyzer's connector**.

Logic Analyzers

E4A10 (D) Which of the following displays multiple digital signal states simultaneously?

A. Network analyzer

B. Bit error rate tester

C. Modulation monitor

D. **Logic analyzer**

Logic analyzers are fancy oscilloscopes with superpowers that make them useful for digital circuits. There are many different types in price ranges that go from not-too-horrible to "I could buy a car for that much money!" Some digital oscilloscopes also offer some logic analyzer functions. As with oscilloscopes and spectrum analyzers, there are also "laptop" logic analyzers that plug into your computer and that cost considerably less than the more powerful stand-alone units.

Logic analyzers display multiple digital signal states, and, depending on the particular analyzer, might display those states in a number of ways for different applications. Some, for instance, will read out the "assembly language" version of the output of a chip. (Assembly language is one step away from the computer's true native tongue, machine language.) Some will test and display the timing of various signal streams, so you can see what is or is not getting sent to where it's supposed to be when it's supposed to be there.

A logic analyzer is not an instrument most of us will ever need, unless we get into designing and testing digital circuits – then, a logic analyzer is right on the verge of absolutely necessary.

Vector Network Analyzers

E4B11 (D) Which of the following can be measured with a vector network analyzer?
  A. Input impedance
  B. Output impedance
  C. Reflection coefficient
  D. **All these choices are correct**

Really, a vector network analyzer is a more sophisticated form of our antenna analyzers – it's looking at forward and reflected power into and out of all the inputs and outputs of the network. From that it can determine **input impedance**, **output impedance**, and the **reflection coefficient**, which is another way of expressing SWR.

Vector network analyzers (sometimes known as network vector analyzers) are all about measuring S parameters.

E4B07 (A) What do the subscripts of S parameters represent?
  A. **The port or ports at which measurements are made**
  B. The relative time between measurements
  C. Relative quality of the data
  D. Frequency order of the measurements

The question pool contains three questions on S parameters. S parameters are used to describe the behavior of a network when a steady state signal is applied to that network. You could use them for any network, but they're most useful for networks handling microwave frequencies. We're going to take a very light look at S parameters, so you can get the general idea of what they are.

Here's the deal on why we use S parameters. With a DC or low-frequency circuit, we can hook up a voltage or signal source, grab our handy test instrument of choice, and measure what's going on quite accurately. We can vary voltages and frequencies on the input side, and trace through the circuit with, for instance, our multimeter or oscilloscope probe, and know what's going on in our circuit.

Once we get to very short wavelengths – say, in the millimeters – that system goes, to put it technically, totally kaflooey. Now the physical sizes of the components and the lengths of the interconnections between them become very, very important. At these frequencies, just the test leads of our instrument are so gigantically long that they, alone, render our measurements useless, never mind what unknown havoc is wreaked by the circuitry of the test instrument itself.

Well, okay, but we can figure out what's going on from the schematic diagram, right? Nope. Schematics don't give us any dimensions of anything – just what component gets hooked up to what other components through which path or paths.

Okay, then, the blueprints – we can just look at the blueprints, get all the component values and the measurements, and do some math, right? Well ... not really. There are interactions with interactions that then interact with the other interactions and – pretty soon we run out of math before we get anything useful. It's chaos in there!

The thinking behind S parameters goes like this: "Hey, who really cares what's going on *inside* the circuit? What we really care about is what comes out compared to what goes in." So we hook up some jazzy test gear called a vector network analyzer and start firing signals into our circuit and measuring what happens. From those measurements, we end up with a bunch of values that are going to boil down to mathematical formulas – what mathematicians call complex numbers.

Let's say our friend, Bob, who has a lot of time on his hands, has made a box. The box is on legs so it sits at a 45 degree angle. There is a hole in the top of the box that says, "Marble Input." There's another hole in the bottom of the box that says "Marble Output." The box is sealed – we can't see what's inside. Bob even lined the inside with lead foil so we can't x-ray the box. (Bob's a little funny in the head.)

Bob asks us, "Hey, can you figure out the performance of this box for me?" Since we are helpful and community service minded hams, we tell him we'll see what we can do.

We can't look inside the box, but we do happen to have a lot of marbles on hand, so we start putting marbles into Marble Input, and we count what happens at the output. It turns out that when we feed in 100 marbles, 60 marbles come out of the output, 35 never come out, and 5, rather surprisingly, bounce right back out of the input.

Those results are the S parameters for Bob's Box. The holes in the box are called "ports" when we talk about S parameters. The Marble Input is port 1, the Output is port 2.

We could create a matrix of our results, with rows and columns, and the concept of a matrix is key to using S parameters. We could number each result by listing the port numbers – we could, for example, say S_{21}, pronounced "Ess-two-one," represents what happened at port 2 when we put marbles into port 1. S_{11} would be what happened at port 1 when we put marbles into port 1. An antenna analyzer is measuring S_{11}.

We'd look at our matrix and could see that Bob's Box has a "forward gain" (or in this case a loss) of 40 marbles, so S_{21} equals 60. At the input, we got five reflected marbles, so S11 equals 5.

So we have a nice little matrix of numbers. All well and good, but our numbers get much, much more useful if we turn those numbers into formulas: we could say we have forward gain of Input x 0.60, and reflected marbles of Input x 0.05.

Using S parameters for electronic circuits is, at its heart, the same process, only we're using signals instead of marbles, we use more measurements, the connections are a bit more complicated, and the formulas in the matrix tend to have a lot more elements.

(We won't go diving into the wonders of linear algebra and matrix computations, but that's one way S parameters get used. They also can be plotted on Smith charts, about which more later.)

E4B03 (C) Which S parameter is equivalent to forward gain?

    A. S11

    B. S12

    C. **S21**

    D. S22

The subscripts of S parameters indicate the "port numbers." They aren't "S11" or "S22", they're "S 1 – 1" and "S 2 – 2." Those subscripts have standard meanings, so, for instance, the standard subscript for forward gain is S_{21}. That's shorthand for "what happened at port 2 when I put a signal into port 1?"

In a two port network, there are four possible parameters; S_{11}, S_{12}, S_{21} and S_{22}.

E4B04 (A) Which S parameter represents input port return loss or reflection coefficient (equivalent to VSWR)?

    A. **S11**

    B. S12

    C. S21

    D. S22

S_{11} and S_{22} are the reflected power parameters. S_{22} is "reverse reflected power", S_{11} is "forward reflected power" – known to us hams as SWR, for Standing Wave Ratio or, more formally, as

VSWR for Voltage Standing Wave Ratio.

Figure 16.4 shows a tiny vector network analyzer. If you look at the left side of it you can see one of the inputs (Channel 0) is marked S_{11}, and that's the one I'm using to measure an antenna's SWR across a band of frequencies.

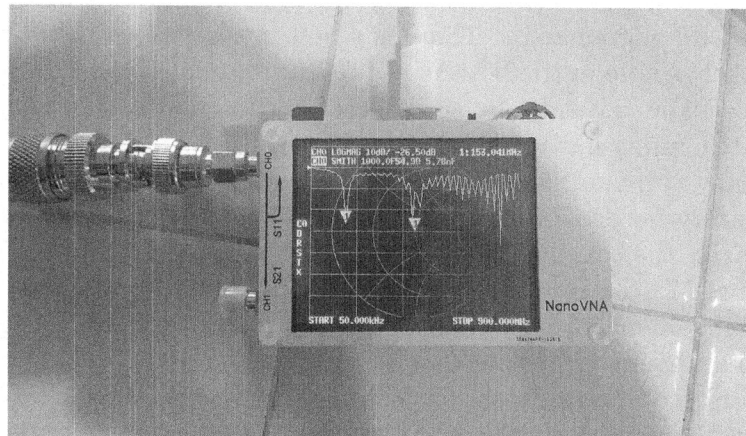

Figure 16.4: NanoVNA vector network analyzer

For S parameters, remember three key facts: The subscripts are the port numbers, S_{21} is forward gain and S_{11} is SWR.

E4B05 (B) What three test loads are used to calibrate an RF vector network analyzer?
    A. 50 ohms, 75 ohms, and 90 ohms
    B. **Short circuit, open circuit, and 50 ohms**
    C. Short circuit, open circuit, and resonant circuit
    D. 50 ohms through 1/8 wavelength, 1/4 wavelength, and 1/2 wavelength of coaxial cable

Vector network analyzers need to be calibrated each time they're used, and the very condensed version of that calibration procedure is that they are tested into a **short circuit, an open circuit, and 50 ohms**. (The actual calibration instructions for some analyzers run about 10 pages, so be glad we got the short version!)

Frequency Counters

E4A05 (D) What is the purpose of the prescaler function on a frequency counter?
    A. It amplifies low level signals for more accurate counting
    B. It multiplies a higher frequency signal so a low-frequency counter can display the operating frequency
    C. It prevents oscillation in a low-frequency counter circuit
    D. **It divides a higher frequency signal so a low-frequency counter can display the input frequency**

We use frequency counters to measure the frequency a particular piece of equipment or circuit is generating. They're often stand-alone pieces of equipment, but some multi-meters include a somewhat limited frequency counter function.

Building frequency counters that measure very high frequencies directly would be painfully expensive and unnecessary. It's easy to build a *prescaler* that **divides a higher frequency signal so a low-frequency counter can display the input frequency**. You'll learn how it's done when we cover flip-flops and decade counters in the chapter on digital logic.

A typical prescaler divides the input frequency by powers of 10, such as 10, 100, 1000, etc. That's how you can buy a frequency counter that will measure up to 2.4 GHz for 40 bucks instead of a few thousand bucks.

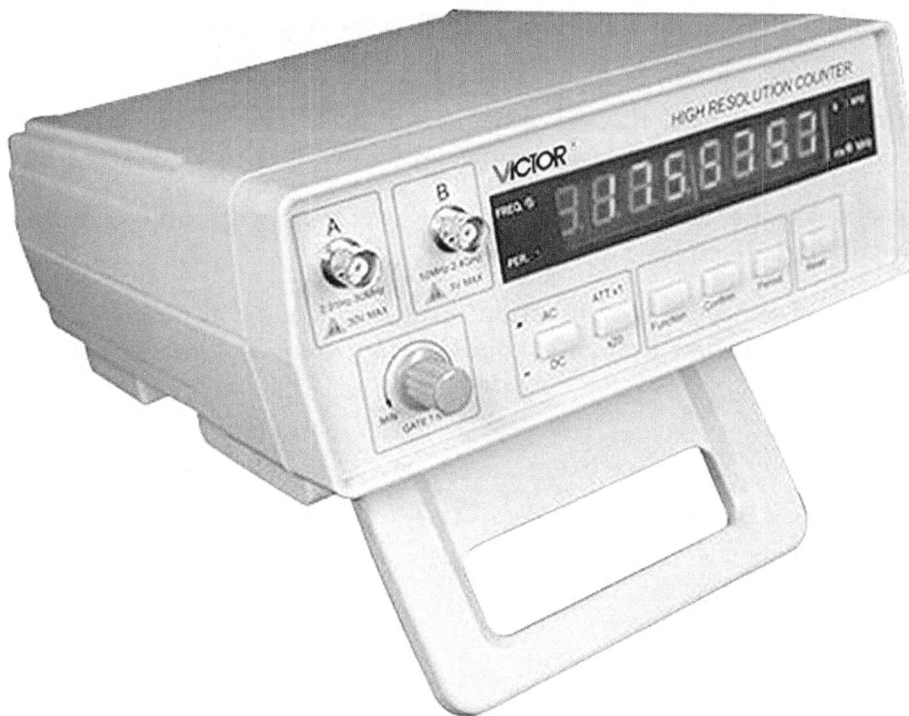

Figure 16.5: Frequency Counter

Key Concepts in This Chapter

- What limits the highest frequency signal that can be accurately displayed on a digital oscilloscope is the sampling rate of the analog-to-digital converter.

- The effect of aliasing on a digital oscilloscope caused by setting the time base too slow is that a false, jittery low-frequency version of the signal is displayed.

- Good practice when using an oscilloscope probe is to keep the signal ground connection of the probe as short as possible.

- To adjust the compensation of an oscilloscope probe, a square wave is displayed and the probe is adjusted until the horizontal portions of the displayed wave are as nearly flat as possible.

- A spectrum analyzer displays RF amplitude on the vertical axis and frequency on the horizontal axis. A spectrum analyzer can be used to display spurious signals and/or intermodulation distortion products in an SSB transmitter.

- To measure intermodulation distortion in an SSB transmitter, modulate the transmitter with two non-harmonically related audio frequencies and observe the RF output with a spectrum analyzer.

- One way to measure SWR is with an antenna analyzer.

- One advantage of using an antenna analyzer compared to an SWR bridge to measure antenna SWR is that antenna analyzers do not need an external RF source. To connect an antenna analyzer, connect the antenna feed line directly to the analyzer's connector.

- The instrument that displays multiple digital signal states simultaneously is a logic analyzer.

- Vector network analyzers can measure input impedance, output impedance, and reflection coefficient.

- The subscripts of S parameters represent the port or ports at which measurements are made.

- The S parameter equivalent to forward gain is S21. The S parameter that represents input port return loss or reflection coefficient (equivalent to VSWR) is S11.

- The three test loads used to calibrate an RF vector network analyzer are short circuit, open circuit, and 50 ohms.

- The purpose of the prescaler function on a frequency counter is to divide a higher frequency signal so a low-frequency counter can display the input frequency.

Chapter 17

Measurement Techniques & Limitations

Frequency Counters

E4B01 (B) Which of the following factors most affects the accuracy of a frequency counter?

    A. Input attenuator accuracy

    B. **Time base accuracy**

    C. Decade divider accuracy

    D. Temperature coefficient of the logic

All a frequency counter does is count "how many pulses happened in how much time." The pulse counting part is reasonably simple – it's that "how much time" part that's challenging, and the accuracy of a frequency counter is almost completely dependent on its **time base accuracy**.

Inexpensive, hand-held frequency counters might be accurate to $\pm$ 4 parts per million, or "ppm." So if you're measuring the output frequency of your transceiver on the 40-meter band at 7 MHz, and the frequency counter says, "Yep, it's exactly 7,000,000 Hz," all you really know is it is somewhere in the range of 7,000,000 plus or minus 28 Hz, since $7 \times 4 = 28$. Spend more money and you can get a lot more accuracy if you need it – to the point where the top-of-the-line, $5,000 and up units don't even specify their accuracy in "parts per million" but "parts per *trillion*."

Directional Power Meters

E4B06 (D) How much power is being absorbed by the load when a directional power meter connected between a transmitter and a terminating load reads 100 watts forward power and 25 watts reflected power?

    A. 100 watts

    B. 125 watts

    C. 25 watts

    D. **75 watts**

There's nothing tricky in this question, so don't overthink it! We're seeing 100 watts of forward power. 25 watts of that power is coming back as reflected power. The law of conservation of energy tells us the rest must be going somewhere, and all that adds up to **75 watts** being absorbed by the load.

Voltmeters

E4B02 (A) What is the significance of voltmeter sensitivity expressed in ohms per volt?

A. **The full scale reading of the voltmeter multiplied by its ohms per volt rating will indicate the input impedance of the voltmeter**

B. When used as a galvanometer, the reading in volts multiplied by the ohms per volt rating will determine the power drawn by the device under test

C. When used as an ohmmeter, the reading in ohms divided by the ohms per volt rating will determine the voltage applied to the circuit

D. When used as an ammeter, the full scale reading in amps divided by ohms per volt rating will determine the size of shunt needed

Well, good luck finding your voltmeter's specification for sensitivity expressed in ohms per volt – not even the upper end multimeters I checked listed that spec. Most of the more affordable meters don't even list their impedance. It has gotten to be a bit of a non-issue with field-effect transistors running most of our voltmeters these days, but if you could find that spec, you could calculate the voltmeter's impedance because **the full scale reading of the voltmeter multiplied by its ohms per volt rating will indicate the input impedance of the voltmeter**. If the full scale reading of the voltmeter is 400 volts and the ohms per volt specification is 10,000, then the impedance of the meter equals $400 \times 10,000 = 4,000,000 \, \Omega$.

There are a lot of words in all the answers, so let's figure out how to keep this one straight. First, we'll take them at their word; this is a *voltmeter*, not a multimeter. You can't use a voltmeter as a galvanometer, nor as an ohmmeter, nor as an ammeter! That eliminates all the wrong answers.

Do keep in the back of your mind, though, that impedance of your voltmeter does become significant if you are measuring a high impedance circuit – say 10 kilohms or more.

RF Ammeters

E4B09 (D) What is indicated if the current reading on an RF ammeter placed in series with the antenna feed line of a transmitter increases as the transmitter is tuned to resonance?

A. There is possibly a short to ground in the feed line

B. The transmitter is not properly neutralized

C. There is an impedance mismatch between the antenna and feed line

D. **There is more power going into the antenna**

This is another question to not overthink! If there's more current going from the transmitter to the antenna, **there is more power going into the antenna**.

Key Concepts in This Chapter

- What most affects the accuracy of a frequency counter is the time base accuracy.

- When a directional power meter is connected between a transmitter and a terminating load and it reads 100 watts forward power and 25 watts reflected power, the load is absorbing 75 watts.

- The significance of voltmeter sensitivity expressed in ohms per volt is that the full scale reading for the voltmeter multiplied by its ohms per volt rating will indicate the input impedance of the voltmeter.

- If an RF ammeter is placed in series with the antenna feed line of a transmitter and the current reading increases, it means there is more power going into the antenna.

Once you finish practicing the Practice Exam for this chapter, you'll be ready to take on Progress Check #4 at fasttrackham.com.

Figure 17.1: Clamp-On Style RF Ammeter

Chapter 18

Receiver Performance Characteristics & Features, Part 1

A sizable amount of the Extra exam is devoted to receiver performance characteristics – so much that we've divided it up across three chapters.

Compared to transmitters, receivers have a *lot* of specifications. Most receivers (or the receiver sections of transceivers) are far more complex than most transmitters or transmitter sections. Compare, for instance, the specs for the transmitter and receiver sections of ICOM's IC-7300 transceiver. They list eight specifications for the transmitter, and that includes the microphone impedance. For the receiver section, they list 27. That's why you'll find almost no mention at all of transmitter performance characteristics in the entire Extra exam, but three chapters worth of receiver performance characteristics.

Phase Noise

E4C01 (D) What is an effect of excessive phase noise in the local oscillator section of a receiver?
    A. It limits the receiver's ability to receive strong signals
    B. It can affect the frequency calibration
    C. It decreases receiver third-order intercept point
    D. It can combine with strong signals on nearby frequencies to generate interference

As we move into Extra Class territory, we have to begin to come to grips with a disturbing fact. Brace yourself.

When we learn about basic electronics, electricity always travels at the speed of light, a 100-ohm resistor is a 100-ohm resistor, a transformer's output is precisely proportional to the ratio of the turns in the primary and secondary, and a 10-kHz oscillator produces a lovely 10-kHz sine wave and nothing else. It's wonderful, and if schematic diagrams and theories were really radios, life would be ever so simple.

In the real world, none of those things is quite true. 100-ohm resistors might be 90 ohms, they might be 110 ohms, and they probably change value as their temperature changes. And, directly relevant to this question, oscillators and filters never generate, block, or pass *only* the specific frequency for which we design them.

Phase noise in an oscillator occurs because every oscillator produces multiple frequencies. Figure 18.1 shows a grossly oversimplified and exaggerated example.

In Figure 18.1 we see two different frequencies being produced by a VFO. The desired frequency is the higher amplitude one, drawn in a solid line. The undesired frequency is the lower amplitude signal, drawn in a dotted line.

As you look at the chart, you can see that at some points, those two frequencies are going to add together, such as at point A. This will not only raise the overall amplitude, it's going

107

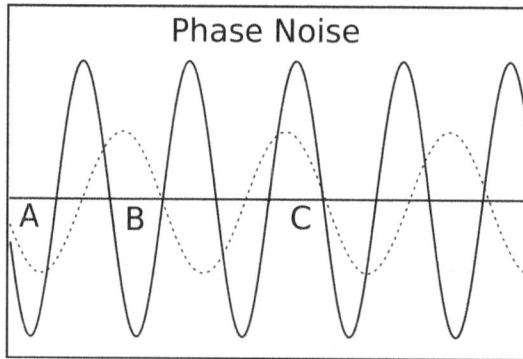

Figure 18.1: Phase Noise

to move that first negative peak slightly to the right ...it will happen at a slightly different time than it "should." At point B, the two signals are trying to cancel each other out, and the negative peak of the desired signal will not only be reduced, it will move just a tiny bit over to the left, happening sooner than it "should." Along about point C, everything is pretty much lined up and the output will be just what it should be for an instant.

That shifting back and forth of the peaks is phase noise, and if you think about it, it's basically a frequency modulation of the VFO.

If you saw a spectrum analysis of a VFO with high phase noise, it would resemble Figure 18.2; rather than a single frequency, it would be spread across a band of frequencies, as though it had side bands.

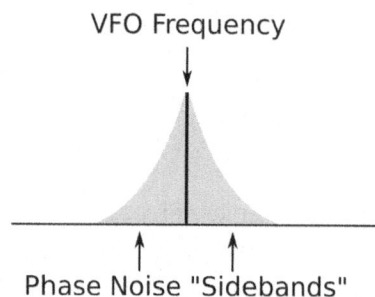

Figure 18.2: VFO Frequency with Phase Noise

Consider that the VFO's frequency is going to mix with the desired frequency, bring it down to the Intermediate Frequency, and send it on through the radio. If, instead, one of the phase noise-created "sidebands" mixes with a nearby strong signal, that strong signal can be brought down to a frequency at or very near the Intermediate Frequency, creating interference.

Phase noise creates inaccuracy and instability of the VFO output. It's like your tuning knob is loose and it's rattling around slightly changing the frequency all the time and causing you to hear stations you don't want to hear. End result – and the answer to this question? Excessive phase noise **can cause strong signals on nearby frequencies to interfere with reception of weak signals.**

E4C15 (D)

What is reciprocal mixing?

A. Two out-of-band signals mixing to generate an in-band spurious signal

B. In-phase signals cancelling in a mixer resulting in loss of receiver sensitivity

C. Two digital signals combining from alternate time slots

D. **Local oscillator phase noise mixing with adjacent strong signals to create interference to desired signals**

Reciprocal mixing is the process of **local oscillator phase noise mixing with adjacent strong signals to create interference to desired signals;** in other words, the process described in the previous question. It is also a parameter than can be used to specify a receiver's immunity to phase noise interference.

Image Response

E4C14 (D) What transmit frequency might generate an image response signal in a receiver tuned to 14.300 MHz and that uses a 455 kHz IF frequency?
   A. 13.845 MHz
   B. 14.755 MHz
   C. 14.445 MHz
   D. **15.210 MHz**

An "image response signal" is a by-product of superheterodyne receiver technology.

You'll recall the superheterodyne receiver works by mixing the original signal with another frequency from the local oscillator – typically one that is 455 kHz away from the original. That mixing, or heterodyning, creates a new copy of the original signal that's at 455 kHz – no matter what frequency the original signal was. That 455 kHz signal is called the Intermediate Frequency. (Some receivers feature a different IF, a variable IF, or even multiple IF stages, but we'll ignore all that for now.)

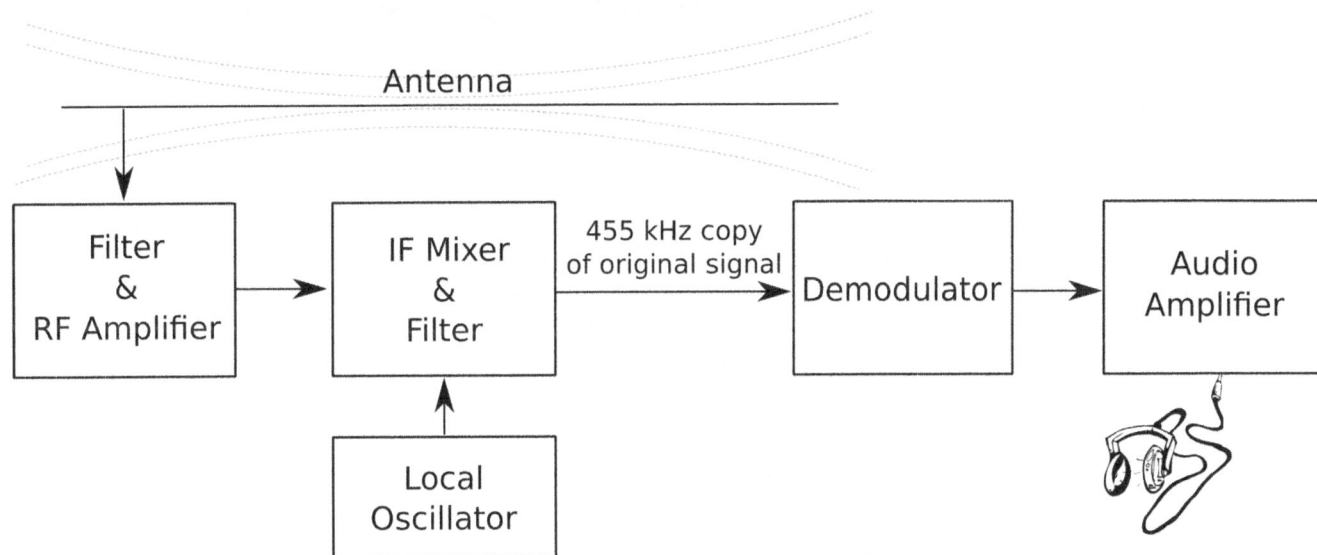

Figure 18.3: Block Diagram of Receiver

When two frequencies heterodyne, we don't just get one frequency. Put simply, we get two. (We actually get more, but the rest are so weak and far away, we can ignore them, at least for the moment.) One is the sum of the two frequencies; the other is the difference between the two frequencies.

The IF stage filters out the original carrier, the VFO (local oscillator), and the sum frequency. All that's left is the Intermediate Frequency, and that gets sent along to the rest of the receiver. So far, so good.

But what if another signal sneaks in there? Well, what happens, is that other frequency ALSO gets mixed with the local oscillator and the original frequency.

Let's do the math. Let's say we're tuned to a station at 14.300 MHz, and there's another station at 15.210. The question tells us the receiver uses a 455 kHz Intermediate Frequency, and that it uses what's called "high side injection", meaning the Intermediate Frequency is 455 kHz above the desired station's frequency. So we have a 14.300 MHz signal, a 15.210 MHz signal, and another signal from the local oscillator at 14.755 MHz, which is 14.300 MHz plus 455 kHz. Let's see if any of the sums or differences present a problem:

Frequency 1	Frequency 2	Sum	Difference
14.300 MHz (Desired Signal)	14.755 MHz (From VFO)	29.055 MHz	455 kHz (Just what we want.)
14.300 MHz	15.210 MHz (Interfering Signal)	29.510 MHz	910 kHz
14.755 MHz (From VFO)	15.210 MHz	29.965 MHz	455 kHz (That's a problem.)

Table 18.1: Image Response Solution

14.300 (the received signal) + 14.755 (from the VFO) = 29.055 MHz. No problem there, that's going to get filtered out in the IF stage.

The received signal + 15.210 (the interfering signal) = 29.510 MHz. No problem.

The received signal – the interfering signal = 910 kHz. Hmmm .... that's two times 455. I wonder if that means anything?

Let's review. Take a look at Table 18.1

The VFO minus the interfering signal = 455 kHz. Uh oh. all of a sudden, we have an extra signal mixed in with the signal we want, and it's going to shoot right through the IF filter because it's right on 455 kHz. The radio thinks that extra station is part of the original signal. It's a ghost station.

That extra signal is called the "image" of the unwanted station, and if the radio doesn't filter out that unwanted station before it gets to the IF mixer, we get "**image signal interference**;" A fancy way to say "I'm hearing that station at 14.300 mixed in with one that is almost a full MHz away!" The ability to reject that unwanted signal is called the receiver's Image Rejection Ratio. The ratio is measured in dB, and a good rule of thumb is anything over 60 dB of Image Rejection is good. That means the unwanted station is going to be 60 dB down from the wanted station, so it's basically lost in other noise.

When you face one of these image response problems, or an intermodulation problem, it's all just a bunch of addition and subtraction (multiplication if you want to get really fancy.) Make a table of the possible combinations and you'll find the culprit.

To solve this particular problem, look for the frequency that is $2 \times 455\,kHz$, or $910\,kHz$ away from the desired signal.

E4C09 (C) Which of the following choices is a good reason for selecting a high frequency for the design of the IF in a superheterodyne HF or VHF communications receiver?

A. Fewer components in the receiver

B. Reduced drift

C. **Easier for front-end circuitry to eliminate image responses**

D. Improved receiver noise figure

In the early days of the superheterodyne receiver, the IF was 20 kHz – they were already operating in the stratosphere of practical frequencies, because of the limitations of available components. By the 1930's, the components to build higher frequency oscillators were available and the IF frequency climbed to the 400 to 600 kHz range. Today, though, we're not limited to what vacuum tubes can produce. Would there be a benefit to bumping up the IF?

Well, let's take the previous problem and instead of a 455 kHz IF frequency, let's go crazy and bump the IF up 1000 times to 455 MHz and see what happens. That means our local oscillator needs to bump up to 469.300 MHz. (14.300 MHz + 455 MHz.) Follow along in Table 18.2.

Well, well! Since the numbers have gotten so large, we see there's a difference from our previous calculations, and the difference frequencies are now in MHz (or nearly MHz) instead of kHz. Building a filter that will filter out a signal that far away – not a problem!

Frequency 1	Frequency 2	Sum	Difference
14.300 MHz (Desired Signal)	469.300 MHz (From local oscillator)	483.600 MHz	455 MHz
14.300 MHz (Desired Signal)	15.210 MHz (Interfering Signal)	29.510 MHz	910 kHz
15.210 MHz (Interfering Signal)	469.300 MHz (From local oscillator)	484.51 MHz	454.090 **MHz**

Table 18.2: 455 MHz IF

Now, my 455 MHz IF example is a bit over-the-top, but the very cool Icom IC-9100 all-band, all-mode, all-singing, all-dancing, DC-to-x-rays radio with multiple IF stages used first-stage IF's between 10 MHz and 243 MHz, depending on the band, so it's not *that* outrageous.

Image response interference occurs in the IF stage of the receiver. Once it happens, the deed is done and nothing can undo it. (It's one reason we're so obsessed with front-end specs on our receivers.)

One great way to prevent image response distortion is to get rid of the frequency that is mixing with the other frequency and creating that distortion before the frequencies get to the IF section. That's what something called a preselector does.

Preselectors

E4D09 (C) What is the purpose of the preselector in a communications receiver?
- A. To store often-used frequencies
- B. To provide a range of AGC time constants
- C. **To increase rejection of signals outside the desired band**
- D. To allow selection of the optimum RF amplifier device

A preselector is an adjustable narrow bandpass filter that is ahead of the RF amplifier in the signal chain. Its purpose is **to increase rejection of signals outside the desired band**. A preselector can be a standalone unit (Figure 18.4) or can be incorporated into the design of a receiver.

E4C02 (A) Which of the following receiver circuits can be effective in eliminating interference from strong out-of-band signals?
- A. **A front-end filter or pre-selector**
- B. A narrow IF filter
- C. A notch filter
- D. A properly adjusted product detector

Strong out-of-band signals can create lots of problems for our reception of desired signals. The best way to solve those problems is to keep those strong out-of-band signals completely out of the radio; that's what a **front-end filter** or **pre-selector** does.

Figure 18.4: MFJ-1048 Standalone Preselector

Capture Ratio

E4C03 (C) What is the term for the suppression in an FM phone receiver of one signal by another, stronger signal on the same frequency?
    A. Desensitization
    B. Cross-modulation interference
    C. **Capture effect**
    D. Frequency discrimination

FM receivers have a challenge AM receivers don't. That challenge is rooted in the very nature of frequency modulation – the information is carried by changes in the carrier's frequency. If two carriers are on the same center frequency, or even on different center frequencies but overlapping, the FM demodulator has a problem – two different variations in frequency! This does not compute! It literally cannot demodulate both signals at once.

For that reason, FM demodulators must discriminate between different carriers and choose one or the other. (This, by the way, is why FM detectors are called discriminators.) Some radios use a limiting circuit ahead of the demodulator to get rid of competing signals, some handle the process in the demodulator itself, but one way or the other, the stronger signal gets demodulated and the weaker gets blanked out. If the signals are close in strength, or both varying in strength, the radio may switch back and forth between them, creating complete chaos. You've probably heard this on your local VHF or UHF repeater when someone is talking and suddenly someone else's stronger signal captures the repeater and the original talker is gone.

This blanking of the weaker signal by a stronger signal is called the **capture effect** and one specification of an FM radio is its capture ratio – essentially, how loud can an adjacent signal be before the demodulator gets captured.

Key Concepts in This Chapter

- Excessive phase noise in the local oscillator section of a receiver can combine with strong signals on nearby frequencies to generate interference-.

- Reciprocal mixing is the term to describe local oscillator phase noise mixing with adjacent strong signals to create interference to desired signals.

- If a receiver is tuned to 14.300 MHz and uses a 455 kHz IF frequency, a transmit frequency that might generate an image response signal is 15.210 MHz.

- A good reason for selecting a high frequency for the design of the IF in a conventional HF or VHF communications receiver is that it is easier for front-end circuitry to eliminate image responses.

- A front-end filter or pre-selector can be effective in eliminating image signal interference. It also helps to increase rejection of signals outside the desired band.

- The term for the suppression of one FM phone signal by another, stronger FM phone signal is capture effect.

Chapter 19

Receiver Performance Characteristics & Features, Part 2

Noise Figure

E4C04 (D) What is the noise figure of a receiver?

    A. The ratio of atmospheric noise to phase noise

    B. The ratio of the noise bandwidth in hertz to the theoretical bandwidth of a resistive network

    C. The ratio of thermal noise to atmospheric noise

    D. **The ratio in dB of the noise generated by the receiver to the theoretical minimum noise**

Since, ideally, we'd have no noise at all in the output of our receiver, the "noise figure" is a rather important measure of receiver performance. But to be meaningful, that figure must be a comparison to some standard.

Our universe is not an electromagnetically quiet place. That was true even before humans started filling our space with purposefully created electromagnetic waves. The Sun, the other stars, our own atmosphere, especially lightning strikes, the earth's core, and more all contribute to a constant wash of electromagnetic noise. No receiver will ever be quieter than this background noise, also known as the "noise floor." Engineers have defined that noise level as, essentially, "zero noise", even though it's not really zero, and that's the standard against which we measure a receiver's noise to get the "noise figure." The noise figure is **the ratio in dB of the noise generated by the receiver to the theoretical minimum noise**. The Noise Figure will always be a positive number – lower is better.

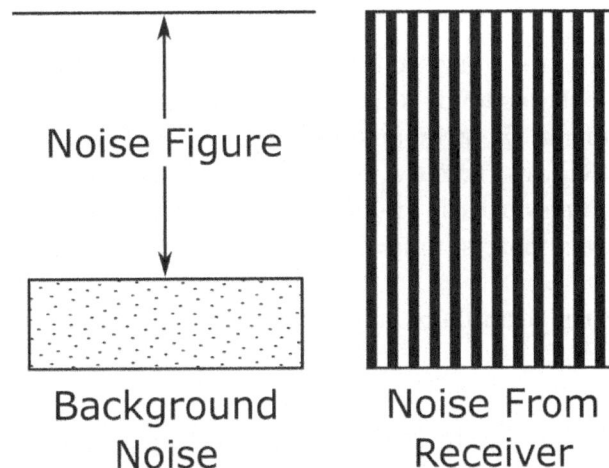

Figure 19.1: Noise Figure

E6E05 (A) Which of the following noise figure values is typical of a low-noise UHF preamplifier?

    A. **2 dB**

    B. -10 dB

    C. 44 dBm

    D. -20 dBm

A noise figure is **the ratio in dB of the noise generated by the receiver to the theoretical minimum noise**. In plain language, it's "how much noisier is this thing than a perfect one of these things?" Lower numbers are better!

Once you have a firm grip on what a noise figure represents, you see that only one answer out of the possible answers is a real noise figure, and that one is **2 dB**. It doesn't even matter that the device in the question is a "low-noise UHF preamplifier."

A -10 dB noise figure would mean the receiver was 10 dB quieter than perfect – that's not in the realm of possibility! There just can't be a negative noise figure.

Noise figures are a ratio – not an absolute value. They're a comparison with a theoretical "perfect device." So the answers in dBm aren't noise figures, they're absolute values of 44 dB higher than a 1 milliwatt signal and 20 dB lower than a 1 milliwatt signal, respectively.

E4C05 (B) What does a receiver noise floor of -174 dBm represent?

    A. The minimum detectable signal as a function of receive frequency

    B. **The theoretical noise in a 1 Hz bandwidth at the input of a perfect receiver at room temperature**

    C. The noise figure of a 1 Hz bandwidth receiver

    D. The galactic noise contribution to minimum detectable signal

The precise value of the noise floor, with regard to receivers, is -174 dBm/Hz – a shorthand way of saying "174 dB below the strength of an input with the strength of 1 milliwatt and a bandwidth of 1 Hz." That's the value of the noise floor, **the theoretical noise at the input of a perfect receiver at room temperature.**

Notice the figure is given as "per Hz of bandwidth." As we make the bandwidth wider, the noise floor is going to rise – in other words, the noise will get louder. At a bandwidth of 400 Hz, for instance, the noise would be 400 times louder.

E4C06 (D) A CW receiver with the AGC off has an equivalent input noise power density of -174 dBm/Hz. What would be the level of an unmodulated carrier input to this receiver that would yield an audio output SNR of 0 dB in a 400 Hz noise bandwidth?

    A. -174 dBm

    B. -164 dBm

    C. -155 dBm

    D. **-148 dBm**

Hoooooooooooooooooooooooooooooooooo boy. What a question! I'll try to translate that into English.

The question is asking, in a very technically precise way, "How loud does a 400 Hz wide signal need to be to be exactly as loud as the noise in the receiver, not counting the noise in the noise floor." You'll recall from the previous question that the noise floor rises – gets louder – proportional to the bandwidth's increase. So a 400 Hz bandwidth is 400 times as noisy as a 1 Hz bandwidth. We have a power ratio of 400 to 1, or, simply, 400.

To put the question even more simply, "If you have a power ratio of 400:1, how many dB is that?"

You might know that the formula to calculate power ratio from dB of change is:

$$Power\,Ratio = 10^{\frac{db}{10}}$$

So, for instance, if we have a 3 dB change, $10^{\frac{3}{10}} \approx 2 \times Power\,Ratio$

But how do we do the reverse? How do we convert a power ratio to decibels? Here's the formula, in simple form:

$$dB = 10_{\log}(Power\,Ratio)$$

Now, strictly speaking, that should be:

$$\frac{P_{Out}}{P_{In}}dB = 10_{log10}\frac{P_{Out}}{P_{In}}$$

However, the simple form works just fine for our purposes, and tells us just what buttons to press on our calculator.

clear 1 0 log 4 0 0) enter

Press "Enter" and you should be looking at 26.02059991 --- but just among us, let's call it 26. That's 26 dB of gain from -174 dB. We're almost there! We *add* 26 to -174, or subtract 26 from +174 and remember to change the sign back to negative, and the answer to the question is **-148 dB**.

E4C13 (C) How does a narrow-band roofing filter affect receiver performance?

A. It improves sensitivity by reducing front end noise

B. It improves intelligibility by using low Q circuitry to reduce ringing

C. **It improves dynamic range by attenuating strong signals near the receive frequency**

D. All these choices are correct

A "roofing filter" is a bandpass filter in the IF section of a receiver. There's nothing particularly special about the circuit that makes it a "roofing" filter; it's wired up like any other bandpass filter. The precise reason we call it a roofing filter is obscure (to say the least) but probably had to do with this filter being like a protective "roof" at the "top" of the receiver, to keep out undesired signals.

Its function is to effectively **improve dynamic range by attenuating strong signals near the receive frequency**. Dynamic range is the difference between the amplitude of the desired signal and the noise floor, so reducing the noise floor enhances dynamic range.

Selectivity

E4C10 (C) What is an advantage of having a variety of receiver IF bandwidths from which to select?

A. The noise figure of the RF amplifier can be adjusted to match the modulation type, thus increasing receiver sensitivity

B. Receiver power consumption can be reduced when wider bandwidth is not required

C. **Receive bandwidth can be set to match the modulation bandwidth, maximizing signal-to-noise ratio and minimizing interference**

D. Multiple frequencies can be received simultaneously if desired

Selectivity is a measure of a receiver's ability to reject all signals except the desired signal. In an ideal world, the selectivity of our receiver would constantly and precisely match the bandwidth of the received signal, since everything outside that bandwidth is, by definition, noise. Of course, that ideal situation is not quite possible, but we want to get as close to it as we can, and having a variety of receiver IF bandwidths from which to select is possible and desirable. With that feature, **receive bandwidth can be set to match the modulation bandwidth, maximizing signal-to-noise ratio and minimizing interference**.

E7C11 (C) Which of the following describes a receiving filter's ability to reject signals occupying an adjacent channel?

    A. Passband ripple

    B. Phase response

    C. **Shape factor**

    D. Noise factor

For maximum ability to reject signals occupying an adjacent channel, a receiving filter's frequency response curve should look like a rectangle; a flat top showing that it is passing the desired band of frequencies perfectly, and straight sides that drop to $-\infty dB$ instantly.

That's never going to happen in analog reality, and we'll only get close in digital reality. We're not terribly concerned with how flat that top is; any reasonably well-designed filter is going to be "close enough" in that area. Our real concern is how steep the sides of that filter are, because that's what's going to keep out the nearby unwanted signals.

We need a handy measurement of the slope of those sides, and that's what the shape factor provides.

To calculate the shape factor of a filter, we need to know the widths of the passband at $-6\,dB$ and at $-60\,dB$. We divide the first value into the second, and that's our shape factor.

$$Shape\ Factor = \frac{Bandwidth\ at\ -60\,dB}{Bandwidth\ at\ -6dB}$$

Let' s try a couple. First, we'll imagine a really terrible bandpass filter. This one has a bandwidth at $-6\,dB$ of 3.5 kHz. At $-60\,dB$ this hunk o' junk has a bandwidth of 350 kHz – it's as wide as the whole 20-meter band.

$$Shape\ Factor = \frac{Bandwidth\ at\ -60\,dB}{Bandwidth\ at\ -6dB} = \frac{350,000}{3,500} = 100$$

That one has a shape factor of 100 or, as the convention goes, "100:1 at 6/60 dB".

Now let's try a really nice SSB filter. This one has a bandwidth of 3.5 kHz at $-6\,dB$. At $-60\,dB$ its bandwidth is 10 kHz.

$$Shape\ Factor = \frac{Bandwidth\ at\ -60\,dB}{Bandwidth\ at\ -6dB} = \frac{10,000}{3,500} = 2.86$$

You can see that when you're looking at shape factors for bandpass filters, lower numbers indicate a steeper slope and, so, are usually better.

Sensitivity/MDS

E4C07 (B) What does the MDS of a receiver represent?

    A. The meter display sensitivity

    B. **The minimum discernible signal**

    C. The multiplex distortion stability

    D. The maximum detectable spectrum

MDS stands for **Minimum Discernible Signal**.

Clearly, for a signal to be detected, it must be louder than the noise floor. The MDS specification of a receiver tells you how faint a signal can be and still be detected *and* come out of the receiver at a specified signal to noise ratio.

Another word for MDS is "sensitivity", and you'll find that's in much more common use in ham radio advertising. Either value can be given as microvolts (μV) or dBm, though the microvolt spec seems, from a quick and decidedly unscientific survey, to be preferred these days, and values in the 0.1 to 0.2 microvolt range are not uncommon for SSB/CW modes. Equivalent dBm values would be in the high 90's, such as -95 dBm to -98 dBm. In either case, lower numbers are better; in other words, -98 dBm is better than -95 dBm.

E4C11 (D) Why can an attenuator be used to reduce receiver overload on the lower frequency HF bands with little or no impact on signal-to-noise ratio?

A. The attenuator has a low-pass filter to increase the strength of lower frequency signals

B. The attenuator has a noise filter to suppress interference

C. Signals are attenuated separately from the noise

D. **Atmospheric noise is generally greater than internally generated noise even after attenuation**

If you've ever tried to produce a high-quality audio recording, or built a high-quality audio system, you know the name of the game is to keep the signal level as high as possible throughout the signal chain without creating distortion. That way the desired signal is as strong as possible relative to the noise in the system.

In the HF reception game, that's not the case. Especially in the lower bands, the incoming signal already contains so much **atmospheric noise** that reducing the input level with an attenuator still leaves the incoming noise much higher than any internal noise in the receiver.

Software Defined Receivers

E4C08 (D) An SDR receiver is overloaded when input signals exceed what level?

A. One-half the maximum sample rate

B. One-half the maximum sampling buffer size

C. The maximum count value of the analog-to-digital converter

D. **The reference voltage of the analog-to-digital converter**

An SDR receiver is a Software Defined Radio.

As digital electronics have gotten faster, better and, let's face it, cheaper, they've become capable of handling more and more radio-related tasks that used to require analog parts. This has opened up vast realms of possibilities, and the once exotic world of SDR's is quickly getting far more accessible to the average ham.

Since radio signals are an analog phenomenon – even when they're carrying digital signals – SDR's will forever be in some way a hybrid of analog and digital technology. As digital technology evolves, SDR's will edge closer and closer to the ideal SDR which would be a computer hooked to an antenna, and we're close to that point now.

An example; ICOM's IC-7300 receiver section is 100% digital starting right after the preselector clear through to the input of the audio amplifier. There's no analog IF section; the signal is digitized with direct sampling and then the internal digital hardware and software takes over. The transmitter side is all digital from the microphone to the final amplifier. All the modulation duties are handled by the computer.

Analog to digital – A/D – conversion is done by sampling the level of the analog signal very rapidly, and converting the instantaneous analog levels to digital form by comparing them to a reference voltage. If those instantaneous levels exceed the reference voltage, the result is a particularly unattractive form of distortion called "clipping."

E4C12 (D) Which of the following has the largest effect on an SDR receiver's dynamic range?

A. CPU register width in bits

B. Anti-aliasing input filter bandwidth

C. RAM speed used for data storage

D. **Analog-to-digital converter sample width in bits**

Dynamic range is the difference between the loudest signal a device can process without a specified level of distortion and the noise floor of the device.

We're concerned about the dynamic range of a receiver because the nature of communications signals is that they have a wide range of signal strengths; we want to be able to hear the very weak stations as well as the very strong ones, so we want a receiver with wide dynamic range.

The largest effect on an SDR receiver's dynamic range is the **analog-to-digital converter sample width in bits**, because that's the limit on the number of different levels the converter can represent.

If you have a converter with an 8-bit sample width, you can represent 256 (2^8) different values. That works out to a dynamic range of about 24 dB; the weakest signal it can receive is 24 dB below the strongest signal it can receive. With a 16-bit sample width, the dynamic range expands to 48 dB.

Key Concepts in This Chapter

- The noise figure of a receiver is the ratio in dB of the noise generated by the receiver to the theoretical minimum noise.

- A typical noise figure value of a low-noise UHF preamplifier is 2 dB.

- The theoretical noise at the input of a perfect receiver at room temperature is -174 dBm/Hz.

- If a CW receiver with the AGC off has an equivalent input noise power density of -174 dBm/Hz, an unmodulated carrier input to this receiver at -148 dBm would yield a signal to noise ratio of 0 dB in a 400 Hz noise bandwidth.

- Narrow-band roofing filters improve dynamic range by attenuating strong signals near the receive frequency.

- The term that describes a receiving filter's ability to reject signals occupying an adjacent channel is shape factor.

- The MDS of a receiver is the Minimum Discernible Signal. It's a measure of sensitivity.

- It's possible to use an attenuator to reduce receiver overload on the lower frequency HF bands with little or no impact on signal-to-noise ratio because atmospheric noise is generally greater than internally generated noise, even after attenuation.

- An SDR receiver is overloaded when input signals exceed the reference voltage of the analog-to-digital converter.

- The analog-to-digital converter's sample width in bits has the largest effect on an SDR receiver's dynamic range.

> Just the act of writing a simple note like "SDR overloaded — reference voltage" can have a powerful positive effect on your memory for that answer.

Chapter 20

Receiver Performance Characteristics & Features, Part 3

Blocking Dynamic Range & Desensitization

E4D01 (A) What is meant by the blocking dynamic range of a receiver?

A. **The difference in dB between the noise floor and the level of an incoming signal that will cause 1 dB of gain compression**

B. The minimum difference in dB between the levels of two FM signals that will cause one signal to block the other

C. The difference in dB between the noise floor and the third order intercept point

D. The minimum difference in dB between two signals which produce third order intermodulation products greater than the noise floor

The blocking dynamic range figure for a receiver is the answer to the question, "How loud does an interfering signal have to be before it blocks the signal I want?" It is defined as **the difference in dB between the noise floor and the level of an incoming signal which will cause 1 dB of gain compression**.

The compression referred to is not purposeful compression, such as that from an AGC, but compression created by the RF amplifier running out of headroom. Every amplifier has a limit for how much input it can handle. Once it hits that limit, it can't "turn up the volume" any more, so the signal gets compressed. As the signal gets compressed, the dynamic range is reduced and the noise floor, consequently, rises. Say bye bye to that weak signal you were trying to hear, it just sank into the noise floor.

Figure 20.1 is a screen shot of the waveforms of some background noise here in my studio (the window was open to the outdoors...) at normal levels. In the middle you see the waveforms of me saying, "dynamic range", then some more background noise. The highest peak is set to exactly 0 dB – just before the system would go into overload.

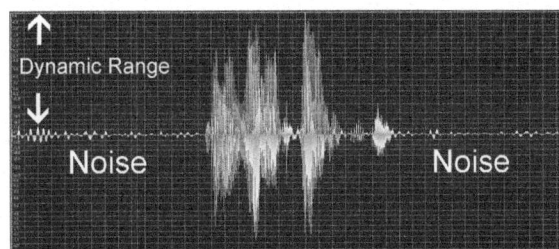

Figure 20.1: Dynamic Range without Clipping

There's about 30 dB of dynamic range – the difference between the peak of 0 and the background noise. Figure 20.2 is the same recording, except I've raised the level of everything 20 dB.

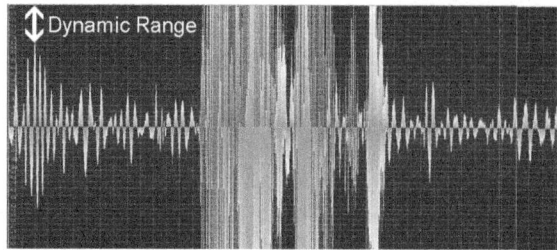

Figure 20.2: Dynamic Range with Severe Clipping

The peak is still 0dB; everything above that is just getting clipped off, just as it would in your amplifier. There was some sort of noise spike early in the recording, which, of course, also got cranked up 20 dB. Now the dynamic range at that moment is about 10 dB. Any signal below -10 dB is going to be covered by the noise.

You could say that the signal in Figure 20.2 has far exceeded the system's blocking dynamic range because it is so compressed and the noise floor has risen.

E4D02 (A) Which of the following describes problems caused by poor dynamic range in a receiver?

A. **Spurious signals caused by cross-modulation and desensitization from strong adjacent signals**

B. Oscillator instability requiring frequent retuning and loss of ability to recover the opposite sideband

C. Cross-modulation of the desired signal and insufficient audio power to operate the speaker

D. Oscillator instability and severe audio distortion of all but the strongest received signals

Remember that the sensitivity of a receiver is directly related to the dynamic range of that receiver. More sensitivity – in other words, more ability to detect faint signals without losing them in the noise floor – is the same thing as more dynamic range. So when a strong adjacent signal pushes the RF amplifier to the point of compression, the noise floor effectively comes up and the sensitivity goes down. We say the receiver gets "desensitized." So **desensitization** is one problem caused by poor dynamic range.

As any amplifier goes into overload, the signals it is carrying start modulating each other, and that's called **cross-modulation**.

So, two problems caused by poor dynamic range in a communications receiver are **spurious signals caused by cross-modulation** and **desensitization from strong adjacent signals.**

E4D12 (A) What is the term for the reduction in receiver sensitivity caused by a strong signal near the received frequency?

A. **Desensitization**

B. Quieting

C. Cross-modulation interference

D. Squelch gain rollback

Desensitization occurs when a strong, unwanted signal near the received frequency overwhelms the receiver and causes the desired signal to be inaudible. This could be the result of an inadequate roofing filter or preselector or the result of excessive phase noise in the VFO, but somehow a strong, unwanted signal is getting into our receiver and wiping out the desired signal.

E4D07 (A) Which of the following reduces the likelihood of receiver desensitization?

A. **Decrease the RF bandwidth of the receiver**

B. Raise the receiver IF frequency

C. Increase the receiver front end gain

D. Switch from fast AGC to slow AGC

If the cause of desensitization is a strong adjacent signal, the cure is to **decrease the RF bandwidth of the receiver** so the unwanted signal never reaches the first stage of the receiver. That's done with a roofing filter or a preselector.

Intermodulation Distortion

E4D06 (D) What is the term for spurious signals generated by the combination of two or more signals in a non-linear device or circuit?
 A. Amplifier desensitization
 B. Neutralization
 C. Adjacent channel interference
 D. **Intermodulation**

You know from your studies of superheterodyne receivers that when two signals with different frequencies mix they produce new signals at frequencies that equal the sum and difference of the original frequencies. (The original frequencies remain, as well.) One term for that process is intermodulation.

First, let's define intermodulation distortion. *Distortion* is any difference between the original signal and the output of the device in question. *Intermodulation distortion* is distortion caused by the mixing of two or more frequencies, resulting in unwanted sum and difference signals. Intermodulation is the result of non-linearity. After all, in a perfectly linear device, there's no distortion of the original signal.

If you've ever tuned a guitar by ear, you've used a difference frequency. For you non-musicians, as two strings on a guitar get closer and closer to playing exactly the same frequency, we start to hear a "beat" frequency – the note sounds like it is getting louder and softer, like "wah, wah, wah." That beat is produced by intermodulation. Its frequency is precisely the difference between the frequencies of the two strings – the closer they are, the slower the beat frequency, until when they perfectly match, the beat is gone.

The same sort of thing can and does happen with radio frequencies; depending on the frequencies involved this intermodulation may be no problem at all or a very significant problem that makes communication on a particular frequency impossible. When intermodulation causes a problem, we call the problem intermodulation distortion or intermodulation interference. When intermodulation is something we intended to create, we call it heterodyning. Image response distortion is sort of a "special case" type of intermodulation, but it is still created by that same underlying process of mixing frequencies.

Frequencies can get mixed in receivers or transmitters. If the intermodulation occurs in your receiver, you have a problem. If the intermodulation occurs in a transmitter, everybody listening to that transmitter has a problem.

Most cases of serious intermodulation distortion are a result of very strong nearby signals.

As noted, when signals combine, they create new signals at frequencies that are the sum and difference of the originals. So, for example, say we somehow mix a signal at 146.52 MHz with one at 146.34 MHz, as we will in question E4D05. Maybe those are frequencies of two repeaters that share a common antenna site, and one repeater signal is getting into the final amplifier of the other repeater. We end up with the sum of 292.86 MHz, and a difference of 0.18 MHz.

Ah, but wait, there's more!

Every transmitter produces harmonics. Those are signals that are double, triple and even quadruple the frequency of the intended frequency. We call harmonics that are double the frequency of the original signal "second order" harmonics. The triples are third order, the quadruples are fourth order.

As a sort of shorthand, we also call sums *and* differences of two frequencies that are near each other "third order" frequencies, since the sums are close to triple one of the original frequencies.

There is no way to build an oscillator that produces a pure, single frequency sine wave, any more than there is a way to plunk an open guitar string and produce a pure, single frequency sine wave. A well-designed transmitter system suppresses those harmonics to a point that they don't cause problems, but they're still there – just very weak, so they don't have much range. But let's say we're very close to another transmitter. Say our antenna is hanging on the same tower right next to theirs. That's a very common situation for ham repeaters. Now we have to think about all the harmonics and their combinations, too. We end up doing a lot of multiplying, adding and subtracting, and end up with tables like Tables 20.1 and 20.2.

Frequency Table	
Frequency 1	146.52
Frequency 2	146.34
Sum	292.86
Difference	0.18

Table 20.1: Frequency Table for Question E4D06

And that still isn't even close to all the possible combinations. We haven't, for instance, even started calculating the fourth order intermodulation products or, hold onto your hat, the intermodulation products of the intermodulation products! In reality, though, those are seldom strong enough to do much damage, so in most cases we can safely ignore all those other possibilities.

If someone is trying to listen to any of those frequencies in the sums and differences, they're going to get some weirdness on their radio every time those repeaters at 146.52 and 146.34 try to transmit at the same time.

The real-world version of solving these intermodulation problems is done, one hopes, *before* the problem arises; for instance, during the planning stages for a new repeater.

E4D08 (C) What causes intermodulation in an electronic circuit?
   A. Too little gain
   B. Lack of neutralization
   C. **Nonlinear circuits or devices**
   D. Positive feedback

If a circuit is linear, the waveform that goes in equals the waveform that goes out. That's linearity.

If an amplifier starts acting like a mixer – and that's what's happening in intermodulation – it's no longer a pure linear amplifier, it has become nonlinear.

The root cause of intermodulation in an electronic circuit is **nonlinear circuits or devices**.

Here's the bad news – there's no such thing as a 100% linear circuit or device in the real world. The good news is, we can get close enough to make things work.

Third Order Intermodulation Distortion Products		
$2f1 + f2$	$(2 \times 146.52) + 146.34$	439.38
$2f1 - f2$	$(2 \times 146.52) - 146.34$	**146.70**
$2f2 + f1$	$(2 \times 146.34) + 146.52$	439.92
$2f2 - f1$	$(2 \times 146.34) - 146.52$	**146.16**

Table 20.2: Third-Order Intermodulation Distortion Products for Question E4D06

E4D03 (B) How can intermodulation interference between two repeaters occur?

A. When the repeaters are in close proximity and the signals cause feedback in the final amplifier of one or both transmitters

B. When the repeaters are in close proximity and the signals mix in the final amplifier of one or both transmitters

C. When the signals from the transmitters are reflected out of phase from airplanes passing overhead

D. When the signals from the transmitters are reflected in phase from airplanes passing overhead

A bad case of intermodulation interference can render a repeater absolutely useless. Given that repeaters cost money and a lot of work, it's definitely best to take steps to head off intermod in advance.

E4D10 (C) What does a third-order intercept level of 40 dBm mean with respect to receiver performance?

A. Signals less than 40 dBm will not generate audible third-order intermodulation products

B. The receiver can tolerate signals up to 40 dB above the noise floor without producing third-order intermodulation products

C. A pair of 40 dBm input signals will theoretically generate a third-order intermodulation product that has the same output amplitude as either of the input signals

D. A pair of 1 mW input signals will produce a third-order intermodulation product which is 40 dB stronger than the input signal

You might recognize the term "third-order intercept" from some of the earlier wrong answers. At last we find out what it means!

If we had an ideal, perfectly linear amplifier we could feed in two closely spaced tones at the input and we'd get the same two tones and nothing else out of the output.

In real amplifiers, when we feed in those two tones we get out the two tones plus other stuff that is a bunch of those sum and difference frequencies we covered earlier. The third-order sums and differences are $2f_1 \pm f_2$ and $2f_2 \pm f_1$, since the sums are pretty close to the value of $3f_1$. We're usually not terribly concerned about the sums, just the differences. The reason we're concerned with those differences is $2f_1 - f_2$ and $2f_2 - f_1$ are going to fall perilously close to our desired frequency when f_1 and f_2 are close to each other in frequency.

The more we crank up the level of the two tones at the input, the more unwanted stuff we get out the other end. At some point, the level of the two tones at the output equals the level of all that unwanted stuff. That's the third-order intercept point. If you imagine the graph of what's happening, it will look something like Figure 20.3.

The third-order intercept figure for a receiver really tells us, "how well does this receiver resist intermodulation distortion?" Higher numbers are better! So if a receiver boasts a 40 dB third-order intercept figure, it means **a pair of 40 dBm signals will theoretically generate a third-order intermodulation product with the same output amplitude as the input signals.** The key phrase to remember in the correct answer is "same output amplitude."

E4D11 (A) Why are odd-order intermodulation products, created within a receiver, of particular interest compared to other products?

A. Odd-order products of two signals in the band of interest are also likely to be within the band

B. Odd-order products overload the IF filters

C. Odd-order products are an indication of poor image rejection

D. Odd-order intermodulation produces three products for every input signal within the band of interest

This question is asking, "Why are we so concerned about odd-order intermodulation products, as opposed to, say, fourth-order intermodulation products?" The way the numbers work out, **the odd-order product of two signals which are in the band of interest is also likely**

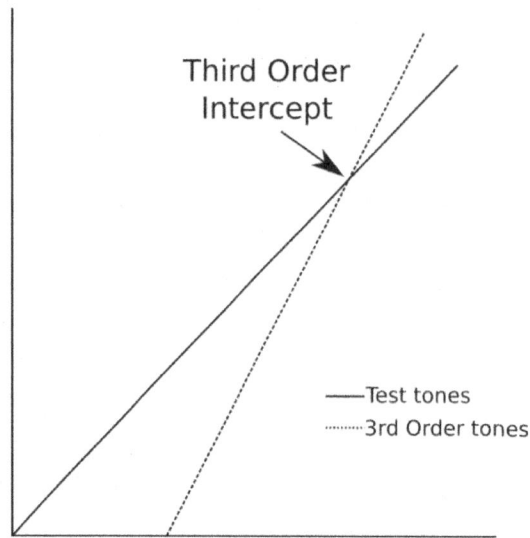

Figure 20.3: Third Order Intercept

to be within the band. Assuming f_1 and f_2 are close to the same frequency, $2f_1 - f_2$ and $2f_2 - f_1$ are always going to fall somewhere nearby. So will the fifth order product, $3f_1 - 2f_2$ and $3f_2 - 2f_1$, etc., though our primary concern is that first one, the third-order intermodulation product.

E4D05 (A) What transmitter frequencies would cause an intermodulation-product signal in a receiver tuned to 146.70 MHz when a nearby station transmits on 146.52 MHz?

    A. **146.34 MHz and 146.61 MHz**
    B. 146.88 MHz and 146.34 MHz
    C. 146.10 MHz and 147.30 MHz
    D. 173.35 MHz and 139.40 MHz

This question requires some careful reading. Here's what it's asking: You have a receiver tuned to 146.70 MHz. You're not transmitting, just receiving. Whenever a nearby station transmits on 146.52 MHz, it messes up your reception on 146.70. Because you are a smart ham, you recognize this as intermodulation distortion. What are the possible frequencies for that *other* station that's garbling the one at 146.70?

Every transmitter produces spurious emissions – harmonics of the main signal. Ideally, these are sufficiently suppressed that they come out of the antenna at very low – and legal -- power, but they're still there. Maybe the intermodulation is coming from the second harmonic of the 146.52 station plus some unknown station; or maybe it's coming from the second harmonic of some unknown station plus the 146.52 station. We don't need to know which it is to figure this out, though; we just need to know whatever is happening, it's ending up at 146.70.

The formulas we want are:

$$f_{imd} = 2f_1 - f_2$$

and

$$f_{imd} = 2f_2 - f_1$$

" f_{imd}" means "the frequency where the intermodulation is appearing," in other words, the frequency to which your receiver is set, in this case 146.70. f_1 is the frequency of the transmitting station, 146.52, and f_2 is the frequency of the unknown station.

With a little algebraic rearrangement, we get:

$$f_2 = 2f_1 - f_{imd}$$

or ...

$$f_2 = 2(146.52) - 146.70 = 146.34$$

One possibility for the frequency that's causing the intermodulation, then, is 146.34 MHz. The other possibility is:

$$f_{imd} = 2_{f2} - f_1$$

We rub a little algebra on that formula to get this:

$$f_2 = \frac{f_{imd} + f_1}{2}$$

...or ...

$$f_2 = \frac{146.70 + 146.52}{2} = 146.61$$

There's our other possible answer; 146.61 MHz. The correct answer for this question is **146.34 MHz and 146.61 MHz**.

Beware the answer that contains 146.88 MHz. Yes, you can come up with that number by doubling your receiver frequency then subtracting 146.52, but that makes no sense – your receiver isn't creating any harmonics that would come close to matching the strength of a received signal.

E4D04 (B) Which of the following may reduce or eliminate intermodulation interference in a repeater caused by another transmitter operating in close proximity?

A. A band-pass filter in the feed line between the transmitter and receiver
B. A properly terminated circulator at the output of the repeater's transmitter
C. Utilizing a Class C final amplifier
D. Utilizing a Class D final amplifier

Circulator

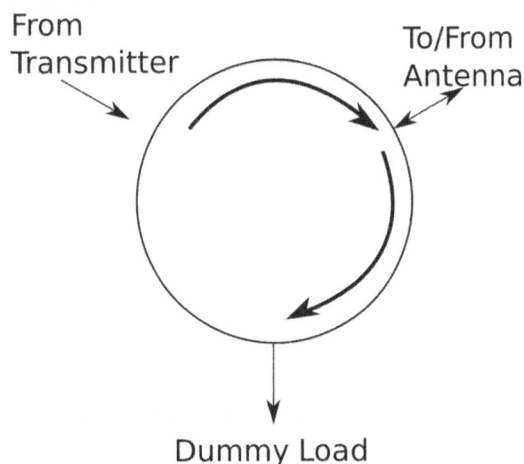

Figure 20.4: Circulator

When we install a repeater transmitter, it's reasonably easy to electrically shield that box from stray RF. We just need a well-grounded metal case around it, and we're all set. Except then we rather recklessly hook it up to an antenna which wants to bring all the RF in the world right into our transmitter, creating tons of intermodulation. Worse, that antenna is often in a location that includes other operating transmitters, meaning the RF in the vicinity is very strong. How do we protect our transmitter from our own antenna?

One device we can use is called a **circulator**.

127

From the outside, a circulator (figure 20.4) looks for all the world like a heavy-duty version of something you probably have somewhere in your home's TV system, which is a splitter. It has, usually, three connections.

However, it's not a splitter. Inside a circulator is a hefty chunk of ferrite and a strong magnet. The carefully adjusted physical relationship of the magnet and the ferrite turns the circulator into a one-way valve for radio waves.

We hook up the input port of the circulator to our transmitter's output. We hook up the output port to the antenna, and we hook the third port to a dummy load. Signals from the transmitter go to the antenna, signals coming back down from the antenna get shunted to the dummy load and never get mixed with the intended signal, nor do they reach the innards of the transmitter, so they can't create intermodulation.

Key Concepts in This Chapter

- The blocking dynamic range of a receiver is the difference in dB between the noise floor and the level of an incoming signal which will cause 1 dB of gain compression.

- Two problems caused by poor dynamic range in a communications receiver are spurious signals caused by cross-modulation and desensitization from strong adjacent signals.

- The term for reduction in receiver sensitivity caused by a strong signal near the received frequency is desensitization.

- A way to reduce the likelihood of receiver desensitization is to decrease the RF bandwidth of the receiver.

- The term for spurious signals generated by the mixing of two or more signals is intermodulation.

- The cause of intermodulation in an electronic circuit is nonlinear circuits or devices.

- With respect to receiver performance, a third-order intercept level of 40 dBm means a pair of 40 dBm signals will theoretically generate a third-order intermodulation product with the same output amplitude as the input signals.

- Odd-order intermodulation products, created within a receiver, are of particular interest because odd-order products of two signals in the band of interest are also likely to be within the band

- The transmitter frequencies that would cause an intermodulation product signal in a receiver tuned to 146.70 MHz when a nearby station transmits on 146.52 MHz are 146.34 MHz and 146.61 MHz. The basic formulas for calculating this are:

$$f_{imd} = 2f_1 - f_2$$

and

$$f_{imd} = 2f_2 - f_1$$

For the problem on the exam, we transpose those formulas to:

$$f_2 = 2f_1 - f_{imd}$$

and

$$f_2 = \frac{f_{imd} + f_1}{2}$$

- A properly terminated circulator at the output of the transmitter is one way to reduce or eliminate intermodulation interference in a repeater caused by another transmitter operating in close proximity.

Chapter 21

Controlling Noise

Noise Blankers

E4E03 (B) Which of the following signals might a receiver noise blanker be able to remove from desired signals?

    A. Signals that are constant at all IF levels

    B. **Signals that appear across a wide bandwidth**

    C. Signals that appear at one IF but not another

    D. Signals that have a sharply peaked frequency distribution

Noise blankers are looking for short bursts of what most people call white noise; **signals which appear across a wide bandwidth**. For example, a noise blanker is very useful in a mobile HF rig to eliminate the "tick tick tick" interference sometimes generated by fuel injectors opening and closing, but useless for eliminating alternator whine, which is a continuous sound.

E4E09 (C) What undesirable effect can occur when using an IF noise blanker?

    A. Received audio in the speech range might have an echo effect

    B. The audio frequency bandwidth of the received signal might be compressed

    C. **Nearby signals may appear to be excessively wide even if they meet emission standards**

    D. FM signals can no longer be demodulated

The question committee has thrown you a big hint by calling it an "IF noise blanker." Noise blankers usually are part of the IF section; putting them farther down the chain, after the filters, would make them less effective.

Noise blankers work by detecting wide bandwidth, broad spectrum signals, since that's the nature of the impulse noises they are designed to reduce. They are like automatic switches that interrupt the signal each time they "hear" a noise pulse. Each interruption, then, produces a dreaded square wave in the IF section. Of course, square waves are rich in harmonics, so now we have a whole lot of frequencies in our IF when all we wanted was that one frequency from the VFO.

Those extra frequencies created by the noise blanker can mix with nearby signals and bring them right into the IF section, making them seem much wider in bandwidth than they actually are. It's something like the effect of reciprocal mixing we covered in Chapter 18 in the section about phase noise; those other stations aren't "talking too wide," you're "listening too wide!"

Noise Sources & Suppression Techniques

E4E04 (D) How can conducted and radiated noise caused by an automobile alternator be suppressed?

    A. By installing filter capacitors in series with the DC power lead and a blocking capacitor in the field lead

    B. By installing a noise suppression resistor and a blocking capacitor in both leads

    C. By installing a high-pass filter in series with the radio's power lead and a low-pass filter in parallel with the field lead

    D. **By connecting the radio's power leads directly to the battery and by installing coaxial capacitors in line with the alternator leads**

When we wire up a mobile ham radio we run the power leads directly from the battery terminals to the radio – inserting, of course, an in-line fuse of the proper value in both leads. It's also acceptable to connect the ground lead to the engine block, if that's where your battery connects to the chassis. That's the main key to reducing alternator whine in your radio. Why does this work? Because the battery acts somewhat like a big capacitor between the positive and negative sides, sucking up voltage surges and at least somewhat shunting high frequency noise to the ground side.

The correct answer says to also install coaxial capacitors "in line" with the alternator leads. I find this puzzling. I have searched the internet high and low for a "coaxial capacitor" and found no reference at all save for a few obscure university class physics problems, and this question on the Extra exam. You can't buy a coaxial capacitor, I can't even find a definition of one, though there are cylindrical capacitors, some of which are capable of handling the high current output of an automobile alternator, and they are constructed coaxially. Still, if we're trying to eliminate high frequencies with an in-line device, we'd want an inductor, which tends to pass low frequencies while suppressing high frequencies. Think snap-on ferrite beads. To use a capacitor, we'd hook it up in *parallel* as a shunt to ground, not "in line" where it would block the flow of DC from the alternator, and most alternators already contain just such a part.

Nevertheless, the correct answer to this question is; conducted and radiated noise caused by an automobile alternator can be suppressed **by connecting the radio's power leads directly to the battery and by installing coaxial capacitors in line with the alternator leads**. Those other answers are way, way off track.

E4E05 (B) How can radio frequency interference from an AC motor be suppressed?

    A. By installing a high pass filter in series with the motor's power leads

    B. **By installing a brute-force AC-line filter in series with the motor leads**

    C. By installing a bypass capacitor in series with the motor leads

    D. By using a ground-fault current interrupter in the circuit used to power the motor

Electric motors create EMI – ElectroMagnetic Interference – through a couple of mechanisms. First, even a brushless motor works by constantly switching magnetic fields, and we know whenever we cause a magnetic field to move, it creates an electromagnetic wave. Motors with brushes and commutators also have at least a little bit of continuous arcing going on, and that too will contribute to EMI. Both of these mechanisms generate something close to a square wave, and the inherent nature of square waves is that they create great galloping gobs of harmonics. (Musical synthesizer players say square waves are "harmonically rich", hams would say those waves are "harmonically nasty." Take your pick.)

Fortunately, an electric motor makes a terrible antenna, and the arrangement of the fields plus the shielding typically around the motor both work in our favor. Unfortunately, the motor is hooked up to wires that are much better transmitting antennas for that noise. We can keep the noise out of the power lines **by installing a brute-force AC-line filter in series with the motor leads**. "Brute force" isn't a type of filter, it just means the filter is heavy-duty since it's going to be passing some heavy loads. What we want is a low-pass filter, so only the 60 Hz power gets through the filter, and the higher frequencies of the EMI never make it out of the motor.

E4E10 (D) What might be the cause of a loud roaring or buzzing AC line interference that comes and goes at intervals?

    A. Arcing contacts in a thermostatically controlled device

    B. A defective doorbell or doorbell transformer inside a nearby residence

    C. A malfunctioning illuminated advertising display

    D. **All these choices are correct**

Be flexible in your thinking about what "comes and goes at intervals" means, and you'll see that for this question the correct answer is **all these choices are correct**. Arcing contacts in a thermostatically controlled device can create some buzzing interference as the contacts open and close. A defective doorbell or doorbell transformer might create some interference when the doorbell is pressed. A malfunctioning illuminated advertising display might only create interference when it's dark outside.

E4E06 (C) What is one type of electrical interference that might be caused by a nearby personal computer?

    A. A loud AC hum in the audio output of your station receiver

    B. A clicking noise at intervals of a few seconds

    C. **The appearance of unstable modulated or unmodulated signals at specific frequencies**

    D. A whining type noise that continually pulses off and on

Personal computers can be nasty sources of interference unless they're well shielded, but that interference most likely will not be a clicking noise at intervals of a few seconds, nor a whining type noise that continually pulses off and on. (If you're hearing a whining noise that continually pulses off and on, you should check your home for the presence of a teenager.) Most likely, if it's a computer making the noise, you'll find evidence of **unstable modulated or unmodulated signals at specific frequencies**.

If you've had the experience of placing a laptop's power supply too close to your gear, you might jump on that answer about "loud AC hum", but for this question just assume the computer is a desktop, because the test authors say "loud AC hum" is a wrong answer.

E4E11 (B) What could cause local AM broadcast band signals to combine to generate spurious signals in the MF or HF bands?

    A. One or more of the broadcast stations is transmitting an over-modulated signal

    B. **Nearby corroded metal joints are mixing and re-radiating the broadcast signals**

    C. You are receiving skywave signals from a distant station

    D. Your station receiver IF amplifier stage is defective

The correct answer to this one might seem like the least likely, but it's the right one. **Nearby corroded metal joints are mixing and re-radiating the broadcast signals**.

Every piece of metal is an antenna. Corroded joints of all sorts can create crude diodes, and diodes can act as detectors and mixers. A couple of local AM stations, especially if they're nearby, can be detected and mixed by that corroded junction, then the sums and differences we learned about in intermodulation can be re-radiated by whatever metal the joint is attached to.

Digital Signal Processors

E4E02 (D) Which of the following types of noise can often be reduced with a digital signal processing noise filter?

    A. Broadband white noise

    B. Ignition noise

    C. Power line noise

    D. **All these choices are correct**

There aren't a lot of things we can do to reduce noise. Really, we can switch the signal off when noise is present, or we can filter the noise frequencies, leaving more of the desired frequencies. Noise blankers and noise gates switch off the signal, and, of course, filters filter whatever frequencies they're designed to filter. That's about it for broad strategies.

If we had really good ears that could accurately measure frequencies and amplitudes and *really* fast fingers working really expensive equipment, we could accomplish noise reduction manually. In fact, back in ancient times, that's exactly how old or noisy recordings were brought back to some semblance of life and how hams in the pre-digital age managed to hear each other. Unfortunately, there are limits to the accuracy of our hearing, the speed of our fingers, and our budgets.

Computers can do it, though! Enter the DSP, the Digital Signal Processor. These are fast becoming standard equipment in most HF rigs, and for good reason. They can accomplish near miracles in pulling an intelligible signal out of a hash of noise. (If your radio does not have DSP, you can find software that will process the audio through your computer or use an outboard filter such as the one shown in Figure 21.1), which connects between your radio's audio output and an external speaker.

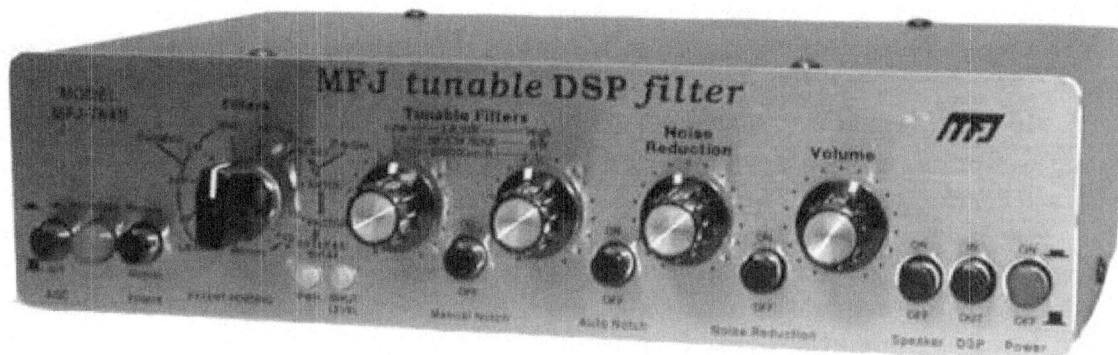

Figure 21.1: MFJ Outboard Digital Signal Processor

A DSP works by converting the analog audio signal to digital form. Once the signal is digitized, the DSP goes to work analyzing it, using complex algorithms to determine what is "signal" and what is "noise." It filters out what it decides is noise, and converts the now transformed digits back into analog, then sends the analog signal along to the audio amplifier.

Because they can deploy as many noise reduction strategies and variations on those strategies as their programming and hardware will permit, DSP's can be effective reducing broadband white noise, ignition noise, and power line noise. If it's noise, a DSP can probably reduce it. So for this question the correct answer is **all of those choices are correct**.

E4E01 (A) What problem can occur when using an automatic notch filter (ANF) to remove interfering carriers while receiving CW signals?

    A. **Removal of the CW signal as well as the interfering carrier**

    B. Any nearby signal passing through the DSP system will overwhelm the desired signal

    C. Received CW signals will appear to be modulated at the DSP clock frequency

    D. Ringing in the DSP filter will completely remove the spaces between the CW characters

A notch filter is a filter that makes a very narrow and deep cut in a band of frequencies. An *automatic* notch filter is part of a digital signal processor and its purpose is to eliminate narrow bands of interfering frequencies.

One disadvantage of using some types of automatic DSP notch filters when attempting to copy CW signals is they can be just too darned powerful! They can get, shall we say, a little carried away with the filtering and filter out everything. After all, they're designed to "notch out" narrow interfering signals; they often can't tell the difference between the narrow CW signal you want to hear and the narrow CW signal that's interfering – so you end up with

removal of the CW signal as well as the interfering carrier. Automatic notch filters, or ANF's, or "notchers" are more suited to wider bandwidth signals, such as SSB or RTTY.

Common-Mode Currents

E4E08 (B) What current flows equally on all conductors of an unshielded multi-conductor cable?
    A. Differential-mode current
    B. **Common-mode current**
    C. Reactive current only
    D. Return current

"Differential mode current" means "current is flowing from a negative source, through the negative side conductor to the device, then from the device through the positive side conductor to a positive source, as in Figure 21.2. Things are operating the way we'd expect, just the way the schematic diagram showed it would.

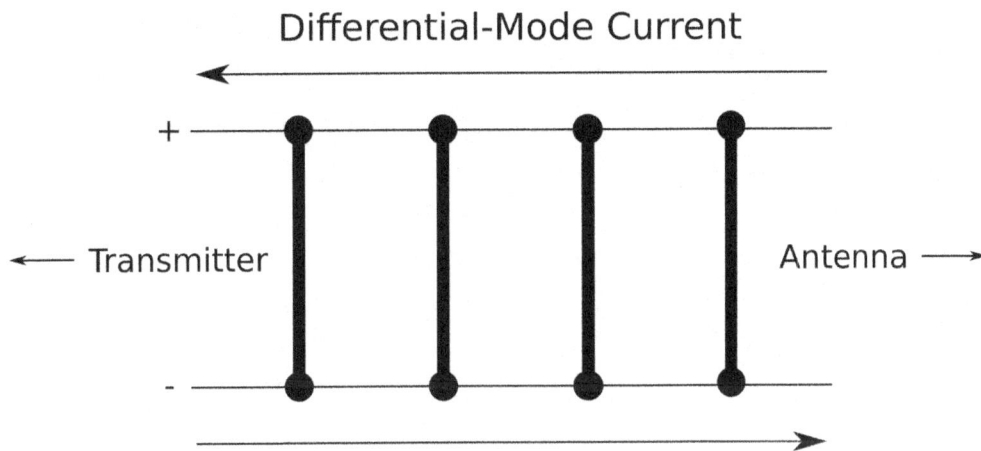

Differential-Mode Current

Figure 21.2: Differential-mode Current on Ladder Line

Common-mode current is a little difficult to imagine. Common-mode current is flowing to or from the device through both conductors as in Figure 21.3.

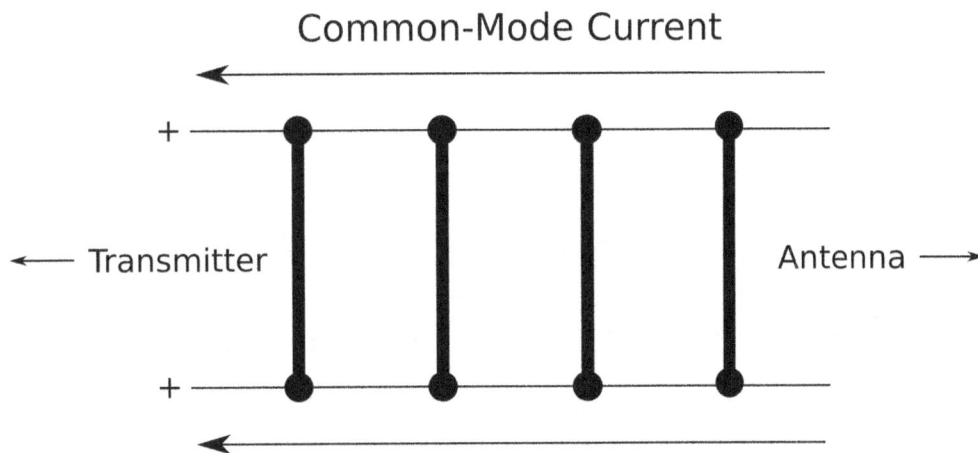

Common-Mode Current

Figure 21.3: Common-mode Current on Ladder Line

You might think the currents would cancel out, and there would be a net zero current, so no effect. Remember, though, current flows in an antenna even though it's an open circuit, and this is very much like that. Depending on their phase relationship, the common-mode current either adds to or subtracts from the desired signal, in effect modulating the signal. The inevitable result is noise and distortion.

The cures for common-mode currents include good RF grounding through wide copper straps and the installation of ferrite beads.

E4E07 (B) Which of the following can cause shielded cables to radiate or receive interference?
    A. Low inductance ground connections at both ends of the shield
    B. **Common-mode currents on the shield and conductors**
    C. Use of braided shielding material
    D. Tying all ground connections to a common point resulting in differential-mode currents in the shield

You've just studied common-mode currents, but we can get this one through process of elimination without knowing a thing about common-mode currents. "Low inductance ground connections at both ends of the shield?" Nope – that's good practice. So is the use of braided shielding material, and so is tying all ground connections to a common point.

Common-mode currents on the shield and conductors can cause shielded cables to radiate or receive interference.

Key Concepts in This Chapter

- Noise blankers can reduce noise signals that appear across a wide bandwidth.

- One undesirable effect that can occur when using an IF noise blanker is that nearby signals may appear to be excessively wide even if they meet emission standards.

- Conducted and radiated noise caused by an automobile alternator can be suppressed by connecting the radio's power leads directly to the battery and by installing coaxial capacitors in line with the alternator leads.

- Noise from an electric motor can be suppressed by installing a brute-force AC-line filter in series with the motor leads.

- All sorts of things can cause a loud roaring or buzzing AC line interference that comes and goes at intervals. For instance, arcing contacts in a thermostatically controlled device; a defective doorbell or doorbell transformer inside a nearby residence, or; a malfunctioning illuminated advertising display.

- One type of electrical interference that might be caused by the operation of a nearby personal computer is the appearance of unstable modulated or unmodulated signals at specific frequencies.

- If you are hearing combinations of local AM broadcast signals within one or more of the MF or HF ham bands, the most likely cause is that nearby corroded metal joints are mixing and re-radiating the broadcast signals.

- Many types of noise can often be reduced with a digital signal processing noise filter. They include broadband white noise, ignition noise, and power line noise.

- A problem that can occur when using an automatic notch filter or ANF to remove interfering carriers while recieving CW signals is removal of the CW signal as well as the interfering carrier.

- Common-mode current flows equally on all conductors of an unshielded multi-conductor cable. Common mode currents on the shield and conductors of shielded cables can cause those cables to radiate or receive interference.

Chapter 22

Principles of Resonant Circuits

In this chapter, we deal with resonant circuits, and we'll be dealing with them over the course of several chapters.

We wouldn't have much in the way of electronics without resonant circuits. We use them to create or to emphasize signals we want, and we use them to block signals we don't want. They're in our transmitters, our antenna tuners, and our receivers. Even our antennas are a form of resonant circuit.

Resonant Circuit Overview

Let's start with a quick review of resonant circuits. In the most basic resonant circuit, we have an inductor, a capacitor, and a source of alternating current. Figure 22.1 shows a schematic of one. It's an "LC" circuit, consisting of some "L" – inductance – and some "C" – capacitance. And, of course, there's a power source.

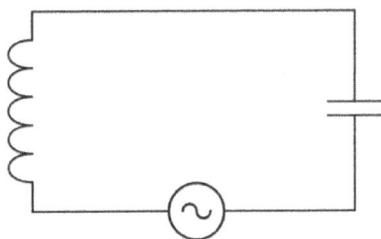

Figure 22.1: Simple Series LC Resonant Circuit

I doubt you'll ever find that exact circuit in the wild. There needs to be some resistance in there somewhere. For one thing, without a resistor the thing's really a short circuit at the resonant frequency. So let's throw in a resistor. Real circuits will look something like Figure 22.2.

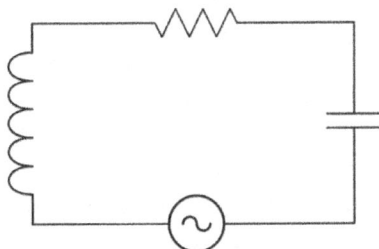

Figure 22.2: Series RLC Circuit

Now we have a series RLC circuit – inductance, capacitance, and resistance. There's only one path through, so it is a series circuit. If the frequency is low, the capacitor blocks it. If the frequency is high, the inductor blocks it. If the frequency is the Goldilocks frequency – the resonant frequency – it can get through both.

When we apply that alternating current, the coil and the capacitor alternately charge and discharge. You'll recall from your earlier studies that inductors and capacitors store and release energy.

Now we have a resonant circuit. The capacitor discharges, charging the coil, which then discharges and recharges the capacitor, and back and forth the current goes. The frequency at which the circuit resonates will be determined by the values of the coil and the capacitor. If we had a zero-resistance circuit, the current would just echo around in there at that resonant frequency forever.

The electronic value that will determine the resonant frequency is reactance, or "X." We'll take a deeper dive into reactance before we're done, but for now think of it as being about the same as resistance in a DC circuit, with an added dimension of frequency, since the "resistance" of an inductor or capacitor changes with frequency.

A circuit is resonant when the reactances of the inductance (X_L) and the capacitance (X_C) match; in other words, when $X_L=X_C$. For a visual representation, see Figure 22.3. That graph shows inductive reactance (X_L) as a solid line. It increases in a straight line as frequency increases. Capacitive reactance (X_C) is shown as a dotted line; it's curved because of the formula for capacitive reactance, and it is decreasing as frequency increases. Where the two lines cross is the resonant frequency.

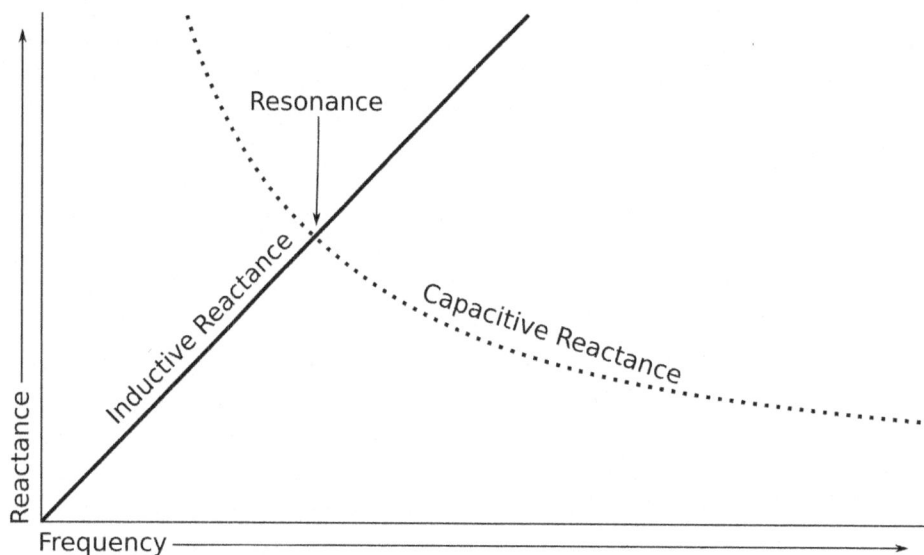

Figure 22.3: Resonance and Reactance

At the resonant frequency of the circuit, a series resonant circuit "looks like" nothing but a resistor to a power source. The inductive and capacitive reactances cancel each other out, and all that's left is whatever resistance is in the circuit.

The impedance of the circuit – its opposition to a particular frequency – is at minimum at the resonant frequency. At frequencies below the resonant frequency, the series circuit "looks like" a capacitor. At frequencies above the resonant frequency it "looks like" an inductor. You know capacitors block low frequencies and inductors block high frequencies, so frequencies above and below the resonant frequency get blocked. In a broad sense, you can think of series resonant circuits as bandpass filters. They tend to let one specific frequency go shooting through while blocking the rest. Indeed, in a series resonant circuit the circuit not only lets that frequency through, the voltages across the components – the "reactances" -- can be higher than the voltage of the input, because of that resonance, much the way a guitar string's rather feeble volume level is raised by the resonance of the guitar's body.

136

Resonant circuits can also be parallel circuits. The topic of parallel resonant circuits is much more complex than the topic of series parallel circuits, but the Extra exam takes mercy on us and keeps things very basic.

Figure 22.4 shows the most basic parallel LC resonant circuit possible.

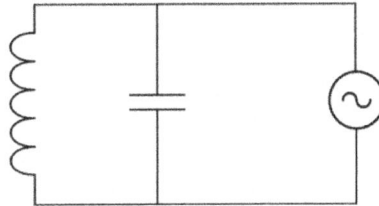

Figure 22.4: Simple Parallel Resonant Circuit

But, of course, we know we most likely need some resistance in there, so let's throw in a resistor. Figure 22.5 represents something you'd see in reality.

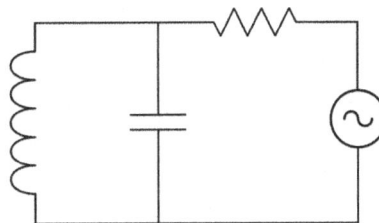

Figure 22.5: Parallel LRC Circuit

Notice the current now has two paths to follow. If it's a high frequency, it can pass through the capacitor. If it's a low frequency, it can pass through the inductor. If it's the resonant frequency, though, the reactances match and the current can't get through either one.

A parallel resonant circuit like the one in Figure 22.5 reverses the behavior of the series resonant circuit. The parallel circuit's impedance is highest at the resonant frequency. At frequencies below the resonant frequency, the circuit looks like an inductor, and inductors pass low frequencies. At frequencies above the resonant frequency, it looks like a capacitor, and capacitors pass high frequencies. Think of the parallel resonant circuits in Figures 22.4 and 22.5 as bandstop filters. They block a certain frequency – the resonant frequency -- and let everything else through. For purposes of the exam, it's important for us to examine the circuit in Figure 22.6.

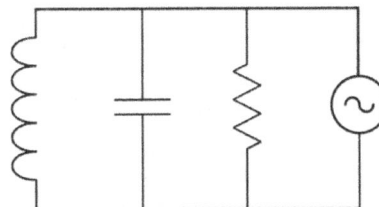

Figure 22.6: Parallel Resonant Circuit with Three Paths

The question pool includes a question about this circuit. It's a parallel resonant circuit with three paths. The rule that its impedance will be highest at the resonant frequency still applies to the LC part of the circuit. When we feed the resonant frequency to that circuit the circuit will look like a purely resistive load because the current will all be going through the resistor – the LC part will be blocking the frequency.

Calculating the Resonant Frequency of a Circuit

Given the complexity of what's going on in a resonant circuit, the formula for calculating the resonant frequency of a circuit is really pretty simple!

$$f = \frac{1}{2\pi\sqrt{L \times C}}$$

To find the resonant frequency in Hz, also known as simply "the resonance," we multiply the inductance in henries times the capacitance in farads, take the square root of that, multiply that square root by 6.28 -- -- then divide that answer into one. Let's try one. What is the resonant frequency of a series RLC circuit if R is 22 ohms, L is 50 microhenries and C is 40 picofarads? We plug in our values. Notice the "R" value isn't used in these calculations, we're just going to work with L and C for the resonant frequency.

Before we get going on this, a reminder – and some Very Good Things To Remember for the exam; milli = 10^{-3}, micro = 10^{-6}, and pico = 10^{-12}.

$$f = \frac{1}{(2\pi\sqrt{(50 \times 10^{-6}) \times (40 \times 10^{-12})})} \approx 3.56\, MHz$$

Calculating Reactance

There are at least two more values we need to calculate if we want to do something useful with a resonant circuit. To get to those, we need to know the reactances of the components. Capacitors and inductors both oppose alternating currents, and this opposition varies with the value of the component and with frequency. We call this opposition Reactance, and in formulas it gets the letter X. It is measured in Ohms. For inductors, as the frequency gets higher, the reactance gets higher; at the lowest frequency, DC, the inductor looks, essentially, like a straight wire.

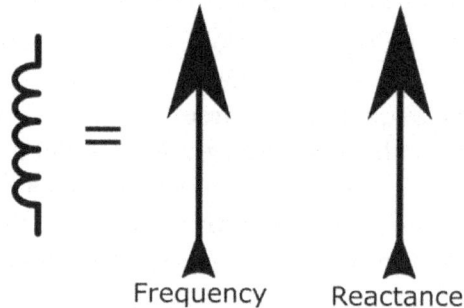

Frequency   Reactance

For capacitors, as the frequency gets higher, the reactance gets lower. At DC, the capacitor looks just like an insulator.

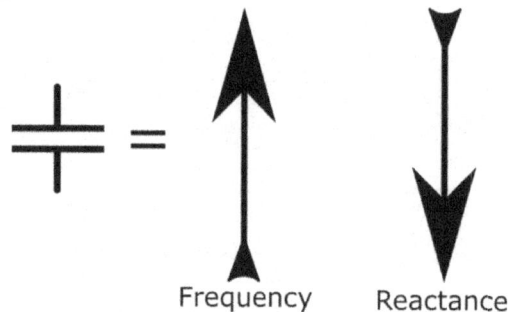

Frequency   Reactance

Since these components behave in opposite ways, we need a formula for inductive reactance and another for capacitive reactance. Let's start with inductive reactance, known in formulas as X_L.

$$X_L = 2\pi f L$$

Inductive reactance equals two times pi times the frequency times the inductance. Notice, if the frequency or the inductance increases, the reactance increases.

We know capacitors work in a way that's the opposite of inductors. Mathematicians would say they work in a way that's the reciprocal of inductors, and so the formula for capacitive reactance is the reciprocal of the formula for inductive reactance. (Reciprocal means "divide that whole thing into one.")

$$X_C = \frac{1}{2\pi fC}$$

Capacitive reactance equals two times pi times the frequency times the capacitance in Farads, all divided into one. Notice in this formula as the frequency or capacitance increases, the reactance decreases.

If you're curious about the 2π that keeps showing up in our formulas, that relates to radians. Radians are an angular measurement used in place of degrees in a lot of scientific work to vastly simplify formulas like this one. There are, of course, 360 degrees in a circle, and there are 2π radians. One cycle of a sine wave "rotates" through 360 degrees or – yep – 2π radians.

The Q Factor

The Q of a resonant circuit relates to the bandwidth of the frequencies affected by the circuit. If you're familiar with high-fidelity terminology, you could think of it as the measure of the shape of the circuit's frequency response. The lower the Q, the wider the bandwidth affected. If we graphed the frequencies passed by a relatively high Q series resonant circuit, it might look something like Figure 22.7

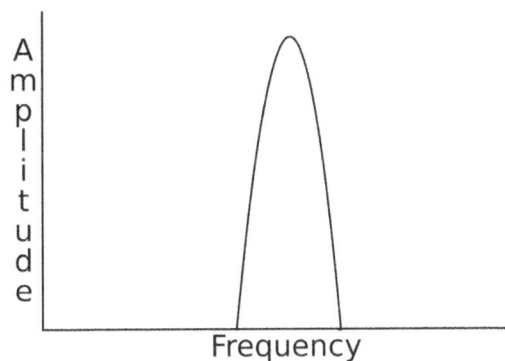

Figure 22.7: High Q Circuit

Another way to describe that high-Q filter, which you'll find in a later question, is that it has "steep skirts." The frequencies passed by a relatively low Q series resonant circuit would look more like Figure 22.8.

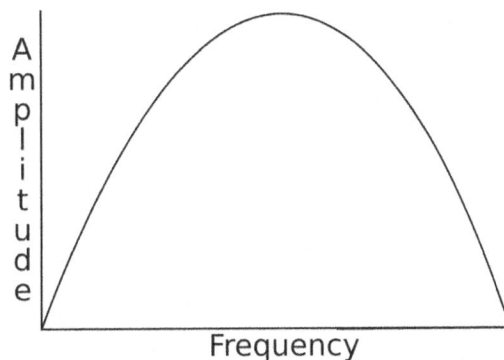

Figure 22.8: Low Q Circuit

(Consider both of those illustrations to be sketches – the actual curves are more complex.)

You can probably see that Q relates directly to *shape factor*, which we covered in Chapter 19.

For parallel resonant circuits, those curves would be upside down relative to the series curves. The calculation for the Q of a resonant circuit is where we use that R number we left out of the resonant frequency calculation. You knew it was going to show up somewhere, right? For series resonant circuits, the formula for Q is:

$$Q = \frac{X}{R}$$

Q equals the reactance at the resonant frequency divided by the resistance in series in the circuit. As reactance increases, the Q increases, and as resistance increases, the Q decreases. We know in a resonant circuit we have X_L and X_C – which X do we use? Either one! At the resonant frequency, X_L and X_C are equal. Let's say we have a reactance of 100 ohms and a resistance of 50 ohms. That gives us a Q of

$$\frac{X}{R} = \frac{100}{50} = 2$$

You might ask, "Two whats? Two ohms, two volts, two Q's.....???" None of those. The Q factor is what they call "a dimensionless term." It's just a factor, a useful number to apply to other calculations, really.

Parallel resonant circuits are a bit more complicated, but for the exam, just flip the formula above upside down.

$$Q = \frac{R}{X}$$

There's one more value relevant to resonant circuits (and so much more), and that's impedance. We will get to it, but for now, just know that impedance is the net result of the interaction of the inductive and capacitive reactances plus the resistance in a circuit. Like reactance, it is also the opposition to AC current flow, also measured in ohms, and for the next few questions, you can think of reactance and impedance as working the same way as resistance in these circuits for these questions.

In the next chapter we cover some questions that use what you've learned so far about resonant circuits.

Chapter 23

Questions About Resonant Circuits

Before we get going on this, let me assure you that if you find it to be a little overwhelming on the first pass through, you are definitely not alone. There's just no avoiding the fact that there is some complexity to resonant circuits, and therefore some complexity to the math that describes them.

One technique I think you will find very helpful is to take the time to actually work the problems yourself. Just seeing the math doesn't do much; but I find when people start writing out the formulas and punching the buttons on the calculator, it all gets a lot less complicated, and how this all works becomes a lot clearer.

Remember your practice exams are open book; you can look up the formulas and even follow along on the illustrations of what buttons to press on your calculator. The idea is to practice coming up with the right answer, not to practice making things as hard as possible!

Resonant Circuit General Characteristics

E5A02 (C) What is resonance in an LC or RLC circuit?
   A. The highest frequency that will pass current
   B. The lowest frequency that will pass current
   C. **The frequency at which the capacitive reactance equals the inductive reactance**
   D. The frequency at which the reactive impedance equals the resistive impedance

When $X_L = X_C$ the circuit will resonate at some particular frequency, and the frequency is called the circuit's "resonance." It's **the frequency at which the capacitive reactance equals the inductive reactance**.

E5A01 (A) What can cause the voltage across reactances in a series RLC circuit to be higher than the voltage applied to the entire network?
   A. **Resonance**
   B. Capacitance
   C. Conductance
   D. Resistance

Resonance can cause the voltage across reactances in series to be higher than the voltage applied to them.

If you're willing to take that on faith, move on to the next question! For those who are either skeptical or curious, here's how it works. Pick up a pencil and your calculator and join in the fun. (This will be good practice.) Let's start with the circuit shown in Figure 23.1.

Let's start crunching numbers.

First, we find the resonant frequency:

$$f = \frac{1}{2\pi\sqrt{L \times C}} = \frac{1}{2\pi\sqrt{(30 \times 10^{-3}) \times (3 \times 10^{-6})}} = 530 Hz$$

141

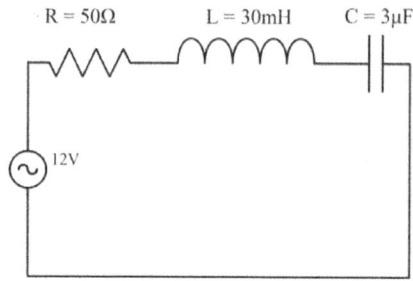

Figure 23.1: Voltage in a Resonant Circuit

To enter that in your TI-30XS calculator, it's:

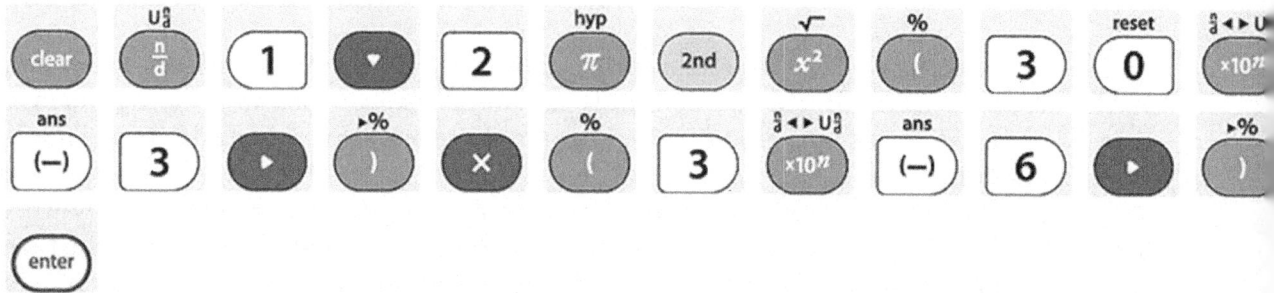

Be sure to use the [(-)] key to make numbers negative – the "subtract" key doesn't work for that.)

Next we need to know the circuit's current at the resonant frequency. This uses Ohm's Law, and all we need to consider is the resistance – at the resonant frequency, XL and XC have effectively cancelled each other out and the coil and capacitor look like a straight wire.

$$I = \frac{E}{R} = \frac{12\,volts}{50\,\Omega} = 0.24A\,(240mA)$$

Now for a little inductive reactance at resonance, X_L.

$$X_L = 2\pi fL = 2\pi \times 530\,Hz \times (30 \times 10^{-3}\,Henries) = 99.9\,\Omega$$

Let's just call it 100 ohms.

We know at resonance, $X_L = X_C$ – the inductive and capacitive reactances are equal, so we must have 100 Ω. of capacitive reactance, too.

Let's look at the voltages across the components at resonance. Since reactance is stated in ohms, we can use Ohm's Law, but it's not the $E = I \times R$ Ohm's Law we know and love. Now we've moved on to an AC version of Ohm's Law, $E = I \times X$.

$X_L = X_C$ at the resonant frequency, so $E_L = E_C$; the voltage across the inductance, L, equals the voltage across the capacitance, C.

$$E = I \times X$$

$$E_L = I \times X_L = 240\,mA \times 100\,\Omega = 24\,volts$$

Wow! The supply voltage is 12V, but at resonance, the voltages across the inductor and the capacitor are 24V peak. It's true, it's true -- **Resonance** can cause the voltage across reactances in series to be higher than the voltage applied to them. Indeed, this is a key principle behind the Tesla coil, which is nothing but a couple of step-up transformers combined with a resonant circuit.

E5A03 (D) What is the magnitude of the impedance of a series RLC circuit at resonance?
    A. High, as compared to the circuit resistance
    B. Approximately equal to capacitive reactance
    C. Approximately equal to inductive reactance

D. **Approximately equal to circuit resistance**

In a series RLC (resonant) circuit the reactance of the LC part drops to, effectively, zero at the resonant frequency. It's like you replaced the capacitor and coil with a straight wire. All that is left in the circuit is the impedance from the resistor. The circuit impedance – its opposition to current flow – is **approximately equal to circuit resistance**.

Figure 23.2: Nikola Tesla

E5A04 (A) What is the magnitude of the impedance of a parallel RLC circuit at resonance?
   A. **Approximately equal to circuit resistance**
   B. Approximately equal to inductive reactance
   C. Low compared to the circuit resistance
   D. High compared to the circuit resistance

This question asks about the impedance of the circuit. We haven't covered any formulas for impedance yet, but we don't need any impedance calculations to solve it; we just need a basic knowledge of resonant circuits.

Remember this circuit?

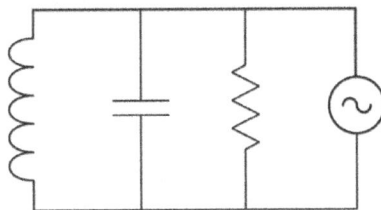

That's the circuit this question is asking about. It's a parallel RLC circuit. The impedance of the LC part on the left will be maximum – as in, "HUGE" -- at the resonant frequency. That means the doors are slammed shut through the inductor and the capacitor, leaving the resistor as the only path back home. You could almost cut the wires to the capacitor and the coil, leaving just the resistor.

That leaves the impedance in this circuit, at the resonant frequency, **approximately equal to circuit resistance.**

This is a really important concept, believe it or not, when you start thinking about tuning antennas.

With this problem, don't get lost in some fog of trying to figure out the parallel resistance/reactance – you can consider the inductor and the capacitor to have been yanked out of the circuit, leaving only the resistor in a series circuit. Think of it this way – let's say that's a 100Ω resistor. The effective "resistance" (really, reactance) of the coil and capacitor are, say, $100,000,000\Omega$ each. If we *do* bother to do the parallel resistance/reactance formula ...

$$\frac{1}{\frac{1}{100,000,000} + \frac{1}{100,000,000} + \frac{1}{100}} = 99.9998\Omega$$

Close enough to 100Ω for radio work, and definitely **approximately equal to circuit resistance!**

Current Flow and Resonant Circuits

E5A07 (A) What is the magnitude of the current at the input of a parallel RLC circuit at resonance?

    A. **Minimum**

    B. Maximum

    C. R/L

    D. L/R

Kirchoff's Law tells us that the current going into a circuit must equal the current going out. At resonance a parallel RLC circuit is allowing the minimum amount of current to pass, so the magnitude of the current at the input is at **minimum.**

E5A06 (B) What is the magnitude of the circulating current within the components of a parallel LC circuit at resonance?

    A. It is at a minimum

    B. **It is at a maximum**

    C. It equals 1 divided by the quantity 2 times pi, multiplied by the square root of inductance L multiplied by capacitance C

    D. It equals 2 multiplied by pi, multiplied by frequency, multiplied by inductance

The *circulating current* in a parallel LC circuit is, put simply, the current that isn't coming out the other end. It's "circulating" back and forth between the L and the C, lost in an eternal electronic echo chamber, never to be seen again. You can think of it as "the stuff that's getting filtered out."

We know ideal, fantasy world parallel LC circuits at resonance pass nothing. The output is at minimum and the circulating current **is at a maximum**.

Resonant Frequencies

E5A14 (C) What is the resonant frequency of an RLC circuit if R is 22 ohms, L is 50 microhenries and C is 40 picofarads?

    A. 44.72 MHz

    B. 22.36 MHz

    C. **3.56 MHz**

    D. 1.78 MHz

Heyyyyyyy .....we've seen this problem before, in the last chapter! To review,

$$f = \frac{1}{(2\pi\sqrt{L \times C})} = \frac{1}{(2\pi\sqrt{(50 \times 10^{-6}) \times (40 \times 10^{-12})})} = 3,558,812.72\,Hz \approx 3.56\,MHz$$

For the exam, you can just break a bunch of rules and simplify it to this:

$$f = \frac{1}{(2\pi\sqrt{5 \times 4})} = 0.035588127 \approx 0.0356$$

Then look for the only answer to this question with 356 in it!

To enter that formula directly in your TI-30XS calculator, press these keys:

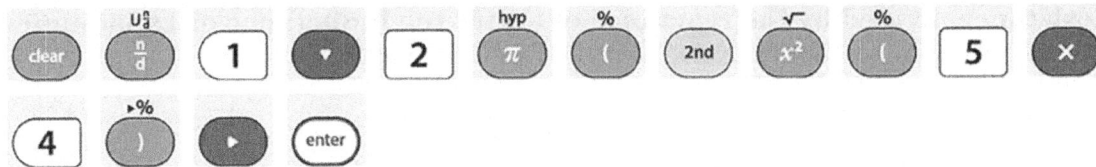

When you press [Enter] you should be looking at 0.035588127.

E5A16 (D) What is the resonant frequency of an RLC circuit if R is 33 ohms, L is 50 microhenries and C is 10 picofarads?

A. 23.5 MHz

B. 23.5 kHz

C. 7.12 kHz

D. 7.12 MHz

Same problem, different numbers to plug in. Remember, when we're calculating the resonant frequency, we don't need the R figure.

$$f = \frac{1}{2\pi\sqrt{L \times C}}$$

$$f = \frac{1}{2\pi\sqrt{(50 \times 10^{-6}) \times (10 \times 10^{-12})}} = 7.117\,MHz \approx 7.12\,MHz$$

Or, the shortcut ....

$$f = \frac{1}{2\pi\sqrt{5 \times 1}} = 0.7117 \approx 0.712$$

On this one, though, notice you need to remember that both of these resonant frequency questions come out in MHz – because they give you 7.12 kHz as an incorrect answer.

Did you notice that 7.12 MHz is precisely double the previous correct answer of 3.56 MHz? Hmmmm. Could be useful to remember....

Q

E5A10 (A) How is the Q of an RLC series resonant circuit calculated?

A. **Reactance of either the inductance or capacitance divided by the resistance**

B. Reactance of either the inductance or capacitance multiplied by the resistance

C. Resistance divided by the reactance of either the inductance or capacitance

D. Reactance of the inductance multiplied by the reactance of the capacitance

Q stands for quality; not as in "how good is this circuit," but "what are the bandwidth characteristics of this circuit."

The formula for calculating the Q of a *series* RLC resonant circuit is:

$$Q = \frac{X}{R}$$

The reactance of either the inductance or capacitance divided by the resistance.

E5A09 (C) How is the Q of an RLC parallel resonant circuit calculated?

A. Reactance of either the inductance or capacitance divided by the resistance

B. Reactance of either the inductance or capacitance multiplied by the resistance

C. **Resistance divided by the reactance of either the inductance or capacitance**

D. Reactance of the inductance multiplied by the reactance of the capacitance

For RLC parallel resonant circuits, where everything is in parallel, the formula for Q is:

$$Q = \frac{R}{X}$$

Resistance divided by the reactance of either the inductance or capacitance.
Think of it this way. Let's take this parallel RLC resonant circuit:

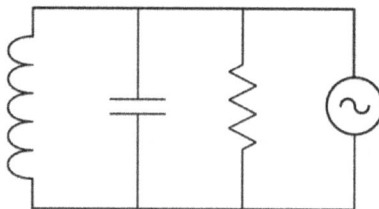

Think of the coil and capacitor on the left being a blockade to the resonant frequency.

Frequencies around the resonant frequency can go through, but those frequencies near resonance face a tough fight. They're going to get filtered out. But if there was a straight wire where that resistor is, the frequencies would just take that route and wouldn't be filtered out at all. As we raise the value of that resistor, we force more and more of the signal to "run the blockade", which means the Q gets higher. We want the R on the top of the fraction so that as R gets bigger, the Q gets higher.

E5A13 (C) What is an effect of increasing Q in a series resonant circuit?
    A. Fewer components are needed for the same performance
    B. Parasitic effects are minimized
    C. **Internal voltages increase**
    D. Phase shift can become uncontrolled

The effect of increasing Q in a resonant circuit is that **internal voltages increase.**

Just think of it this way. High Q means some intense filtering is going on, and that energy being filtered has to go somewhere. It isn't going out the other end, so it must be in the circuit. That's where it is, "circulating" back and forth between the inductance and the capacitance. **As Q increases, internal voltages increase**.

E4B08 (C) Which of the following can be used to measure the Q of a series-tuned circuit?
    A. The inductance to capacitance ratio
    B. The frequency shift
    C. **The bandwidth of the circuit's frequency response**
    D. The resonant frequency of the circuit

High Q = narrow bandwidth in a series resonant, that is, "tuned"circuit. You can use **the bandwidth of the circuit's frequency response** as a relative measure of the Q of a series tuned circuit.

E5A15 (A) Which of the following increases Q for inductors and capacitors?
    A. **Lower losses**
    B. Lower reactance
    C. Lower self-resonant frequency
    D. Higher self-resonant frequency

Since the formula for Q is $Q = \frac{X}{R}$, if we decrease the R we end up with a higher Q. Well, losses in a circuit come from resistance, so if we lower the resistance – say, by using some fancy low resistance wire, or just shortening all the signal paths – we raise the Q.

E5A05 (A) What is the result of increasing the Q of an impedance-matching circuit?
    A. **Matching bandwidth is decreased**
    B. Matching bandwidth is increased
    C. Matching range is increased
    D. It has no effect on impedance matching

The higher the Q, the narrower the bandwidth of frequencies that are either passed, in the case of a series resonant circuit, or blocked, in the case of a parallel resonant circuit.

In the case of an impedance matching circuit, increasing the Q narrows the range of possible impedance matches – the **matching bandwidth is decreased.**

Half-Power Bandwidth

E5A11 (C) What is the half-power bandwidth of a resonant circuit that has a resonant frequency of 7.1 MHz and a Q of 150?

    A. 157.8 Hz

    B. 315.6 Hz

    C. **47.3 kHz**

    D. 23.67 kHz

Here's where that Q number gets put to practical use. This question asks about a parallel resonant circuit. We know a parallel resonant circuit opposes the flow of the resonant frequency. Let's say we have a high Q parallel resonant circuit and its frequency response looks like this:

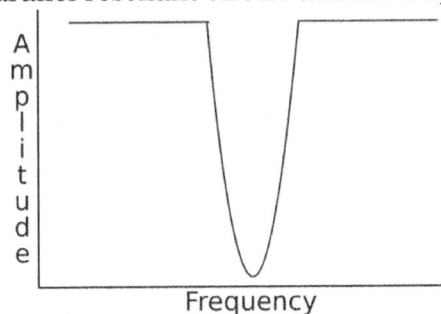

The half-power bandwidth is the frequency difference between the lowest frequency that is dropped to half power, which is – 3 DB and the highest frequency that is dropped to half power.

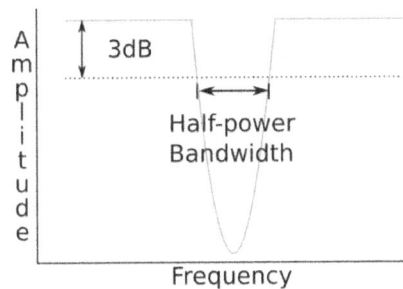

Figure 23.3: Half-power Bandwidth

Now that you understand the concept of half-power bandwidth, here's how to calculate that bandwidth. Happily, it's a simple formula.

$$half\ power\ bandwidth = \frac{f_{resonant}}{Q}$$

The half-power bandwidth equals the resonant frequency divided by the Q.

$$half\ power\ bandwidth = \frac{7,100,000\ Hz}{150} = 47,333\ Hz = 47.3\ kHz$$

E5A12 (C) What is the half-power bandwidth of a resonant circuit that has a resonant frequency of 3.7 MHz and a Q of 118?

    A. 436.6 kHz

    B. 218.3 kHz

    C. **31.4 kHz**

D. 15.7 kHz

Here's the same question as the previous question, just with different numbers to plug into our formula.

$$half\ power\ bandwidth = \frac{3,700,000\,Hz}{118} = 31,355\,Hz \approx 31.4\,kHz$$

Key Concepts in This Chapter

- The resonance of an LC or RLC circuit is the frequency at which the capacitive reactance equals the inductive reactance.

- Resonance can cause the voltage across reactances in a series RLC circuit to be higher than the voltage applied to the entire network.

- The magnitude of the impedance of a series RLC circuit at resonance is approximately equal to circuit resistance.

- The magnitude of the impedance of a parallel RLC circuit at resonance is approximately equal to circuit resistance.

- The magnitude of the current at the input of a parallel RLC circuit at resonance is minimum.

- The magnitude of the circulating current within the components of a parallel LC circuit at resonance is at a maximum.

- The formula to calculate the resonant frequency of an RLC circuit is:

$$f = \frac{1}{2\pi\sqrt{L \times C}} \tag{23.1}$$

- The formula to calculate the Q of an RLC *series* resonant circuit is:

$$Q = \frac{X}{R} \tag{23.2}$$

- The formula to calculate the Q of an RLC *parallel* resonant circuit is:

$$Q = \frac{R}{X} \tag{23.3}$$

- The effect of increasing Q in a series resonant circuit is internal voltages increase.

- The bandwidth of the circuit's frequency response can be used to measure the Q of a series-tuned circuit.

- Lower losses increase the Q for inductors and capacitors.

- The result of increasing the Q of an impedance-matching circuit is that matching bandwidth is decreased.

- The formula to calculate half-power bandwidth of a resonant circuit is:

$$half\ power\ bandwidth = \frac{f_{resonant}}{Q} \tag{23.4}$$

Chapter 24

Time Constants and Phase Relationships

Time Constants

E5B01 (B) What is the term for the time required for the capacitor in an RC circuit to be charged to 63.2% of the applied voltage or to discharge to 36.8% of its initial voltage?

    A. An exponential rate of one

    B. **One time constant**

    C. One exponential period

    D. A time factor of one

The whole key to resonant circuits is that different values of components charge and discharge at different rates. The term for that rate is *time constant*, and it is represented by the Greek letter *tau*; τ.

One time constant is defined as the time required, in seconds, for the capacitor in an RC circuit to be charged to 63.2% of the applied voltage. The formula for calculating it is quite simple:

$$\tau = RC$$

The τ is simply the resistance in the circuit, in ohms, times the capacitance in farads. The bigger the resistance or the capacitance, the bigger the τ, and the longer it will take to charge the capacitor.

But why 63.2%? What does that number have to do with anything? Did they just throw random numbers in a hat and pick one? Nope. Let's take a look at how a capacitor charges. No matter what the value, the voltage across the capacitor is going to create the sort of flattening curve shown in Figure 24.1 as the capacitor charges.

At first, lots of electrons are rushing in, and the voltage is increasing rapidly. Then, as the capacitor fills, it gets harder to push in more electrons – think of blowing up a balloon.

So, let's imagine we're going to charge a capacitor to 100 volts, and let's say the τ of the circuit is 1 second. You can see a graph of what happens in Figure 24.1. After the first τ – one second – the voltage across the capacitor is 63.2 volts. After another second, it has only gone up 23.3 volts to 86.5 volts. We wait another second – now the voltage has only gone up 8.5 volts to 95 volts. Another whole second goes by with a gain of only 3.2 volts to 98.2 volts and yet another full second of charging only gets us to 99.3 volts. If it seems like we're never going to quite get to 100 volts at this rate, you are correct!

It's one of those Zeno's Paradox deals. Zeno was that Greek philosopher who reasoned that one could never walk across a room because before one could get across the room, one had to travel half-way across the room, and before one could travel half-way ... well, you can see where this is going. Or perhaps I should say, where it isn't going! Mathematicians call this sort

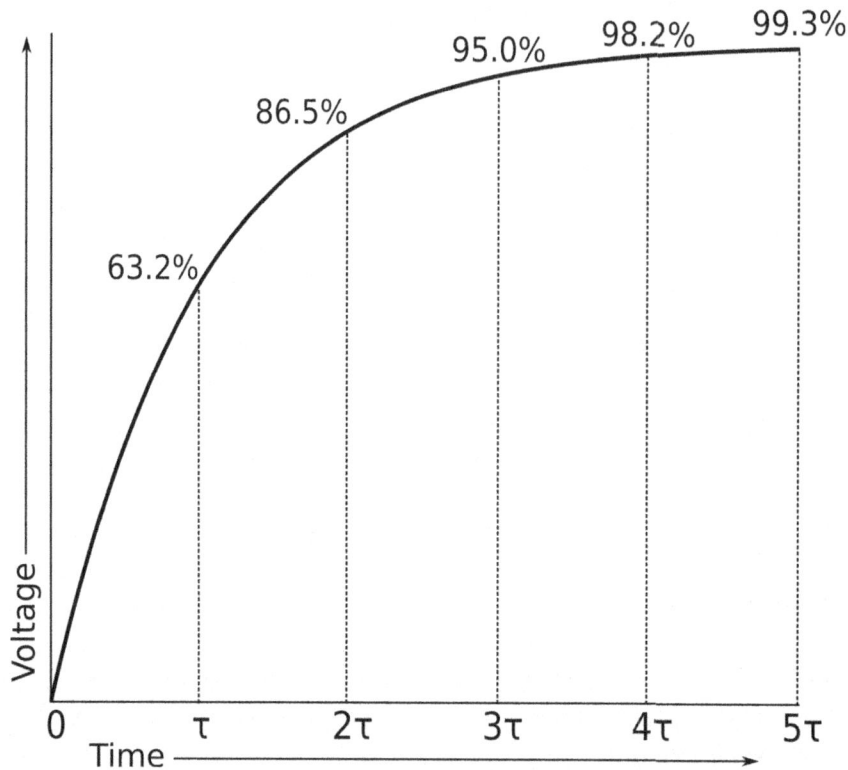

Figure 24.1: Series RC Circuit Capacitor Voltage

of thing a non-summable infinite series. As it turns out, these turn up all over the physical world, not just in electronics, and the math to describe these phenomena is based on what are called *natural logarithms*.

The most common logarithms, such as the ones we use to calculate dB problems, are "base 10 logarithms." They answer the question, "To what power do I need to raise the number 10 (the base number) to equal this number?" The base 10 logarithm of 5 is ≈ 0.698, and $10^{0.698} \approx 5$. The base of natural logarithms is a mathematical constant known as e, or Euler's Constant, which is approximately 2.178. It's a rather strange number that shows up in electronics, physics, geology, and even biology. And $1 - \frac{1}{e} \approx 0.632$ So, aside from its mathematical and engineering significance, it turns out that using 63.2% to define the time constant, τ, spares us from using the natural logarithm (l_n) and e^x keys on our calculator and simplifies our formula down to $\tau = RC$.

Inductors have time constants, too, though none of the questions on the exam deal with time constants of inductors. Just for your reference, the formula is:

$$\tau = \frac{L}{R}$$

E5B04 (D) What is the time constant of a circuit having two 220-microfarad capacitors and two 1-megohm resistors, all in parallel?

    A. 55 seconds

    B. 110 seconds

    C. 440 seconds

    D. 220 seconds

Now we get to use our simple formula for the time constant, $\tau = RC$, after we do some slightly more complicated math to figure out the R and the C.

The question describes the circuit shown in Figure 24.2.

We have two capacitors and two resistors, all in parallel. We need to know the total resistance in the circuit and the total capacitance. There's no special formula for "parallel time

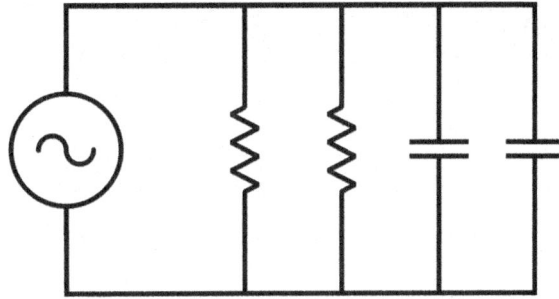

Figure 24.2: Schematic for Question E5B04

constant." (Nice, right?)

The resistors are 1 megohm each, and they're in parallel. Because we wisely and diligently studied the *Fast Track* books, we remember that two equal resistors in parallel add up to one-half the value of one of the resistors, so we have 500 kΩ of resistance, or 500,000 Ω.

Capacitors in parallel add together, just like resistors in series, so we have

$$2 \times 220 \; microfarads = 440 \; microfarads$$

440 $\mu farads$ of capacitance, also known as 440×10^{-6} or 0.00044 farads.

Now we multiply the capacitance by the resistance to find our τ, the time constant.

$$500,000 \; \Omega \times (440 \times 10^{-6} \; farads) = 220 \; seconds$$

Phase Angle

As we advance in electronics, it seems *phase angle* becomes a more and more important concept.

Let's start with the nature of a sine wave. A sine wave is really just a graph of a circle – you could say it's a graph of a spinning circle. Imagine a wheel that is rotating clockwise once per minute. We'll set that wheel next to a sheet of graph paper, so the center of the wheel is lined up with the x axis. You can see a picture of this in Figure 24.3.

Let's pick a point on the circle at what would be 9 o'clock on a clock face. We'll take a ruler and at the "zero seconds" point on the x axis, we make a dot. We wait five seconds. Now the point on the wheel has rotated up to the 10 o'clock position. We find the five seconds mark on the x axis, and the point on the y axis that the point is lined up with now. We make a mark above the x axis, to the right of the first mark. Every five seconds, we plot the position of that point on the wheel on our graph. After the wheel rotates 360°, we'll have a plot of a sine curve.

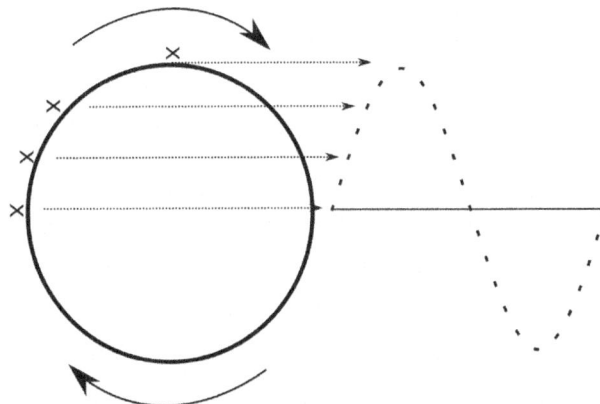

Figure 24.3: Sine Wave

If this is still puzzling, here's a URL for a cool little animation of the process:

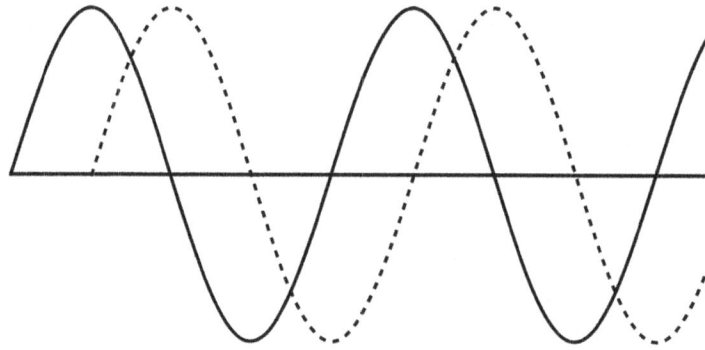

Figure 24.4: 90° Phase Angle

https://upload.wikimedia.org/wikipedia/commons/0/08/Sine_curve_drawing_animation.gif

When we talk about a phase angle, we're talking about comparing one position on that circle to another, in terms of degrees of rotation.

Figure 24.4 shows what a 90 degree phase angle (or 90° phase *difference*) between two sine waves of equal frequency looks like:

Just as the first sine wave – represented by the solid line – reaches the 90° mark, the second sine wave is at the 0 mark and starting up. We call that a "90° phase angle."

If the sine waves matched up at all points, we'd say they're "in phase", but these are "out of phase", and they're out of phase by 90°.

Now things get a little strange. When we learn DC electronics, we learn that voltage and current always go together; put a higher voltage on a circuit, the current instantly gets higher, because of Ohm's Law. I call it the "garden hose" model of electricity, and you've heard lots of garden hose related similes as you've learned DC electricity. The garden hose model works great. For DC.

AC circuits make us set aside that thinking. In AC circuits, current and voltage can get out of phase. We can have spots where there is current with no voltage, or voltage with no current. Current and voltage can be put out of phase by capacitors and inductors.

Let's say in the illustration above, that the dotted line represents voltage and the solid line represents the current. In that illustration the current is *leading* the voltage by 90°, as it does in the case of a capacitor. In other words, the current is "happening" before the voltage. As you'll learn, that means we have a "negative phase angle."

E5B09 (D) What is the relationship between the AC current through a capacitor and the voltage across a capacitor?
    A. Voltage and current are in phase
    B. Voltage and current are 180 degrees out of phase
    C. Voltage leads current by 90 degrees
    D. **Current leads voltage by 90 degrees**

When AC passes through a capacitor, the voltage lags behind the current. As that voltage is on the up cycle, in effect, current passes freely through the capacitor. There's no resistance there at all, so no voltage across the capacitor. When the voltage peaks, the current has dwindled to, effectively, nothing. As the voltage drops on the part of the cycle heading toward zero, the pressure on the incoming side of the capacitor is reduced, and current starts to flow in the reverse direction before the voltage ever becomes negative. Current leads voltage by 90 degrees.

E5B10 (A) What is the relationship between the AC current through an inductor and the voltage across an inductor?
    A. **Voltage leads current by 90 degrees**
    B. Current leads voltage by 90 degrees
    C. Voltage and current are 180 degrees out of phase
    D. Voltage and current are in phase

152

When AC passes through an inductor the current lags behind the voltage; remember, as that voltage is headed up, those magnetic lines of force are generating opposition to current flow. For simplicity's sake, we'll say the inductor looks like an insulator to the current. Then the voltage peaks, and starts heading back toward zero, and those magnetic lines of force start collapsing, creating current flow. The end result is **voltage leads current by 90 degrees.**

The summer of 1885 in London, England, was dreadfully hot. Of course, in those days, there were no refrigerators, but there were ice boxes, and there were commercial ice services that would deliver big five pound blocks of ice to homes to keep food fresh.

During that summer, Oliver Heaviside, was sweating away at his desk, coming up with the equations that would lay the foundation for our understanding of reactances and impedances. Heaviside was, by all accounts, absolutely brilliant, but is all but forgotten these days. Among other innovations, he invented coaxial cable, and his Telegrapher's Equations made trans-Atlantic telegraphy possible.

Figure 24.5: Oliver Heaviside

Every week, Heaviside's meditations would be interrupted by his ice delivery from Eli's Ice Company (Roy G. Biv, prop.) Heaviside was a notorious sourpuss and anything but a social butterfly – he was almost a recluse – and Eli's deliveries always irritated him; until one day, he suddenly saw that Eli was, unwittingly, the perfect mnemonic for the voltage and current relationships Heaviside was discovering. "Gadzooks!", he cried, "Voltage leads current across an inductor and current leads voltage across a capacitor – ELI the ICE man! Oh, jolly good."

Ever since that fateful (and completely fictitious) day, hams and other electronics students have used "ELI the ICE Man" to remember that across an inductance (across an L) E (voltage) comes before I (current); E leads I. Across a Capacitance, I leads E. Good ol' ELI the ICE Man!

When we say the voltage is leading the current, as it does across an inductance, we mean the voltage "happens" before the current. See Figure 24.7.

We could also describe that state of affairs as "the current is *lagging* the voltage."

When we say the current is leading the voltage, as it does across a capacitance, that means the current happens first, followed by the voltage. Of course, we could also say "the voltage is lagging the current." That's the case in Figure 24.8.

E5A08 (C) What is the phase relationship between the current through and the voltage across a series resonant circuit at resonance?

A. The voltage leads the current by 90 degrees

B. The current leads the voltage by 90 degrees

C. **The voltage and current are in phase**

D. The voltage and current are 180 degrees out of phase

Figure 24.6: Eli The Ice Man (Actual Photo)

Voltage Leading Current

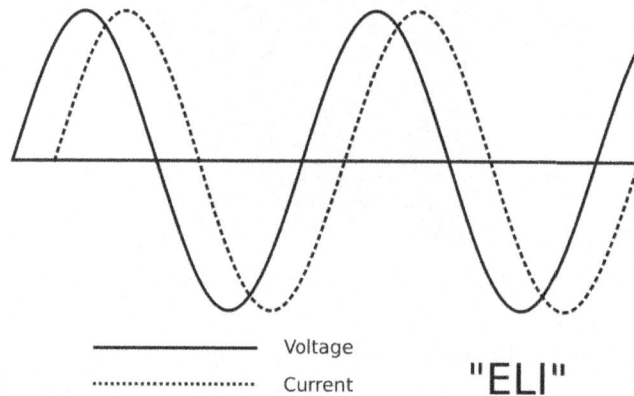

Voltage
Current   "ELI"

Figure 24.7: Voltage Leading Current: ELI

Here's the real heart of how a resonant circuit works. In a series resonant circuit at resonance, **the voltage and current are in phase**.

ELI the ICE man tells us that the inductor in the circuit is causing the voltage to lead the current, and the capacitance is causing the current to lead the voltage. As we approach the frequency where those two effects balance out, the overall effect is to move the current and voltage closer and closer together in phase, until at resonance they match.

Well, so what? Ah! Remember, power always equals voltage times current. When voltage and current are out of phase, we always have either less current or less voltage to multiply together than if they were in phase. And that's how resonant circuits work, and, on a very-much-related note, why we're forever concerned with matching impedances as well.

We are now fast approaching a series of questions that, based on my conversations with hams over the years, has scared a lot of hams away from the Extra exam. DO NOT PANIC. I will get you through this, and, in fact, these are some of the easiest questions on the exam to get right. You'll see.

Seriously, if I can do this, you can do this. I passed high school math mostly on my good looks and winning personality – and probably on the teacher's eagerness to never see me again. So, relax.

Current Leading Voltage

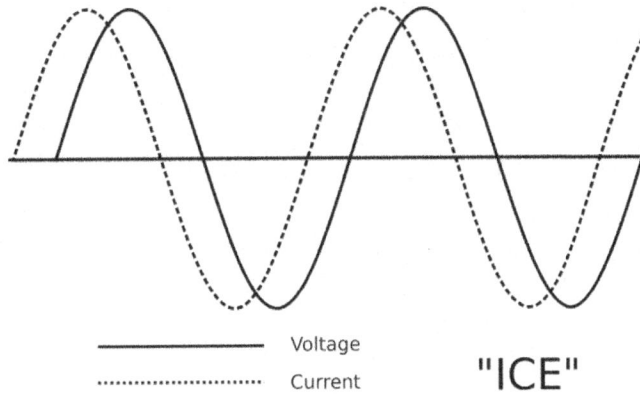

Figure 24.8: Current Leading Voltage: ICE

Besides, the two sections that seem to be the most intimidating account for, at the most, two questions out of 50 on the test – you can completely mess them up and still pass! But – you are a *Fast Track* student! You will not mess them up!

E5B07 (C) What is the phase angle between the voltage across and the current through a series RLC circuit if XC is 500 ohms, R is 1 kilohm, and XL is 250 ohms?
    A. 68.2 degrees with the voltage leading the current
    B. 14.0 degrees with the voltage leading the current
    C. **14.0 degrees with the voltage lagging the current**
    D. 68.2 degrees with the voltage lagging the current

To calculate the phase angle of a series RLC circuit, we need the capacitive reactance (X_C), the inductive reactance (X_L), and the resistance. Lucky for us, they've given us all the values we need in the question, and all we need to do is plug them into the right formula. Here's the right formula:

$$Phase\,Angle = \tan^{-1}\left(\frac{X_L - X_c}{R}\right)$$

The answer comes out in degrees. If the degrees are a negative number, that means the current is leading the voltage or, put another way, the voltage is lagging the current.

So let's plug in our numbers. If you're using the TI-30XS calculator I recommend, look at the "tan" key. Above it, you'll see "tan-1", and you'll be using that key by pressing the "2nd" key in the upper left hand corner – the bright green one – then the "tan" key.

$$Phase\,Angle = \tan^{-1}\left(\frac{250\,\Omega - 500\,\Omega}{1000\,\Omega}\right) = -14.03624347° \approx -14.0°$$

The sign of the phase angle tells us where the voltage is relative to the current. A negative phase angle means the voltage is behind the current.

The answer is the phase angle is **14.0 degrees with the voltage lagging the current.**

I'll give you the exact keys I pressed on my calculator to get this answer.

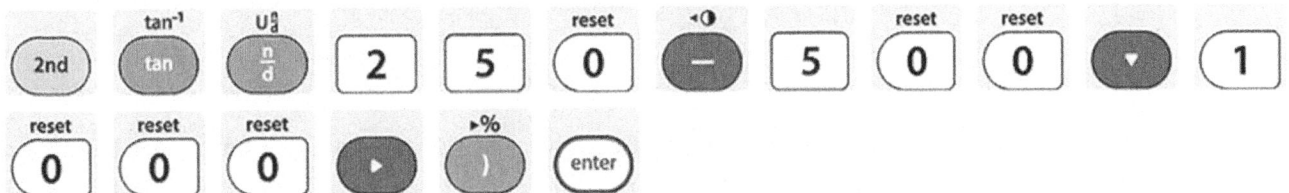

Press "Enter" and you should be looking at that -14.036 number. That's all there is to it. Congratulations, you were just doing trigonometry.

If you can't seem to find the "tan^{-1}" key on your scientific calculator, look for "arctan", or "arctangent." It's the same function, just a different way to write it.

Now let's step back a moment. Does that negative phase angle make sense?

We have more capacitive reactance than inductive reactance, so we have a net capacitive reactance. ELI the ICE man tells us that across a capacitance, voltage lags current, so yes, negative 14° makes sense.

E5B08 (A) What is the phase angle between the voltage across and the current through a series RLC circuit if XC is 100-ohms, R is 100-ohms, and XL is 75 ohms?

    A. **14 degrees with the voltage lagging the current**
    B. 14 degrees with the voltage leading the current
    C. 76 degrees with the voltage leading the current
    D. 76 degrees with the voltage lagging the current

This is the same problem with different numbers. Let's plug in our values ...

$$Phase\,Angle = \tan^{-1}\left(\frac{75\,\Omega - 100\,\Omega}{100\,\Omega}\right) = -14.03624347° \approx -14.0°$$

Hmmmm .... the answer is **14 degrees with the voltage lagging the current**. Seems vaguely familiar.

E5B11 (B) What is the phase angle between the voltage across and the current through a series RLC circuit if XC is 25 ohms, R is 100-ohms, and XL is 50 ohms?

    A. 14 degrees with the voltage lagging the current
    B. **14 degrees with the voltage leading the current**
    C. 76 degrees with the voltage lagging the current
    D. 76 degrees with the voltage leading the current

$$Phase\,Angle = \tan^{-1}\left(\frac{50\,\Omega - 25\,\Omega}{100\,\Omega}\right) = +14.03624347° \approx +14.0°$$

We get **14.0 degrees with the voltage *leading* the current**, because we have a positive value for the phase angle.

All the phase angle problems on the exam work out to either $-14°$ or $+14°$. The formula's pretty simple, but the real key to this, so far as the exam goes, is remembering that in this and other reactance related equations, *capacitive reactance is a negative value*, so we always subtract the X_C from the X_L. As we get into calculating and graphing impedance, capacitive reactance is a negative value there, too, and you'll literally get a picture of that relationship. When you take the exam, you know if they're asking about phase angle and the capacitive reactance (X_C) is a bigger number than the inductive reactance (X_L) the answer has to be $14°$ with the voltage lagging the current. If X_L is bigger, it's $14°$ with the voltage leading the current. Positive number, voltage leading, negative number, voltage lagging. Think of phase angle as "how far ahead is the voltage."

Key Concepts in This Chapter

- The term for the time required for the capacitor in an RC circuit to be charged to 63.2% of the applied voltage is one time constant. The formula for the time constant is:

$$\tau = RC$$

- The time constant of a circuit having two 220 microfarad capacitors and two 1 megohm resistors, all in parallel, is 220 seconds.

- The relationship between the current through a capacitor and the voltage across a capacitor is that current leads voltage by 90°.

- The relationship between the current through an inductor and the voltage across an inductor is that the voltage leads current by 90°.

- The mnemonic for remembering those relationships is ELI the ICE man.

- In a series resonant circuit at resonance, voltage and current are in phase.

- The formula for calculating the phase angle through a series RLC circuit is:

$$Phase\ Angle = \tan^{-1}\left(\frac{X_L - X_c}{R}\right)$$

- The phase angle between the voltage across and the current through a series RLC circuit if XC is 500 ohms, R is 1 kilohm, and XL is 250 ohms is 14.0° with the voltage lagging the current.

- The phase angle between the voltage across and the current through a series RLC circuit if XC is 100-ohms, R is 100-ohms, and XL is 75 ohms is 14° with the voltage lagging the current.

- The phase angle between the voltage across and the current through a series RLC circuit if XC is 25 ohms, R is 100 ohms, and XL is 50 ohms is 14° with the voltage leading the current.

When you are done with the Practice Exam for this chapter, find out how you're doing overall by taking Progress Check #5 at fasttrack-ham.com.

Chapter 25

Coordinate Systems and Phasors

Impedance

In this section, you're going to learn how to calculate and graph the impedance and phase angle of a circuit. We'll also cover some less-well-known electronics terms that relate to those calculations and graphs.

Let's start with a review of impedance basics.

Impedance, Z, is the opposition to current flow in an AC circuit. It is measured in ohms. In an Ohm's Law style calculation, it can be substituted for R, just like reactance, X.

$$\frac{E}{I} = R \; or \; X \; or \; Z$$

That says, in math, that impedance is the *ratio* of voltage and current. A "high impedance" circuit is a high voltage, low current circuit. A "low impedance" circuit has low voltage and high current. Obviously, low and high impedance are relative terms; there's no specific impedance value where "high impedance" begins.

Impedance is created by the interaction of four values.

FREQUENCY: Resistance, capacitance, and inductance are all fixed values, in the sense that those components are what they are, no matter what circuit they are in. Impedance is not a fixed value. You can't just go to the electronics parts store and buy x amount of impedance. Impedance usually varies with frequency.

RESISTANCE: A resistor in an AC circuit behaves just the same way a resistor behaves in a DC circuit. It dissipates electrical power as heat. A 50-ohm resistor in an AC circuit is exactly the same as a 50-ohm resistor in a DC circuit.

CAPACITIVE REACTANCE: A capacitor passes high frequencies, and opposes low frequencies. Its opposition to a particular frequency is its reactance.

INDUCTIVE REACTANCE: An inductor passes low frequencies – remember, it's essentially invisible to direct current, which has a frequency of zero – and opposes high frequencies, just the opposite behavior of a capacitor. The quantification of the inductor's opposition to a specific frequency is its reactance.

It's those characteristics that let us combine resistors, inductors and capacitors in various ways to create resonant circuits.

According to Eli the Ice Man, reactance affects the phase relationship of voltage and current.

By controlling the inductance and capacitance in an AC circuit, we control those phase relationships. When we mathematically combine the Resistance and Reactance in a circuit, we get Impedance, represented in formulas by the letter **Z**.

Obviously, this is a more complex calculation than E = I x R. It takes a couple of steps to complete. First, we have to calculate the Capacitive Reactance and the Inductive Reactance, then use those to calculate the Impedance.

Resistance is, of course, measured in ohms. So is reactance. So is impedance. While this has been the source of confusion for generations of folks learning electronics, it's critical to our calculations that all the units are the same.

Resistive impedance is equal to the ohms of resistance, no matter what the frequency. Simple. There is no "resistive reactance", resistors aren't reactive. They're a "pure" impedance. A 50-ohm resistor has 50 ohms of impedance, period. Put mathematically, $X_R = R$.

Inductive reactance is :

$$X_L = 2\pi f L$$

In the inductive reactance formula, as the frequency or inductance increase, reactance increases. Mathematicians would say that in that formula, frequency, inductance, and reactance have a direct relationship.

Capacitive reactance is the reciprocal of inductive reactance. In a sense, it's the opposite, since capacitors act in a manner that's the opposite of inductors. The formula is:

$$X_C = \frac{1}{2\pi f C}$$

In the capacitive reactance formula, as frequency or capacitance *increase*, the reactance *decreases*. It's what math folks call an inverse relationship.

To calculate the total reactance of a circuit containing X_L and X_C we subtract the X_C from the X_L.

$$X = X_L - X_C$$

It's always "subtract X_C from X_L", never the reverse, because capacitive reactance is a negative value.

On to the last step: This is the formula for Z, impedance:

$$Z = \sqrt{R^2 + X^2}$$

Impedance equals the square root of the resistance squared plus the reactance squared.

With a little basic algebra, we can rewrite that formula as:

$$R^2 + X^2 = Z^2$$

or....

$$A^2 + B^2 = C^2$$

Whaddaya know! It's the good ol' Pythagorean Theorem. Eighth grade math. For all the mystery and mathematics that surround the topic, every single impedance problem comes down to nothing but a "what is the length of hypotenuse C" problem. Not only that, but once you do the math to graph it out as a triangle, the angle between the base and the hypotenuse equals the number of degrees the current is out of phase with the voltage (see Figure 25.1.)

In Figure 25.1, we have the resistance plotted as a point on the X axis of our graph – that's "X" the axis, not "X" for reactance – and reactance plotted on the Y axis. In the example in Figure 25.1, the reactance is more inductive than capacitive, so we end up with a positive value for reactance. We also end up with a positive phase angle when we draw the hypotenuse of this triangle by connecting the zero point of the X axis to the point we plotted on the Y axis above the resistance value on the X axis. The length of that hypotenuse equals the impedance.

If the reactance was more capacitive than inductive, we'd plot a negative number on the Y axis, and end up with a triangle like the one shown in Figure 25.2.

That triangle shows a negative phase angle.

Those diagrams have a technical name – they are called "phasor diagrams." Notice that in both phasor examples above, we're really just identifying a point on a graph, which is the point where the impedance and reactance lines meet. We could specify that point in a couple of ways, and it turns out that each different way comes in handy in different situations. Imagine that

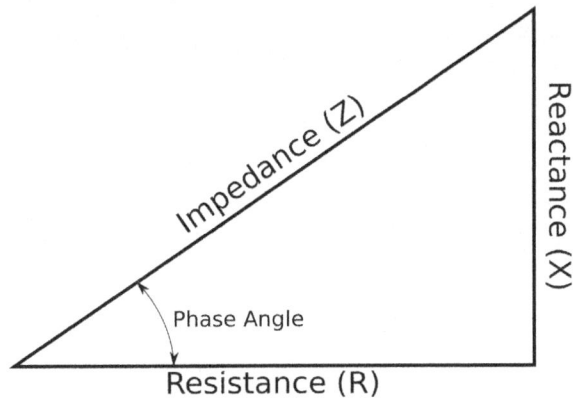

Figure 25.1: Phasor Diagram Showing Inductive Reactance

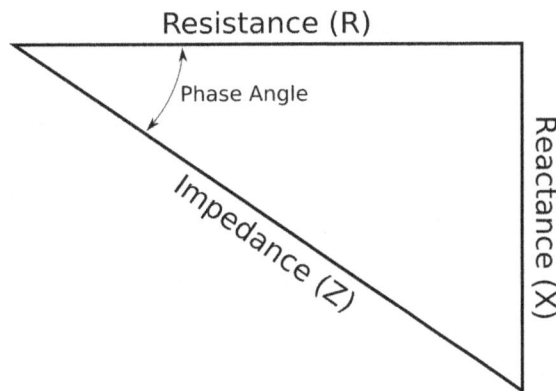

Figure 25.2: Phasor Diagram Showing Capacitive Reactance

point is a destination where you want someone to meet you. You could tell them, "Ralph, you're going to start at the corner of X Street and Y Avenue. You walk straight east on the X road 400 paces, make a 90 degree turn to the left, then walk 300 paces north and you'll be there." If you did that, you'd be using what the Amateur Extra exam calls "rectangular coordinates." Specifically, the 300 and the 400 are the rectangular coordinates.

If Ralph and you know all about this sort of thing, you don't need all those words – you could shorten the whole thing down to "(400, 300)", and that's how you'd write the instructions for Ralph if you were both mathematicians, engineers, or, say, Amateur Extra ham radio operators.

Another way you could get Ralph to the same destination would be to say, "Ralph, stand right here at the corner of X Street and Y Avenue, facing due east. Now turn $36.87°$ to your left, walk 500 paces and you'll be there." Were you to use that system, Ralph ends up at the same spot, but you have used what are called "polar coordinates" and the polar coordinates you used were $36.87°$ and 500. If you had told ol' Ralph to turn $36.87°$ to his right, we'd call that $-36.87°$. If you gave Ralph those directions in math shorthand, you'd just write, "(500, 36.87 °)." Ralph would know those were polar coordinates because one of the values is in degrees.

If we use polar coordinates, we call that line that forms the hypotenuse – in other words, the impedance line – a "vector." Vectors have a magnitude, represented by the length, and an angle which is, in this case, the phase angle.

Because the exam asks about both rectangular and polar coordinates, it's important to get the distinction, which is really a pretty simple one; rectangular coordinates specify the points on the X and Y axes, polar coordinates specify the angle and length of the hypotenuse, also known as the vector. Realize, too, that in practice, no matter which set of coordinates we use, we end up with the same picture and the same values for the reactances, the impedance, and the phase angle. Ralph ends up in the same spot. In practice, if we start with a known (or desired) impedance and phase angle, we'd use polar coordinates.

When we plot polar coordinates, it's handy to use some special polar coordinate graph paper, so the polar coordinate plot shown in Figure 25.3 is on some of that. It has coordinates of (6.5, 35°).

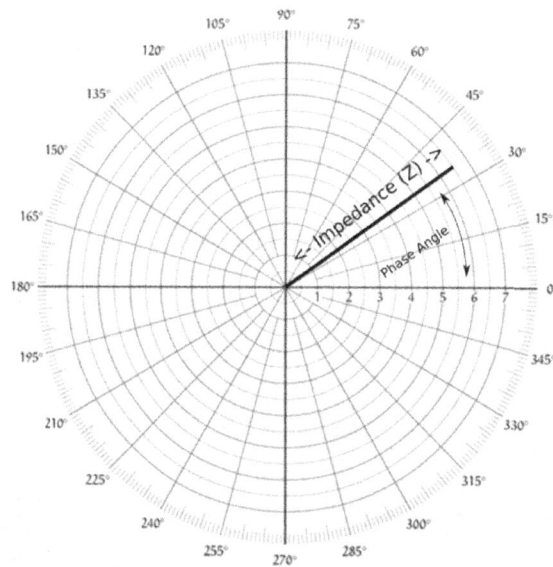

Figure 25.3: Figure 25.1 in Polar Coordinates

From there, we can calculate what we don't know, which is the resistance and reactance. It's just a matter of a tiny bit of trigonometry to calculate the rest of the triangle.

If we know the resistance and reactance, we use the rectangular coordinate system, and we'll give this one, shown in Figure 25.4, coordinates of (4,3).

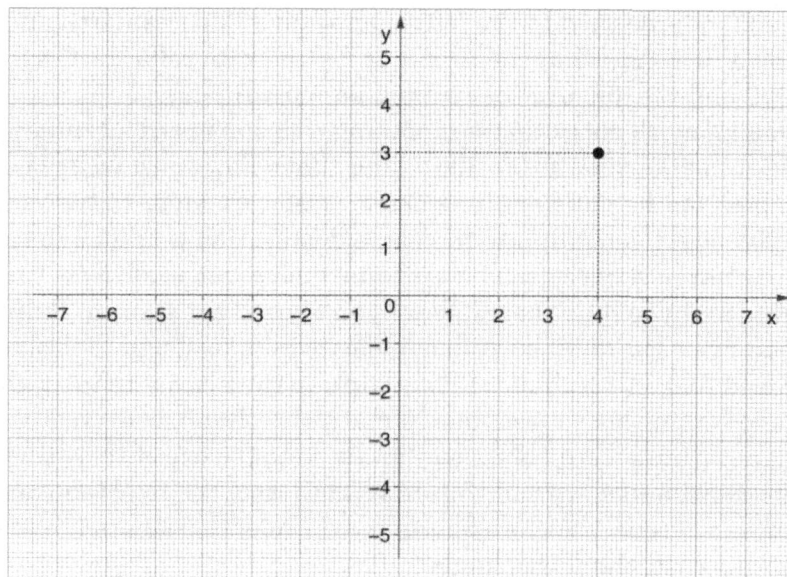

Figure 25.4: Plotting Rectangular Coordinates

From there, we can calculate the length of the hypotenuse, from (0,0) to (4,3) and the phase angle.

We'll get to the questions about coordinate systems and phasors in the next chapter, but first, a couple of questions about electronic terms that relate to impedance.

162

Susceptance and Admittance

For most electronics values, there's an inverse or reciprocal term as well. To take the inverse or reciprocal of a value simply means to divide it into one. For instance, the inverse of resistance is conductance. The symbol for conductance is G. So ...

$$G = \frac{1}{R}$$

You can work any electronics formula with the standard values or the inverse values. Doing so is necessary for a few advanced calculations, handy for other calculations, and simply a matter of personal preference or schooling in most cases.

The extra exam asks you to have some familiarity with a couple of these reciprocal terms that relate to impedance problems, though you won't actually use the values on the exam.

E5B06 (C) What is susceptance?
    A. The magnetic impedance of a circuit
    B. The ratio of magnetic field to electric field
    C. **The imaginary part of admittance**
    D. A measure of the efficiency of a transformer

Susceptance is just the **inverse of reactance**. Mathematically, it's the reactance divided into 1.

$$Susceptance = \frac{1}{Reactance}$$

In fact, susceptance is a lot like conductance, in that it tells us how easily a particular component or circuit passes electricity, but unlike conductance, susceptance changes depending on the frequency we're talking about.

Another way to define susceptance is as **the imaginary part of admittance**. We'll cover what "imaginary" means in this context in the next chapter. For now, just know that the resistance component of a complex impedance is called the "real" part and the reactance is "imaginary." Admittance is the reciprocal of impedance. Impedance can contain resistance and reactance; i.e., the real component and the imaginary component. You could say, then, that admittance can contain conductance and susceptance, with susceptance being the imaginary part.

The symbol for susceptance is B.

That guy Oliver Heaviside came up with the term, susceptance. He also coined the terms admittance, impedance, conductance, inductance, and several more.

E5B05 (D) What happens to the magnitude of a reactance when it is converted to a susceptance?
    A. It is unchanged
    B. The sign is reversed
    C. It is shifted by 90 degrees
    D. **It becomes the reciprocal**

They're asking, "How do we mathematically turn a reactance into a susceptance?"

If we have 100 ohms of reactance, the susceptance is $\frac{1}{100}$ *siemens* (the unit of susceptance and of conductance.) We get that by dividing the value – as the question says, the magnitude – into one. The magnitude of the susceptance is **the reciprocal** of the reactance.

$$B = \frac{1}{X} = \frac{1}{100} = 0.01 siemens$$

E5B02 (D) What letter is commonly used to represent susceptance?
    A. G
    B. X
    C. Y
    D. **B**

The symbol for susceptance is the letter **B**.

E5B12 (A) What is admittance?

    A. **The inverse of impedance**

    B. The term for the gain of a field effect transistor

    C. The turns ratio of a transformer

    D. The inverse of Q factor

Just as the inverse of reactance is susceptance, **the inverse of impedance** is admittance. Again, you take the impedance, divide it into 1, and that's the admittance.

E5B03 (B) How is impedance in polar form converted to an equivalent admittance?

    A. Take the reciprocal of the angle and change the sign of the magnitude

    B. **Take the reciprocal of the magnitude and change the sign of the angle**

    C. Take the square root of the magnitude and add 180 degrees to the angle

    D. Square the magnitude and subtract 90 degrees from the angle

These are the same polar coordinates we went over on page 160, consisting of a phase angle and a magnitude.

Since admittance is the reciprocal of impedance, all we need to do to convert impedance in polar form to an equivalent admittance (still in polar form) is take the reciprocal of the magnitude and change the sign of the angle.

Let's take a circuit in which $R = 1\ k\Omega$, $X_C = 500\ \Omega$, and $X_L = 250\ \Omega$. The impedance of this circuit is $968\ \Omega$ with a phase angle of – this should sound familiar – positive 14 degrees.

$$Z = \sqrt{X^2 + R^2} = \sqrt{(-250\ \Omega)^2 + (1000\ \Omega)^2} = 1030\ \Omega \approx 1000\ \Omega$$

$$Phase\ Angle = \tan^{-1}(\frac{250\ \Omega - 500\ \Omega}{1000\ \Omega}) = -14.03624347° \approx -14.0°$$

In polar coordinates, the impedance of this circuit is $(1000\Omega, -14°.)$ The magnitude is 1000Ω and the angle is $-14°$. The admittance is

$$(\frac{1}{1000}\ siemens, +14°) = (0.001\ siemens, +14°)$$

Perhaps your curious mind is wondering if there is an inverse of capacitance. There is! It's elastance, and it is measured in darafs – farad spelled backward. This slightly amusing information is completely left out of the Amateur Extra exam, a tragic oversight.

Key Concepts in This Chapter

- Susceptance can be described as the imaginary part of admittance.

- The magnitude of a reactance becomes the reciprocal when converted to a susceptance.

- The letter commonly used to represent susceptance is B.

- Admittance is the inverse of impedance.

- To convert impedance in polar form to an equivalent admittance take the reciprocal of the magnitude and change the sign of the angle.

Chapter 26

Questions About Coordinate Systems and Phasors

Phasor Diagrams

E5C05 (C) What is the name of the diagram used to show the phase relationship between impedances at a given frequency?

    A. Venn diagram

    B. Near field diagram

    C. **Phasor diagram**

    D. Far field diagram

When we graph the relationship of resistance and reactance, we end up with a diagram called a **phasor diagram**. It might be in polar coordinates, or it might be in rectangular coordinates, but either way, it's a phasor diagram.

E5C04 (D) What coordinate system is often used to display the resistive, inductive, and/or capacitive reactance components of impedance?

    A. Maidenhead grid

    B. Faraday grid

    C. Elliptical coordinates

    D. **Rectangular coordinates**

Very formally speaking, the polar coordinate system only *displays* the phase angle and the magnitude of the impedance; we'd have to figure out the resistance and reactance. Besides, the polar coordinate system is not listed as a possibility here, so for this question, it's **rectangular coordinates.**

E5C08 (D) What coordinate system is often used to display the phase angle of a circuit containing resistance, inductive and/or capacitive reactance?

    A. Maidenhead grid

    B. Faraday grid

    C. Elliptical coordinates

    D. **Polar coordinates**

You just knew this question was coming, right? It has to be either polar or rectangular coordinates. Since "rectangular coordinates" doesn't appear among the answers, **polar coordinates** is the winner!

E5C09 (A) When using rectangular coordinates to graph the impedance of a circuit, what do the axes represent?

A. **The X axis represents the resistive component and the Y axis represents the reactive component**

B. The X axis represents the reactive component and the Y axis represents the resistive component

C. The X axis represents the phase angle and the Y axis represents the magnitude

D. The X axis represents the magnitude and the Y axis represents the phase angle

In rectangular coordinates, **the X axis represents the resistive component** of the impedance **and the Y axis represents the reactive component.**

E5C01 (A) Which of the following represents capacitive reactance in rectangular notation?

A. **–jX**

B. +jX

C. Delta

D. Omega

This requires a bit of explanation. This is formal engineering math stuff, but in the end it's really simple. It will come up again when we cover Smith charts. Besides, this is one of those parts of the test prep process where a lot of hams' eyes glaze over and they decide that bowling looks like a much more attractive hobby. You don't need to do that, we promise.

For every simple formula we use on the practical level, there are acres of dense, complex mathematical proofs. Georg Ohm didn't just finish his breakfast, jot down $E = I \times R$, say "Yeah, that should work" and start idly wondering if he could work in a round of golf before lunch. He wrote a whole treatise backing it up, and the whole thing had to be rigorously accurate. We just never see all his proofs because, hey, we know it works.

In the same way, there are reams and reams of calculus behind the simple formulas we use for impedance calculations, as well, and they involve some pretty esoteric math, very little of which has trickled down to us common folk. Stuff that looks like this:

$$\frac{2}{\sqrt{-g}}\partial_\alpha(\sqrt{-g}g^{\alpha\mu}g^{\beta\upsilon}\sigma_{[\mu_0 A_\upsilon]} = 2\nabla_a\left(\nabla^{[\alpha}A^{\beta]}\right) = \mu_0 J^\beta$$

(If you're curious, and if I've gotten every little Greek letter in the right spot, that's real. It's a form of one of Maxwell's Equations that led to pretty much everything we know about electromagnetic waves. We don't need to know it.)

Formally speaking, then, we represent capacitive reactance in rectangular notation with what's called $-jX$; for instance, $-j25$ would be "25 ohms of capacitive reactance." "$-j$"? Where'd the "$-j$" come from?

In electrical engineering, a lower case "j" is used as the symbol for The Imaginary Number. Not <u>an</u> imaginary number, but <u>The</u> Imaginary Number which is $\sqrt{-1}$. It's imaginary because there is no square root of negative one.

$$1 \times 1 = 1$$
$$-1 \times -1 = 1$$
$$?? \times ?? = -1$$

Despite the fact that "j" is not what mathematicians call a "real" number, it is quite useful in some forms of math and a phasor diagram is one of those forms. (Sharp-eyed students will note that the Maxwell's Equation above contains not one but two square roots of negative numbers.)

Why do we call it "j" – well, mathematicians actually don't. They call it "i", but electrical engineers were already using I for current, so to avoid confusion, they took the next available letter which happened to be "j."

When we represent an impedance that contains reactance, we're representing what mathematicians call a "complex value." It's not a simple number like 12 or 1,432, it's a value with those parts we mentioned when we covered polar coordinates; it has a magnitude and an angle. This is why an impedance that contains reactance is called a complex impedance. Mathematicians would say the value is "on the complex plane" of the graph, and complex planes have a *real* axis and an *imaginary axis*. To us, the real axis looks just like an X axis and the imaginary axis looks just like a Y axis.

Because of all this, then, we say the resistance is the "real" component of the impedance, while the reactance is the "imaginary" component. These terms have absolutely nothing to do with the physical reality or unreality of the components, they just have to do with the math.

Now, the good news. In practice, here in the real world, you don't have to solve the mystery of $\sqrt{-1}$. You don't even have to enter it on your calculator, which will only give you a cryptic message like "DOMAIN Error." **You just stick a plus sign in front of the inductive reactance value and a minus sign in front of the capacitive reactance value and carry on.** If you want to be very, very proper, and get that coveted gold star from the teacher, you write a $+j$ or a $-j$ in front of the value, but it won't change your graph in the slightest.

E5C06 (B) What does the impedance $50-j25$ represent?
   A. 50 ohms resistance in series with 25 ohms inductive reactance
   B. **50 ohms resistance in series with 25 ohms capacitive reactance**
   C. 25 ohms resistance in series with 50 ohms inductive reactance
   D. 25 ohms resistance in series with 50 ohms capacitive reactance

Those are rectangular coordinates. (There should be a comma between them and they should be in brackets, $(50, -j25)$, but we work with what we get, right?) The 50 tells us the position on the X axis, which for phasor diagrams is always the resistance axis. So right away, we know we have 50 ohms of resistance. The –j25 part tells us where to plot on the Y axis, and since it is negative we know it is capacitive reactance. The correct answer is **50 ohms resistance in series with 25 ohms capacitive reactance**.

The exam will only ask about series circuits in this section, by the way.

E5C02 (C) How are impedances described in polar coordinates?
   A. By X and R values
   B. By real and imaginary parts
   C. **By phase angle and magnitude**
   D. By Y and G values

When we use polar coordinates, we use two coordinates. One is an **angle** – in this case, the phase angle – and the other is a length, which the test writers have called here the **magnitude**.

E5C03 (C) Which of the following represents an inductive reactance in polar coordinates?
   A. A positive magnitude
   B. A negative magnitude
   C. **A positive phase angle**
   D. A negative phase angle

The sign of the phase angle tells you where the voltage is relative to the current. If the phase angle is positive, the voltage is leading the current, and that can only happen in an inductive reactance. A reactance that is more inductive than capacitive is represented in polar coordinates by **a positive phase angle**.

E5C10 (B) Which point on Figure E5-1 best represents the impedance of a series circuit consisting of a 400-ohm resistor and a 38-picofarad capacitor at 14 MHz?
   A. Point 2
   B. **Point 4**

Figure E5-1

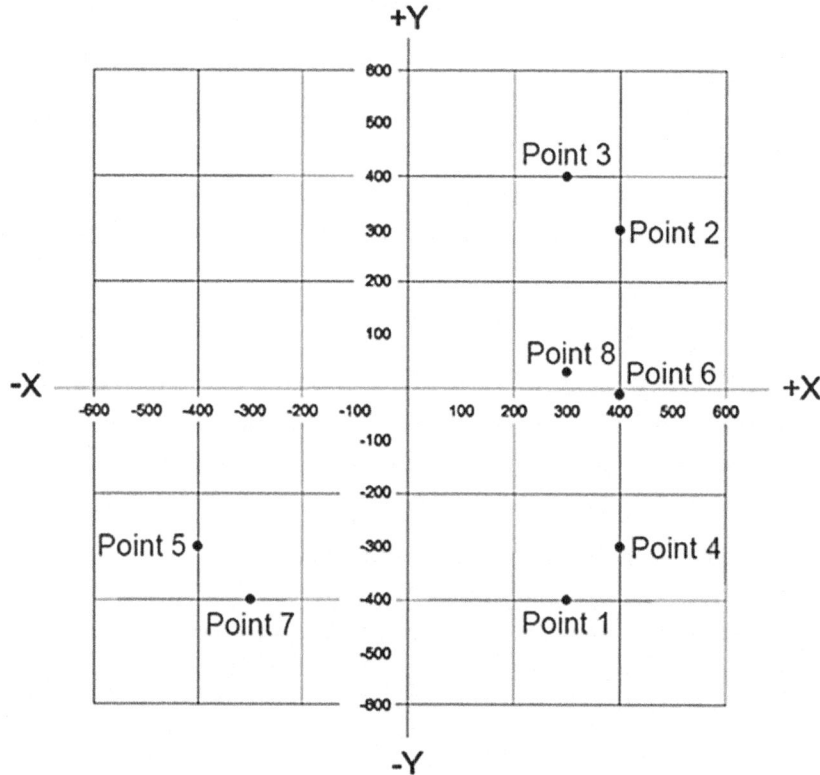

Figure 26.1: Figure E5-1 in the question pool

C. Point 5

D. Point 6

There are four possible questions on the exam that ask you to plot a phasor diagram for a particular circuit. It's the same circuit in every question, but the component values are different. All these questions use the rectangular coordinate system.

Let's get this one figured out.

All we need is the resistance in the circuit and the sum of the reactances. We'll plot the resistance on the X axis. In this case, we have 400 ohms of resistance, and since resistance is always a positive number, right away we can narrow down the possible correct points on Figure E5-1 to those lined up with the +400 on the X axis, and those are, from top to bottom, points 2, 6, and 4. Wouldn't you know it, they included all of those points in the answers, so there's no shortcut available ...yet.

Ah, but wait! There's no inductor in this circuit, only a capacitor. That tells us the correct answer *must* be on the negative side of the Y axis, the part below the X axis. There's only one point at 400 on the X scale with a negative value on the Y scale and it is **point 4**. We're done!

E5C11 (B) Which point in Figure E5-1 best represents the impedance of a series circuit consisting of a 300-ohm resistor and an 18-microhenry inductor at 3.505 MHz?

A. Point 1

B. **Point 3**

C. Point 7

D. Point 8

This circuit has 300 ohms of resistance, so go to 300 on the X axis. We have narrowed down the possible correct answers to three points, points 3, 8 and 1.

There's only inductive reactance in this circuit. That tells us the correct answer must be on the positive half of the Y axis – uh, oh, points 3 and 8 are both positive. We're not done. We'll need to calculate the inductive reactance.

The formula for inductive reactance is $X_L = 2\pi f L$. We know the f, which is 3.505 MHz, or $3.505 \times 10^6\ Hz$, and we know the L, 18 $microhenries$, or $18 \times 10^{-6}\ henries$.

$$X_L = 2\pi \times (3.505 \times 10^6\ Hz) \times (18 \times 10^{-6} Henries) = 396.406161\ \Omega \approx 400\ \Omega$$

Call it 400 ohms of inductive reactance. In a very scientific and precise fashion, we put our finger on the 300 on the X axis and go straight up until we're even with the 400 on the Y axis and whaddaya know, there's **point 3**.

E5C12 (A) Which point on Figure E5-1 best represents the impedance of a series circuit consisting of a 300-ohm resistor and a 19-picofarad capacitor at 21.200 MHz?

    A. **Point 1**
    B. Point 3
    C. Point 7
    D. Point 8

In this problem, we have 300 ohms of resistance. That puts us at 300 on the X axis. Once again, we have three possible points. This circuit has no inductor in it, only a resistor and a capacitor, we know that our correct answer must be in that lower right-hand quadrant where we have positive values of X and negative values of Y. There's only one point at 300 on the X axis and a negative number on the Y axis, and that's **point 1**.

If you've gotten your mind wrapped around rectangular coordinates, this next one should be a snap.

E5C07 (D) Where is the impedance of a pure resistance plotted on rectangular coordinates?

    A. On the vertical axis
    B. On a line through the origin, slanted at 45 degrees
    C. On a horizontal line, offset vertically above the horizontal axis
    D. **On the horizontal axis**

If there's no inductive reactance and no capacitive reactance, the impedance is located **on the horizontal axis**.

Are you still with me? Good! You've made it through the section of the exam that, based on my completely unscientific survey, seems to have run off more would-be Extra Class hams than any other. We have about 90% of the math on the exam behind us now, and the remaining 10% is fairly concrete and simple. You just plug some numbers into a formula and the answer comes chugging out the other end. No more graphing, no more weird numbers that don't actually exist, I promise.

Key Concepts in This Chapter

- The diagram used to show the phase relationship between impedances at a given frequency is a phasor diagram.

- The coordinate system often used to display the resistive, inductive, and/or capacitive reactance components of impedance is known as rectangular coordinates.

- The coordinate system often used to display the phase angle of a circuit containing resistance, inductive and/or capacitive reactance is known as polar coordinates.

- When using rectangular coordinates to graph the impedance of a circuit, the X axis represents the resistive component and the Y axis represents the reactive component.

169

- In rectangular coordinates, $-jX$ represents capacitive reactance.

- An impedance described as 50, $-j25$ represents 50 ohms resistance in series with 25 ohms capacitive reactance.

- In polar coordinates, impedances are described by phase angle and magnitude.

- In polar coordinates, a positive phase angle represents an inductive reactance.

- The impedance of a pure resistance plotted on rectangular coordinates is on the horizontal axis.

Chapter 27

AC and RF Energy in Real Circuits

Real-World Conductors

E5D01 (A) What is the result of skin effect?

    A. **As frequency increases, RF current flows in a thinner layer of the conductor, closer to the surface**

    B. As frequency decreases, RF current flows in a thinner layer of the conductor, closer to the surface

    C. Thermal effects on the surface of the conductor increase the impedance

    D. Thermal effects on the surface of the conductor decrease the impedance

As frequency increases, RF current flows in a thinner layer of the conductor, closer to the surface. That's the result of skin effect.

At DC, there is no skin effect. The effective diameter of a conductor is equal to the actual diameter of the conductor.

As the frequency increases, the alternating current starts inducing other currents inside the conductor and these tend to cancel each other out, leaving more and more of the current outside the center of the conductor.

At 60 Hz – household electricity – the skin effect is about 1/3 inch – so, no need to take it into account until you start to get to some pretty beefy wires more than about 2/3 inch thick which will, one presumes, be carrying equally beefy amounts of current.

If we multiply that 60 Hz frequency by 1000 to 60 kHz, the skin effect in copper wire is about 1/100 of an inch; we have virtually all of the current flowing in the outer 1/100 inch of the wire. That's not a lot of wire if we plan on it carrying a lot of current. Our wire is already turning into a resistor.

At 6 MHz, almost all the current is flowing in the outer 1/1000 of an inch of a copper conductor. Now we really need to think about skin effect and how it affects the resistance of the wire at that frequency.

As we keep going up, that skin gets thinner and thinner. There are some ways of working with this. Up to about a megahertz, you may need to use "Litz wire." It's cable made of braided, insulated strands. If you've ever seen what's inside a computer network Cat5 cable, that's a modified form of Litz wire. Or, you might use copper pipe as a conductor – a big, hollow conductor. This is part of why when you see a high-power VHF or higher frequency installation, it looks more like plumbing than wiring. That stuff that looks like plumbing is commonly referred to as "waveguide."

E5D02 (B) Why is it important to keep lead lengths short for components used in circuits for VHF and above?

A. To increase the thermal time constant

B. To avoid unwanted inductive reactance

C. To maintain component lifetime

D. All these choices are correct

Any conductor has some amount of self-inductance. It doesn't matter if it is perfectly straight – not even vaguely coil shaped – it still has some amount of self-inductance. In fact, when Faraday discovered inductance, he wasn't using coils at all, just straight wires. Coils came later.

The longer the conductor, the greater the self-inductance, and since inductive reactance is proportional to frequency, the greater the frequency the greater the self-inductance.

At low frequencies, and reasonable conductor lengths, we don't concern ourselves much with self-inductance, since it is negligible. (Our transmission lines are an exception to that.) But as frequencies increase, it becomes not so negligible, and somewhere near the VHF range it becomes quite significant indeed. **To avoid unwanted inductive reactance** we keep lead lengths short for components used in circuits for VHF and above.

E5D04 (B) Why are short connections used at microwave frequencies?

A. To increase neutralizing resistance

B. To reduce phase shift along the connection

C. To increase compensating capacitance

D. To reduce noise figure

We know from the question above that we want short connections at VHF and higher frequencies so that we reduce unwanted inductance. But they don't seem to have given us any answers that include inductance here.

Ah, but we also know inductance affects the phase angle, so by reducing unwanted inductance we automatically **reduce phase shift along the connection**.

E5D03 (D) What is microstrip?

A. Lightweight transmission line made of common zip cord

B. Miniature coax used for low power applications

C. Short lengths of coax mounted on printed circuit boards to minimize time delay between microwave circuits

D. Precision printed circuit conductors above a ground plane that provide constant impedance interconnects at microwave frequencies

We can use copper pipe, or "waveguide" to conduct VHF and higher frequencies. But running big copper pipes inside a cell phone or a handheld radio just doesn't work! That's where microstrip comes in. Microstrip is waveguide on the scale of a printed circuit board. Microstrips are **precision printed circuit conductors above a ground plane that provide constant impedance interconnects at microwave frequencies.**

E5D06 (D) In what direction is the magnetic field oriented about a conductor in relation to the direction of electron flow?

A. In the same direction as the current

B. In a direction opposite to the current

C. In all directions; omni-directional

D. In a circle around the conductor

You might have, once upon a time, learned this as the right-hand rule. The most basic version goes like this: Hold a conductor in your right fist with your thumb extended in the direction of the current flow; your fingers are showing you the orientation of the magnetic field. **The magnetic lines of force form a circle around the conductor.**

Reactive Power & the Power Factor

E5D09 (B) What happens to reactive power in an AC circuit that has both ideal inductors and ideal capacitors?

A. It is dissipated as heat in the circuit

B. It is repeatedly exchanged between the associated magnetic and electric fields, but is not dissipated

C. It is dissipated as kinetic energy in the circuit

D. It is dissipated in the formation of inductive and capacitive fields

The key word in that question is "ideal." "Ideal" components of any kind are convenient fantasies. They don't exist in the real world, just on paper. So if we had an ideal inductor and an ideal capacitor hooked up with ideal wires and we charged the circuit, the charge would go back and forth between the coil and the capacitor forever.

The question asks about "reactive power." Reactive power is almost a contradiction in terms. It is "wattless" power. It does no work, it is nonproductive.

Recall our diagram of two sine waves 90° out of phase.

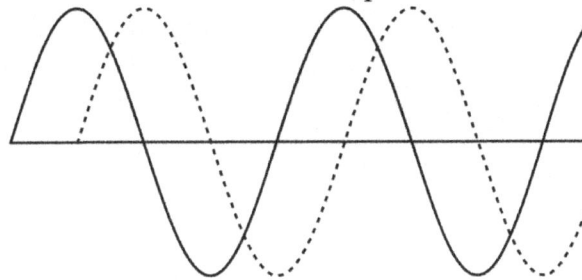

90° Phase Difference

We'll say the solid wave is voltage and the dotted wave is current. Notice how each time the voltage hits maximum positive or negative, the current is at 0. Since power is volts times amps that gives us zero watts at those moments. There are even large parts of the cycle where the current is positive while the voltage is negative, and vice versa.

In fact, when we look at each point of time on that graph and calculate the real power, it comes out something like the bold line curve on the graph in Figure 27.1.

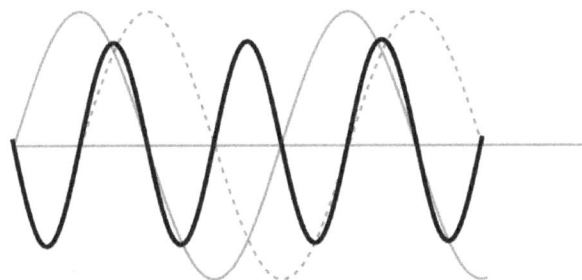

Figure 27.1: Power with 90 Degree Phase Angle

Understand, that bold line in Figure 27.1 represents real power, and half the time it's actually negative – it's pushing power back to the source! When we average it all out, we have no real power in this circuit at all.

At other phase angles, we get more real power and less reactive power. For instance, at 60° of phase angle, (see Figure 27.2) it's a 50/50 split between productive power and reactive power.

Notice in Figure 27.2, there are many more points where the voltage and current are cooperating and making real power.

If we multiply the volts and amps going into the circuit, we get a figure called *apparent power*. But if we factor in how much of that is trapped in reactive power, we get the *real power*.

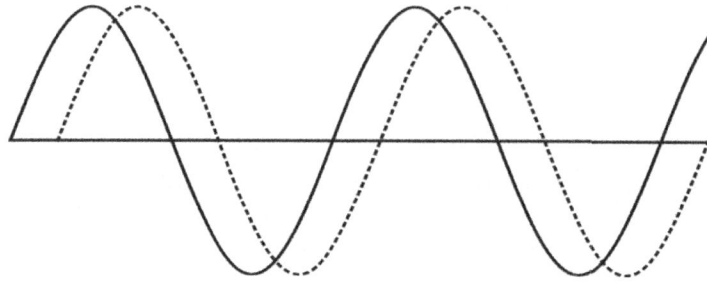

Figure 27.2: 60 Degree Phase Angle

The ratio of real power to apparent power is called the *power factor*, and once we know the phase angle it is easy to calculate that power factor.

$$Power\,Factor = \cos(Phase\,Angle)$$

The power factor is equal to the cosine of the phase angle. If you're using your trusty TI-30XS calculator, you just press the "cos" key, enter the phase angle, add a ")", press enter, and there's the power factor. Unfortunately, if you just enter, say, "30" for the phase angle, the 30-XS will tell you the answer is $\frac{\sqrt{3}}{2}$ which is perfectly accurate and perfectly useless for the exam, but if you enter the angle as "30.0" it will tell you the power factor for a 30 degree phase angle is 0.866.

E5D14 (A) What is reactive power?

  A. **Wattless, nonproductive power**
  B. Power consumed in wire resistance in an inductor
  C. Power lost because of capacitor leakage
  D. Power consumed in circuit Q

Reactive power is **wattless, nonproductive power**.

E5D10 (A) How can the true power be determined in an AC circuit where the voltage and current are out of phase?

  A. **By multiplying the apparent power by the power factor**
  B. By dividing the reactive power by the power factor
  C. By dividing the apparent power by the power factor
  D. By multiplying the reactive power by the power factor

Apparent power is the power we would calculate by multiplying the voltage applied to the circuit by the amps going into the circuit. In a DC circuit, apparent power is real power. Once we have an AC circuit that contains some reactance, apparent power and real power are not necessarily equal. In this question, they call real power "true power" – it's the same thing.

We determine the true power in an AC circuit **by multiplying the apparent power times the power factor.** If we apply 100 volts to a circuit and it draws 50 amps, that's 5,000 watts of apparent power. But if the power factor is 0.866, the true power is only $5000 \times 0.866 = 4,330$ watts.

E5D11 (C) What is the power factor of an RL circuit having a 60-degree phase angle between the voltage and the current?

  A. 1.414
  B. 0.866
  C. **0.5**
  D. 1.73

$$Power\,Factor = \cos(Phase\,Angle) = \cos(60.0) = 0.5$$

E5D15 (D) What is the power factor of an RL circuit having a 45-degree phase angle between the voltage and the current?

A. 0.866

B. 1.0

C. 0.5

D. **0.707**

For this one, you could just memorize the fact that a 45-degree phase angle always yields a 0.707 power factor, or you can calculate it with the cosine of the phase angle:

$$Power\ Factor = \cos{(Phase\ Angle)} = \cos{(45.0)} = 0.707$$

Does that 0.707 number sound vaguely familiar? It's the same number you used on the General Exam to calculate the RMS voltage from a peak voltage, and vice versa. Coincidence? Nope.

E5D05 (C) What is the power factor of an RL circuit having a 30-degree phase angle between the voltage and the current?

A. 1.73

B. 0.5

C. **0.866**

D. 0.577

This problem uses the same calculation as the previous problem. Grab that TI-30XS, press the "cos" key, then "30.0", a ")" and "Enter." Boom! A 30-degree phase angle yields a power factor of 0.866.

E5D12 (B) How many watts are consumed in a circuit having a power factor of 0.2 if the input is 100 VAC at 4 amperes?

A. 400 watts

B. **80 watts**

C. 2000 watts

D. 50 watts

We can easily calculate the apparent power of this circuit by multiplying 100 volts by 4 amps to get 400 watts. The question asks us how many watts are consumed, though, and in an AC circuit with reactance that's a different number than the apparent power.

To get the *real power*, which is the amount of power actually consumed or dissipated in the circuit, we multiply the apparent power by the power factor.

$$Real\ Power = Apparent\ Power \times Power\ Factor$$

We have a power factor of 0.2, so to calculate the real power consumed in the circuit, we multiply 400 watts by 0.2 to get **80 watts.**

E5D08 (D) How many watts are consumed in a circuit having a power factor of 0.6 if the input is 200VAC at 5 amperes?

A. 200 watts

B. 1000 watts

C. 1600 watts

D. **600 watts**

They're asking, "What's the real power in this circuit?" VAC stands for "Volts of Alternating Current."

Joule's Law tells us that the power – the apparent power – of the circuit is 1000 watts; $200\ watts \times 5\ amperes$. Multiply the 1000 watts by the power factor of 0.6 to get the real power figure of **600 watts.**

E5D07 (B) How many watts are consumed in a circuit having a power factor of 0.71 if the apparent power is 500VA?

A. 704 W

B. 355 W

C. 252 W

D. 1.42 mW

They use "VA" instead of "W" – VA stands for "volts times amps" and is often used rather than "watts" in AC circuits, because "watts" means "power consumed" or "real power." VA implies "apparent power."

$$500VA \times 0.71 = \textbf{355 W}$$

E5D13 (B) How many watts are consumed in a circuit consisting of a 100-ohm resistor in series with a 100-ohm inductive reactance drawing 1 ampere?

A. 70.7 watts

B. 100 watts

C. 141.4 watts

D. 200 watts

In this question, they give us neither the power factor, nor the phase angle, nor even the applied voltage, because we don't need them.

All we need for this one is Ohm's Law because we know the actual current flow through the circuit and the resistance in the circuit. (Only resistance consumes power, so we can ignore the inductance.) We just treat the circuit as though it was a DC circuit containing only the resistance, since inductance doesn't affect DC.

We can calculate the power consumed in one step or two. In one step, it's twinkle, twinkle little star, power equals I squared R:

$$P = I^2 \times R = 1^2 \times 100 = 100 \, Watts$$

Those one step power formulas tend to somehow fall out of my head at the worst possible time, though, so here's a review of how to get there in two steps:

$$E = I \times R = 1 \times 100 = 100 \, Volts$$

$$P = E \times I = 1 \times 100 = 100 \, Watts$$

Either way, we have **100 watts** of power consumed.

Key Concepts in This Chapter

- The result of skin effect is that as frequency increases, RF current flows in a thinner layer of the conductor, closer to the surface.

- It is important to keep lead lengths short for components used in circuits for VHF and above to avoid unwanted inductive reactance.

- Short connections are used at microwave frequencies to reduce phase shift.

- Microstrip is what we call precision printed circuit conductors above a ground plane that provide constant impedance interconnects at microwave frequencies.

- The magnetic field about a conductor forms a circle around the conductor.

- In an AC circuit that has both ideal inductors and ideal capacitors, reactive power is repeatedly exchanged between the associated magnetic and electric fields but is not dissipated.

- Reactive power is wattless, nonproductive power.

- The formula to find the power factor of a circuit is

$$Power\ Factor = \cos(Phase\ Angle)$$

- The power factor of an RL circuit having a 60-degree phase angle between the voltage and the current is 0.5.

- The power factor of an RL circuit having a 45-degree phase angle between the voltage and the current is 0.707.

- The power factor of an RL circuit having a 30-degree phase angle between the voltage and the current is 0.866.

- To find the real power consumed in a circuit, multiply the apparent power (E x I) by the power factor.

$$Real\ Power = Apparent\ Power \times Power\ Factor$$

- The power consumed in a circuit having a power factor of 0.2 if the input is 100 VAC at 4 amperes is 80 watts.

- The power consumed in a circuit consisting of a 100-ohm resistor in series with a 100-ohm inductive reactance drawing 1 ampere is 100 watts.

Chapter 28

Principles of Semiconductor Devices

In this chapter and the next two, you'll learn about semiconductors of several types.

We use the term "semiconductor" for any material or component that is "sort of a conductor." It might be a device that conducts electricity in only one direction. In that case we call it a diode. It might be a component that conducts electricity in proportion to an input signal. That's a transistor.

The simplest semiconductor devices are diodes, but we had diodes before we mastered semiconductor technology. Thomas Edison almost discovered the vacuum tube diode some 150 years ago, while he was trying to invent the electric light bulb. Among the many observations, there was noted a blue glow around one pole of the filament of an early light bulb prototype, with a blackening effect near the opposite pole. Something seemed to be flowing from one of the poles of the filament to other places. Now we know this was caused by the light bulb acting as a primitive diode. For a few, brief shining moments, this effect was known as "Hammer's Phantom Shadow", after William Hammer, the technician who noticed it, but with characteristic modesty, old Tom quickly renamed it "The Edison Effect". He promptly filed a patent on it and just as promptly forgot all about it since it had, in his mind, no practical application and he had plenty of other useful problems to solve.

Some twenty years later, in 1904, John Ambrose Fleming realized The Edison Effect could be put to use to detect radio signals and patented the vacuum tube diode. It had a cathode, a hot filament to heat the cathode, and a plate. Just a couple of years later, Lee de Forest added a charged grid between the cathode and the plate, and that was the triode. By varying the charge on the grid, the triode allowed the control of a large voltage with a small voltage, so it could serve as an amplifier. Aside from some fine tuning through the addition of extra grids in tetrode and pentode tubes, and various refinements that made them smaller and more durable, vacuum tubes were the peak of technology for rectification, switching, and amplification for the next 50 years.

Solid-state diodes were discovered even before the Edison Effect, but they were notoriously finicky and unreliable. Maybe somewhere along the way you built a crystal radio with a cat's whisker of wire that, if you poked it in just the right spot on a chunk of galenium, would work. Well, that was about the state of the art for solid-state until the 1940's when we could make semiconductor materials pure enough to work consistently. With access to those new, pure materials, a team at Bell Labs started experimenting with controlling the current through a solid-state diode with a voltage. Eventually, they came up with a scheme that worked and the transistor was born. It took a few years for transistors to become reliable and affordable enough to be practical, but once they were, they created previously unimaginable possibilities for electronics. Atlantic Magazine named the transistor the fourth most important technological breakthrough "since the wheel", behind the printing press, electricity and penicillin.

The Technician exam basically asks, "Do you know what a diode is? How about a transistor? Do you have some idea of how those things work? You know there's more than one type? Good!" At the General level, we went into a bit more detail, and looked at some more varieties of solid-state devices. The Amateur Extra exam asks you to really delve into what I'd call the anatomy and physiology of solid-state components and the practical applications of the various kinds.

This "Bias" Business

There's a term that will come up several times in the course of the exam, and many more times if you read about or work with semiconductors of any sort, and it's one that many folks new – or even not so new -- to this sort of thing have told me they are puzzled by. It's "bias", and its sub-categories, "forward bias" and "reverse bias." So let's clear those up here. You can think of bias as "positivity" or "negativity" that's intentionally added to the signal in a circuit. That signal could be as simple as the voltage in a power supply or as complex as an incoming RF signal, but it still has some positivity or negativity added to it to make things function.

For example, Figure 28.1 shows an unbiased sine wave. We've seen these over and over.

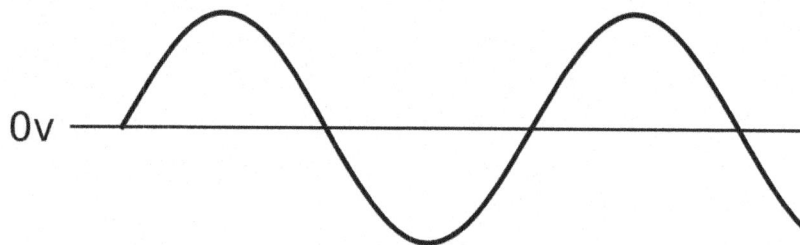

Figure 28.1: Unbiased Sine Wave

It's oscillating happily along, swinging up to the positive side, and down to the negative side. If we add in some DC voltage, though, we end up with a "biased" sine wave. Figure 28.2 shows one with positive bias.

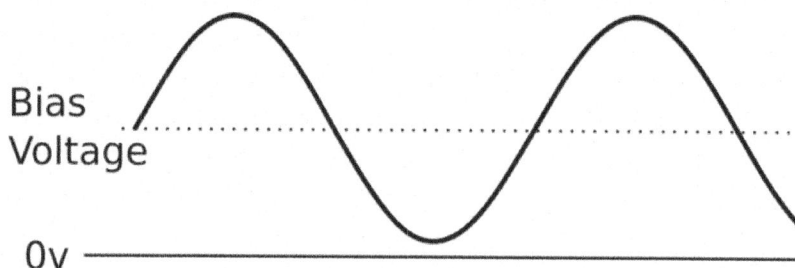

Figure 28.2: Positively Biased Sine Wave

It's still oscillating, but now, it's a varying positive voltage. We could also have a negatively biased signal, as shown in Figure 28.3.

So, why would we do that? For the moment, forget about whether the signal is biased positive or negative. We'll get to that. The reason we bias is that every semiconductor has a threshold voltage. Semiconductors don't conduct anything below a certain voltage. They're what we call non-linear; what goes in doesn't match what comes out, until we get up into the operating voltages of the component. We have to hit that threshold voltage before the semiconductor starts conducting, and it will only conduct linearly up to a higher voltage known as the saturation voltage – then it just keeps conducting at the same level. It's as "on" as it

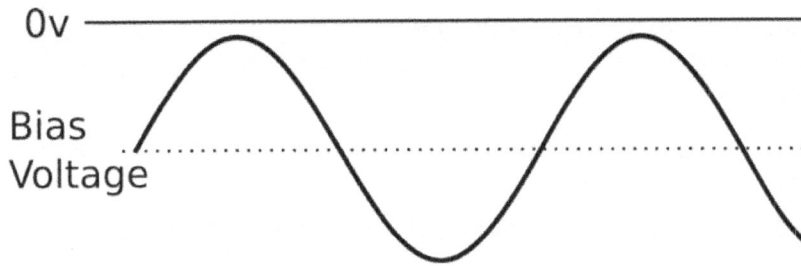
Figure 28.3: Negatively Biased Sine Wave

can get. For a diagram of this, see Figure 28.4. (The threshold voltage is sometimes called the *cutoff* voltage of the device.)

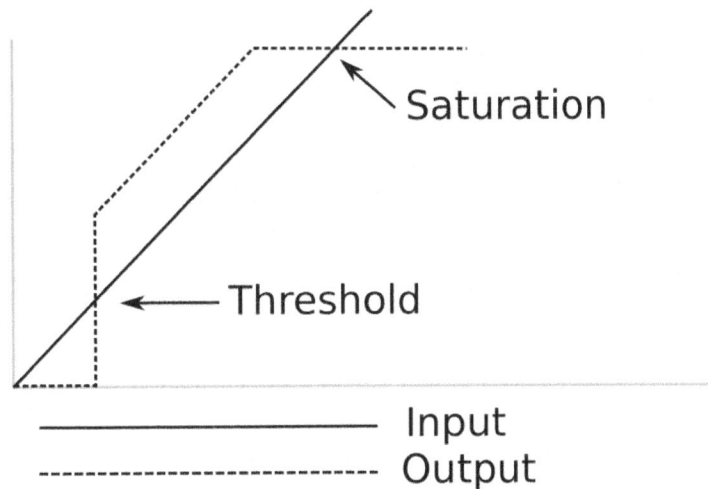
Figure 28.4: Threshold and Saturation

If we fed that unbiased sine wave into a transistor that we wanted to act as an amplifier, the amplifier would only be operating less than half the time. Let's say it's an NPN transistor, so it conducts when there's a positive voltage on the base. That transistor would only allow current to pass between the emitter and the collector when the voltage was above, say, a half volt positive. Any other time, it's just like a switch that's turned off.

In the case of a diode, we say it is "forward biased" when the voltage we are applying is negative on the cathode side and positive on the anode side. In the case of an "ordinary" diode, that would make the diode conduct, so long as the voltage was over its threshold voltage. Figure 28.5 shows a forward biased diode on the left, with a negative voltage being applied to the cathode.

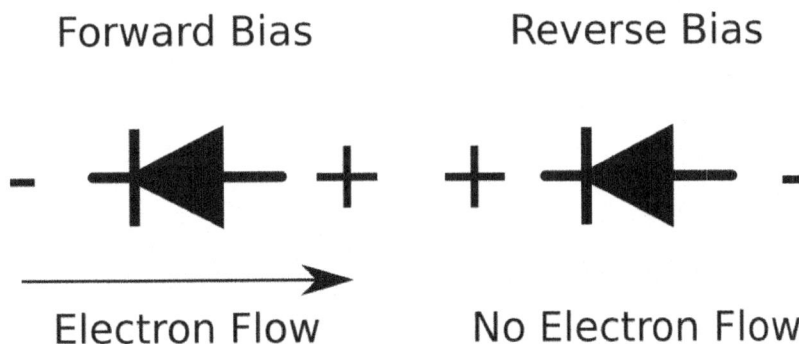
Figure 28.5: Forward and Reverse Biased Diodes

When an ordinary diode is "reverse biased", like the diode on the right in Figure 28.5, it means we're applying a positive voltage to the cathode and a negative voltage to the anode. The diode does not conduct.

With transistors, the positives and negatives can switch around, depending on whether the transistor is an NPN or a PNP, but the principles of bias remain the same – "forward bias" is the bias that makes the semiconductor conduct, "reverse bias" makes it non-conductive.

Chapter 29

Questions About Semiconductors and Semiconductor Devices

Semiconductor Materials

E6A02 (A) Which of the following semiconductor materials contains excess free electrons?
- A. **N-type**
- B. P-type
- C. Bipolar
- D. Insulated gate

Semiconductor materials are characterized as either "N-type" or "P-type." **N-types** have an excess of free electrons, so localized regions of the material carry a **N**egative charge. P-types have a deficiency of electrons, so tiny areas of them carry a positive charge. In some atoms in P-type material, there's no electron where an electron is "supposed" to be, so we call those areas with a deficiency of electrons, "holes."

Both types of material start as pure semiconductor material, such as silicon or gallium arsenide. Pure semiconductor material is really just an insulator, and not much of one at that. The magic of semiconductors depends on a process called "doping." A miniscule amount of a doping compound – phosphorous, arsenic, or any number of others -- is added to the molten semiconductor material, and it becomes either N or P type material, depending on the doping compound. The amount of doping compound is tiny. Imagine you have a wheelbarrow of sand. Add one grain of sugar – that's roughly the right proportion.

E6A04 (C) What is the name given to an impurity atom that adds holes to a semiconductor crystal structure?
- A. Insulator impurity
- B. N-type impurity
- C. **Acceptor impurity**
- D. Donor impurity

Some impurities added to semiconductors create holes, or positively charged areas, in the crystal structure. These holes "accept" electrons, so an impurity that creates holes is called an **acceptor impurity**.

E6A01 (C) In what application is gallium arsenide used as a semiconductor material?
- A. In high-current rectifier circuits
- B. In high-power audio circuits
- C. **In microwave circuits**
- D. In very low frequency RF circuits

The history of electronic invention is often the history of the quest to manage higher and higher frequencies. Each time we have made a leap into a new frequency range, more usable spectrum has opened up and new applications have become possible. For instance, without components capable of operating in the 2 GHz band, your smart phone would be a lot bigger thing to carry around.

Advances in solid-state devices typically depend on new manufacturing techniques, new designs and either the availability of new materials or discoveries of previously unknown characteristics of available materials.

The most common diode material is silicon. It is ridiculously common – it's, essentially, purified sand. It works quite well, well up into the UHF band and beyond.

Once we get WAY up there, though, up in the microwave band, the actual speed at which electrons flow through the semiconductor material – the material's *electron mobility* – begins to be very significant and limiting. Silicon's electron mobility is lower than gallium arsenide's, so gallium arsenide is used **in microwave circuits**.

E6E01 (B) Why is gallium arsenide (GaAs) useful for semiconductor devices operating at UHF and higher frequencies?
  A. Higher noise figures
  B. **Higher electron mobility**
  C. Lower junction voltage drop
  D. Lower transconductance

As we covered in the last question, gallium arsenide is useful for semiconductor devices operating at UHF and higher frequencies is that it has **higher electron mobility.**

Semiconductor Functions & Applications

E6A03 (C) Why does a PN-junction diode not conduct current when reverse biased?
  A. Only P-type semiconductor material can conduct current
  B. Only N-type semiconductor material can conduct current
  C. **Holes in P-type material and electrons in the N-type material are separated by the applied voltage, widening the depletion region**
  D. Excess holes in P-type material combine with the electrons in N-type material converting the entire diode into an insulator

A PN-junction diode is simply a diode that has a P side and an N side. It's a single crystal, but half is rich in electrons and the other half is rich in holes. All the useful properties of solid-state diodes and transistors come from the junction between those two materials. That junction creates what is called the **depletion region**.

Both electrons and holes can move around within the crystal lattice, but whether they can cross over to the other side is determined by the state of that depletion region. In the case of a diode, the state of the depletion region is determined by the polarity of the charge applied to the diode.

Let's start with a diode that has no charge applied to it at all. It's just sitting in its package on your workbench. Because there is no charge applied, the holes and electrons have diffused through the crystal lattice, much like smoke might diffuse into the air in a closed room.

Now we apply a negative voltage to the N side and a positive voltage to the P side. This is called "forward biasing" the diode. The negative voltage is pushing the electrons over to the P side and the positive voltage is pushing holes toward the N side. The depletion region gets narrower, squeezing down toward that PN-junction. When the depletion region squeezes down to, effectively, nothing the diode conducts electricity almost like a straight wire.

In reverse bias, shown in Figure 29.2, just the opposite occurs. Now electrons and holes are both being pulled away from the depletion region. The electrons *could* flow, but only if they could be replaced by holes, and the holes are going in the opposite direction. **The holes**

Forward Biased PN-junction Diode

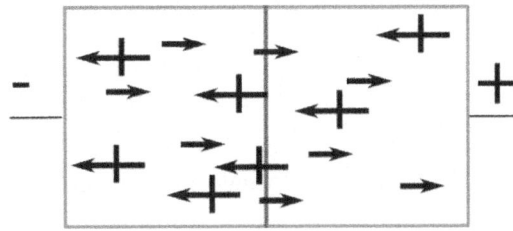

Figure 29.1: PN-junction Diode in Forward Bias

in P-type material and electrons in the N-type material are pulled away from the center by the applied voltage, widening the depletion region.

Reverse Biased PN-junction Diode

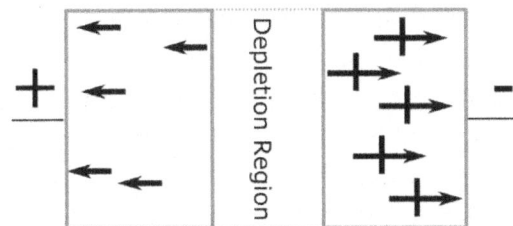

Figure 29.2: PN-junction Diode in Reverse Bias

(Disclaimer: The above is a workable but simplistic explanation of the phenomenon. A really detailed explanation would require at least a couple of courses in quantum physics.)

E6F07 (B) What is a solid-state relay?

A. A relay using transistors to drive the relay coil

B. **A device that uses semiconductors to implement the functions of an electromechanical relay**

C. A mechanical relay that latches in the on or off state each time it is pulsed

D. A semiconductor delay line

A solid-state relay is **a device that uses semiconductors to implement the functions of an electromechanical relay.**

Relays are often used to control high power circuits with low power control signals. Your car's lights work off a relay because they draw a lot of amps and putting a switch that would handle that kind of amperage on your control panel (or in the handle of your turn signal stalk) would be difficult. By using a low current through the switch to control a relay, everything's handled.

An electromechanical relay is just a switch that is run by an electromagnet. We hook a low voltage control signal to the electromagnet and a high voltage circuit we want to control to the switch contacts. Energize the electromagnet, it pulls the switch closed, and the controlled circuit turns on. When we switch off the control voltage, the electromagnet de-energizes and a spring pulls the switch open. (Or vice versa if the relay is normally closed.)

A solid-state relay just uses a transistor as a really dumb amplifier. Let's say we want to use a bipolar junction transistor as a relay. We know when we energize the base, the transistor will conduct electricity between the emitter and the collector. We hook up a control signal to the base, then run the circuit we want our relay to control through the emitter/collector and we have our relay.

185

Bipolar Junction Transistors

E6A06 (B) What is the beta of a bipolar junction transistor?
- A. The frequency at which the current gain is reduced to 0.707
- B. **The change in collector current with respect to base current**
- C. The breakdown voltage of the base to collector junction
- D. The switching speed

The *beta* of a transistor is the ratio of **the change in collector current with respect to base current.** In most applications beta is roughly equal to the gain of the transistor.

E6A07 (D) Which of the following indicates that a silicon NPN junction transistor is biased on?
- A. Base-to-emitter resistance of approximately 6 to 7 ohms
- B. Base-to-emitter resistance of approximately 0.6 to 0.7 ohms
- C. Base-to-emitter voltage of approximately 6 to 7 volts
- D. **Base-to-emitter voltage of approximately 0.6 to 0.7 volts**

When a voltage is applied to the base of a bipolar transistor, current can flow between the emitter and collector. If zero volts is applied to the base, nothing flows. The nature of transistors is that if, say 0.00001 volts is applied to the base, they still won't allow current to flow. It takes some particular minimum voltage, which varies by the transistor, to start current flowing from emitter to collector. Think of it as sort of a trigger point, a voltage that "turns the transistor on." That voltage is called the bias voltage, and there are lots of clever tricks that can be done by varying how a transistor is biased.

So how do we know if we have biased a particular transistor to an "on" state? We can measure the **base-to-emitter voltage** and it should measure **approximately 0.6 to 0.7 volts.** In fact, that's the entire range for the base-to-emitter voltage of a silicon transistor – the minimum is 0.6 volts and the maximum is 0.7 volts.

We will come across that range of voltage values again when we cover silicon junction diodes. Other materials and other methods of constructing semiconductors can change that value, but it's a constant when it comes to silicon to silicon junctions.

E6A08 (D) What term indicates the frequency at which the grounded-base current gain of a transistor has decreased to 0.7 of the gain obtainable at 1 kHz?
- A. Corner frequency
- B. Alpha rejection frequency
- C. Beta cutoff frequency
- D. **Alpha cutoff frequency**

Every transistor has a frequency range. Some can only work effectively in the audio frequency range, some can go clear on up into microwaves. One measure of that frequency response is the **alpha cutoff frequency**.

The alpha of a transistor is the ratio of the collector current to the emitter current. A simpler way to put it is, alpha tells us "how much current will get through the transistor." Under normal conditions, alpha is usually pretty close to 1.0; what goes in comes out, with just a little loss in the base.

As frequencies rise though, the size of the gap between the emitter and the collector – in other words, the amount of base between them – starts to become significant, because of the time it takes the signal to make it through the base. As frequency rises, the alpha will start to drop. Once the frequency is high enough to make the alpha drop to about 0.7 – in other words, 70% – we call that the **alpha cutoff frequency**. The precise number, by the way, should be strangely familiar to you by now, it's our old friend 0.707. Not a coincidence.

Field Effect Transistors

E6A05 (C) How does DC input impedance at the gate of a field-effect transistor compare with the DC input impedance of a bipolar transistor?

    A. They are both low impedance

    B. An FET has lower input impedance

    C. **An FET has higher input impedance**

    D. They are both high impedance

An FET has high input impedance; a bipolar transistor has low input impedance.

As hams, we're typically very concerned with matching impedances for maximum power transfer. But maximum power transfer isn't *always* desirable in a circuit. For an extreme example, think of a voltmeter. Setting safety issues aside for a moment, we still don't want the power in the circuit we're measuring to transfer into the voltmeter. The voltmeter would become part of the circuit and we wouldn't get accurate readings. In that case, we want a huge impedance mismatch between the meter and the circuit being measured.

In certain amplifier designs, we also want that sort of mismatch between one circuit element and another, so that the next element doesn't adversely affect the performance of the previous element.

In other designs, we want one element tightly coupled to another element – in other words, we do want the impedances to closely match.

The FET's **high input impedance** gives designers more choices about that impedance matching.

E6A12 (D) Why do many MOSFET devices have internally connected Zener diodes on the gates?

    A. To provide a voltage reference for the correct amount of reverse-bias gate voltage

    B. To protect the substrate from excessive voltages

    C. To keep the gate voltage within specifications and prevent the device from overheating

    D. **To reduce the chance of static damage to the gate**

There are two main families of field-effect transistors, JFETs and MOSFETs. JFETs are Junction Field Effect Transistors. MOSFETs are Metal Oxide Semiconductor Field Effect Transistors – say that three times quickly and you'll know why everyone calls them MOSFETs.

Both sorts of FETs are constructed like a sandwich; a channel with a source at one end and a drain at the other, flanked by gates on the sides. Unlike JFETs, though, MOSFETs have a tiny layer of insulation – the "oxide" part of the name – between the metal gate pieces and the semiconductor channel. So you have a layer of Metal, then a layer of Oxide, and finally a layer of Semiconductor. M-O-S.

One of the main advantages of MOSFETs is that they require almost no current to the gates in order to function; just voltage. In other words, the input has a very high impedance. That's a result of that layer of insulation.

That oxide insulator is very, very thin – just a few molecules is all it takes. It doesn't take much voltage to ruin that insulation – and that static discharge from your finger when you shuffle across a carpet on a dry day can be tens of thousands of volts; far more than enough to ruin a MOSFET. That's why many MOSFET devices have a Zener diode built right into them, hooked up in reverse bias from the gates to ground; normal charges stay on the gates but excess charges get conducted away to a safe place. The purpose of the Zener diode is **to reduce the chance of static damage to the gate**.

E6A09 (A) What is a depletion-mode FET?

A. **An FET that exhibits a current flow between source and drain when no gate voltage is applied**

B. An FET that has no current flow between source and drain when no gate voltage is applied

C. Any FET without a channel

D. Any FET for which holes are the majority carriers

"Depletion mode" is the fancy way of saying that a transistor **exhibits a current flow between the source and drain when no gate voltage is applied**. It's a switch that's "always on" until a voltage is applied – the opposite behavior from what we usually expect of a transistor.

Transistor Schematic Symbols

We have two and only two questions about figure E6-1, which you'll find on page 188.

E6A10 (B) In Figure E6-1, what is the schematic symbol for an N-channel dual-gate MOSFET?

A. 2

B. **4**

C. 5

D. 6

Figure E6-1

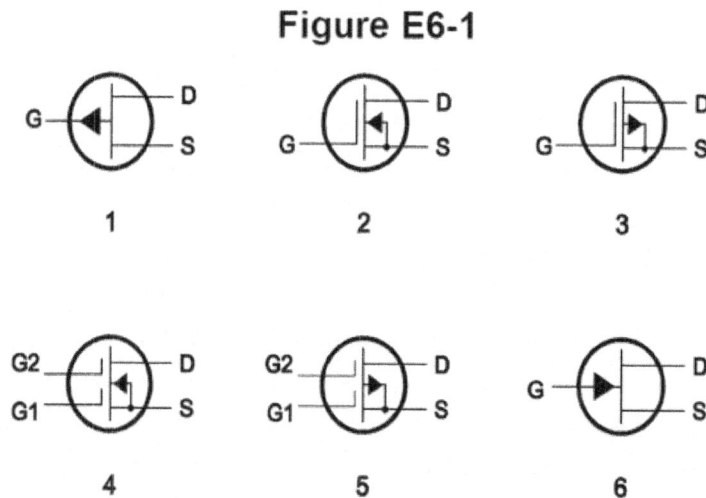

Figure 29.3: Figure E6-1 in the question pool

This question asks about an N-channel dual gate MOSFET. MOSFETs have insulated gates, so we're looking for a symbol where the gate parts are separated from the source and drain. Symbols 2, 3, 4 and 5 qualify, so we've eliminated 1 and 6.

There are two schematic symbols that show two "G" connections, which stands for "gate." Good, we're down to two out of six.

Now, remember that for FET's, the symbol for **N**-channel devices has the arrow pointing "**N**", and you can spot figure **4** as the correct answer for this question.

E6A11 (A) In Figure E6-1, what is the schematic symbol for a P-channel junction FET?

A. **1**

B. 2

C. 3

D. 6

For this question we're just looking for a simple FET. We want a symbol with no separation between source, gate and drain. Symbols 1 and 6 fill the bill. Symbol 6 has the arrow pointing IN, so it's an N channel device. That leaves us with symbol **1**.

Did you notice both of the correct symbols for Figure E6-1 are in the left hand column? The diagram on the actual exam has to match the one you see here, so that might come in handy at test time.

Key Concepts in This Chapter

- The type of semiconductor material that contains excess free electrons is N-type.

- An impurity atom that adds holes to a semiconductor crystal structure is called an acceptor impurity.

- Gallium arsenide is used as a semiconductor material in preference to germanium or silicon in microwave circuits. The reason it is useful in those applications is that it has higher electron mobility than silicon.

- A PN-junction diode does not conduct current when reverse biased because holes in P-type material and electrons in the N-type material are separated by the applied voltage, widening the depletion region.

- A solid-state relay is a device that uses semiconductors to implement the functions of an electromechanical relay.

- The beta of a bipolar junction transistor is the change in collector current with respect to base current.

- A silicon NPN junction transistor is biased on when base-to-emitter voltage is approximately 0.6 to 0.7 volts.

- The term that indicates the frequency at which the grounded-base current gain of a transistor has decreased to 0.7 of the gain obtainable at 1 kHz is the alpha cutoff frequency.

- Compared to a bipolar transistor, the DC input impedance at the gate of a field-effect transistor is higher.

- Many MOSFET devices have internally connected Zener diodes on the gates to reduce the chance of the gate insulation being punctured by static discharges or excessive voltages.

- A depletion-mode FET is an FET that exhibits a current flow between source and drain when no gate voltage is applied.

- In Figure E6-1, the schematic symbol for an N-channel dual-gate MOSFET is symbol number four.

- In Figure E6-1, the schematic symbol for a P-channel junction FET is symbol number one.

When you complete the Practice Exam for this chapter satisfactorily, you'll be ready to take Progress Check #6 at fasttrackham.com.

Chapter 30

Principles of Diodes

We have had solid-state diodes for much longer than we had any use for solid-state diodes. German physicist Ferdinand Braun first made note of the "one-way valve" characteristics of certain crystals in 1874. Like so many other discoveries in electricity, at the time it amounted to, "Huh. Interesting, I guess," and that was about it. We should note that Braun went on to make major contributions to the development of radio and shared the Nobel Prize in Physics with Marconi for his contributions.

Once we started to really understand how semiconductors work, in the 1950's, new variations on the basic diode were invented – and invented, and invented. I found one source that listed 21 distinct types of diodes, with hundreds of applications. That odd effect Braun noticed in 1874 turns out to be one of the most versatile, useful tools in our electronic tool kit.

Here's an overview of the types of diodes that appear on the Extra exam.

Zener Diodes

Given how many times the topic is covered in the Technician and General exams, by now you certainly know that a typical, silicon junction diode is a one-way valve for electricity. Zener diodes have also been covered along the way, at least briefly.

To review; a Zener diode operates like a normal diode, conducting electricity when the diode is forward biased. Forward biased means the cathode is negative and the anode is positive. The diode blocks current flow when it is reverse biased. With the Zener diode, though, there's an "until." The until part is, it blocks the flow of electricity when reverse biased until the reverse bias voltage rises to the "breakdown voltage," also known as the "Zener voltage", and then it conducts in the reverse direction. Now, normal diodes will do that too, but the voltage at which they'll do it is unstable, and just a little excess reverse bias will let the magic smoke out of the diode and destroy its usefulness. That's not so with Zeners, which can be built to "Zener" at very precise voltages, and can withstand higher levels of reverse current. The Zener voltage can range from a handful of volts to hundreds of volts, so Zener diodes are available for a wide range of purposes.

If a normal diode is a one-way valve for electricity, a reverse biased Zener diode is like a spring loaded pressure-relief valve.

An important characteristic of diodes in general is that they all create a particular voltage drop in forward bias, which is equal to the threshold voltage – not a coincidence. It's not quite like a resistor, where the voltage drop equals the current times the resistance. A diode always provides about the same voltage drop, no matter the current – within rather narrow limits. A Zener diode happens to create a very stable voltage drop across a wide range of currents.

Figure 30.1 shows a simple voltage regulator circuit using a Zener diode. The input is on the left, the output on the right.

Somewhere in the device being fed power by the circuit shown in Figure 30.1 is a component that needs a nice, stable 5v supply. We grab some power from the power supply that's equal

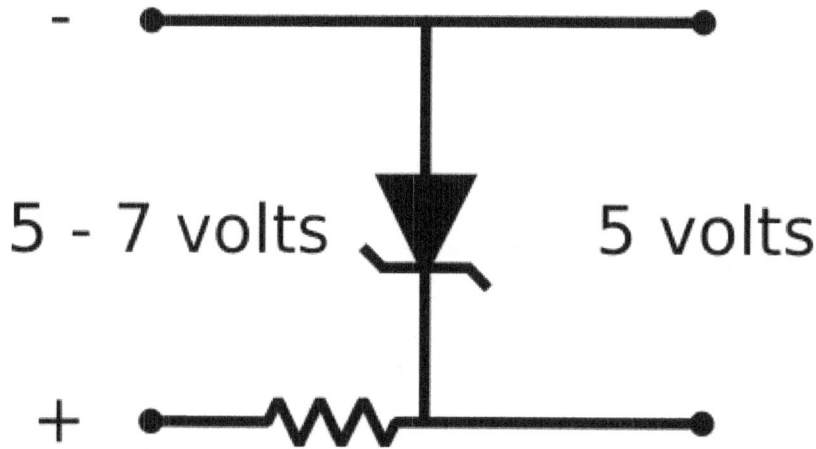

Figure 30.1: Simple Voltage Regulator

to or a little over 5v. In our example, I've made it an unstable 5 to 7 volts. We put a resistor in the line – important, partly because otherwise it's about to become a short circuit – and put a Zener diode with a Zener voltage of 5v in parallel, operating in reverse bias. So long as that incoming voltage is 5v, the Zener might as well not even be there – it's just an insulator between two wires. When the voltage wanders up over 5 volts, though, the Zener diode starts bleeding off that excess voltage and keeps the output right at 5v. (In practice, it would be Zenering all the time, but this works better for getting an idea of how it works.) That **Zener diode provides a very stable voltage drop over a wide range of currents**; a key phrase to remember for the exam.

Schottky Diodes

Normal silicon diodes have an inherent limitation. No matter what we do, they require about 0.6 volts of forward bias before they start conducting. That voltage required for the diode to conduct is called the forward voltage, or the forward voltage drop.

Because there's no such thing as an instantaneous voltage change, that 0.6 volts becomes a time issue, and therefore a linearity issue. For however much time it takes a waveform to go from 0.0 to 0.6 volts, the diode is not conducting and that part of the waveform has vanished, so what goes in no longer matches what comes out.

Diodes can also be used as switches. If we're using one as a switch, instead of a linearity issue, we have a response time issue. If we could somehow make that diode start conducting at a lower voltage, we could have a switch that responded faster. We could build higher frequency devices, or faster computer circuits.

The Schottky diode, (pronounced "SHOT-key") also known as a *hot carrier diode*, has a much lower forward voltage drop of 0.15 to 0.45 volts. Schottky diodes also have inherently lower noise than junction diodes, making them quite useful in applications like VHF and UHF mixers and detectors.

The magic of a normal diode occurs at the semiconductor-to-semiconductor junction – that point in the middle where the P material meets the N material. In a Schottky diode, the magic happens at a metal-semiconductor junction; the point where a metal layer meets a semiconductor layer. There's a very thin layer of metal sandwiched between the P and N material layers. The metal is usually something like paladium, chrome, or platinum.

Figure 30.2 shows the symbol for a Schottky diode:

Notice the shape of the cathode symbol in Figure 30.2. If you hold your head right, it looks something like an "S" for Schottky.

Figure 30.2: Schottky Diode Symbol

PIN Diodes

The PIN in PIN diode stands for Positive-Intrinsic-Negative. Pure semiconductor material, with no impurities added, is neither P nor N material. It is called "intrinsic" semiconductor material. A PIN diode consists of a layer of P material and a layer of N material separated by a layer of "I" material; undoped semiconductor material. (In practice, it isn't quite pure – it is very lightly doped.)

This makes an absolutely lousy normal diode. It doesn't work well in a basic diode function, rectifying waveforms. However, that large intrinsic region makes it a great variable attenuator or switch for radio frequencies, because the resistance of the diode is directly proportional to the forward DC bias voltage applied to it. It works like a mechanical potentiometer, but mechanical potentiometers are pretty much useless at radio frequencies.

PIN diodes also have low junction capacitance; any capacitor loses capacitance as the plates get farther apart, and the intrinsic material in the middle of the PIN sandwich separates the "plates."

The most common symbol for a PIN diode is simply a "normal" diode symbol with a P next to it, shown in Figure 30.3

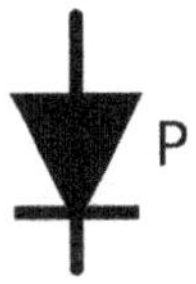

Figure 30.3: PIN Diode Symbol

Varactor Diodes

We know a PIN diode can act as a variable attenuator. Wouldn't it be handy if we could also have an electronically variable capacitor? It would, indeed, and that's what a varactor is.

Varactors may take the prize for "Electronic Component With The Most Names." You'll see them called varactors, variable capacitance diodes, variable reactance diodes, varicaps, and even tuning diodes. All the names mean the same thing. Every diode is a capacitor. A capacitor is two charged conductors separated by a non-conductor. In a diode we have two charged "sort-of conducting zones" separated by a depletion layer which is a "less conducting" zone – that's a capacitor. Every diode operating in reverse bias is even a variable capacitor – the narrower we make the depletion zone, the higher the value of the capacitance. Varactors are just diodes designed to exploit and enhance this inherent quality of diodes and make it useful.

Varactors can be used in tuning circuits, in oscillators, in filters – really, anywhere we need a variable capacitor, so long as the value of capacitance we need is fairly low. The most common varactors can manage a range of 1 to 10 pF. For special applications, varactors are available up to the 100 – 500 pF range.

The common symbol for a varactor, shown in Figure 30.4, mashes together the symbols for a diode and for a capacitor.

Figure 30.4: Varactor Diode Symbol

Tunnel Diodes

The tunnel diode is so-named because it takes advantage of a quantum mechanics effect called "tunneling." We're not going to go down the rabbit hole of the bizarre logic of quantum mechanics – it's fascinating, but we don't need it for our purposes.

Tunnel diodes have the unique characteristic that as the forward bias voltage increases, the current passed decreases. It's almost like Ohm's Law gets turned upside down inside the tunnel! Well, any device that lets us control current with an applied voltage (or vice versa) can be used as an amplifier, and tunnel diodes can be used as amplifiers. Tunnel diodes can also be used as oscillators, in part because their breakdown voltage, the reverse bias voltage where they start conducting "backwards," is zero. Figure 30.5 shows the tunnel diode symbol.

Figure 30.5: Tunnel Diode Symbol

Point-Contact Diode

Behold the point-contact diode in Figure 30.6!

Figure 30.6: Point Contact Diode

That one dates from about 1906. It was the first commercially available diode, and it was sold for the purpose of being a detector. Today, we'd probably just use a five-cent junction diode, but you have to admit that apparatus is one glorious piece of antique gear.

Figure 30.7: Close-up of Point Contact Diode

It should be reasonably apparent why it is called a point-contact diode. There's a piece of silicon or germanium in the one pictured – that's the flat plate under the point. The container holding the silicon can be moved back and forth so the operator can find a spot that acts as a diode. Point contact diodes can also be made from other semiconducting materials, including galenium, germanium, even a lightly rusted razor blade! (Yep, that rusty razor blade is a Metal Oxide Semiconductor.)

Point contact diodes are a form of Schottky diode – the junction of the metal point and the semiconductor material forms a Schottky barrier, and these diodes exhibit most of the characteristics of what we usually think of as a Schottky diode.

The unit illustrated is obsolete, but they do make more practical versions that look like modern components and that fit on printed circuit boards. They're for a few specialized applications, especially for low level microwave signals where they handily outperform their cousins, the Schottky diodes.

I chose this illustration of an antique for a reason, though – look at the spot where the "point contact", which is the anode, and the silicon, which is the cathode, meet. You can see a magnified view in Figure 30.7.

Now check out the schematic symbol for a diode.

So THAT'S where they got it!

LED

LED's are Light Emitting Diodes. They have all sorts of marvelous characteristics from extraordinary efficiency to low cost and long life, and most of us literally own millions of them in our computer and television screens.

In fact, every diode is a light emitting diode. Every time an electron falls into a hole in semiconducting material, a photon is emitted, because the electron is going from a high energy state to a lower energy state. LED's are just optimized to accent this feature of diodes and packaged so the photons can escape.

Silicon diodes emit light in the infrared region, so we can't see that light. Using different materials in the LED lets us create visible light in all the colors across the visible light spectrum.

To emit light, LED's must be connected in forward bias.

The symbol for an LED, Figure 30.8 is a standard diode symbol with two arrows pointing away from it to indicate the light being emitted.

Figure 30.8: Light Emitting Diode Symbol

Chapter 31

Diode Applications

E6B01 (B) What is the most useful characteristic of a Zener diode?
  A. A constant current drop under conditions of varying voltage
  B. **A constant voltage drop under conditions of varying current**
  C. A negative resistance region
  D. An internal capacitance that varies with the applied voltage

Zener diodes are often used as voltage regulators because they provide **a constant voltage drop under conditions of varying current.**

E6B02 (D) What is an important characteristic of a Schottky diode as compared to an ordinary silicon diode when used as a power supply rectifier?
  A. Much higher reverse voltage breakdown
  B. More constant reverse avalanche voltage
  C. Longer carrier retention time
  D. **Less forward voltage drop**

Schottky diodes are used in applications where high speed is essential. They get that rapid responsiveness because, compared to junction diodes, they have **less forward voltage drop** – another way to say they have a lower threshold voltage. Though the question doesn't specify the type of power supply, that rapid responsiveness would be particularly useful in switching power supplies. Schottky diodes are not, however, useful in high voltage applications.

E6B06 (D) Which of the following is a common use of a Schottky diode?
  A. As a rectifier in high current power supplies
  B. As a variable capacitance in an automatic frequency control circuit
  C. As a constant voltage reference in a power supply
  D. **As a VHF/UHF mixer or detector**

The last question was asking about what important characteristic of a Schottky diode would be useful if it was used as a power supply rectifier. In reality, though, you'd be more likely to find a Schottky diode being used **as a VHF/UHF mixer or detector**.

E6B04 (A) What type of semiconductor device is designed for use as a voltage-controlled capacitor?
  A. **Varactor diode**
  B. Tunnel diode
  C. Silicon-controlled rectifier
  D. Zener diode

The semiconductor device that can be used as a voltage-controlled capacitor goes by many names. Among them is **varactor diode**.

E6B05 (D) What characteristic of a PIN diode makes it useful as an RF switch?
  A. Extremely high reverse breakdown voltage
  B. Ability to dissipate large amounts of power

C. Reverse bias controls its forward voltage drop

D. **Low junction capacitance**

Recall that as the value of a capacitor decreases, reactance increases; as frequency increases, reactance decreases.

What makes a PIN diode useful as an RF switch is that because it has very **low junction capacitance**, it has very high reactance – and thus, very high impedance – at high frequencies. "Normal" diodes, with low junction capacitance, can't function well as switches at high frequencies because they can't present a high impedance to a signal.

E6B11 (A) What is used to control the attenuation of RF signals by a PIN diode?

A. **Forward DC bias current**

B. A sub-harmonic pump signal

C. Reverse voltage larger than the RF signal

D. Capacitance of an RF coupling capacitor

When we use PIN diodes as RF attenuators, we control the attenuation by changing the **forward DC bias current**.

E6B07 (B) What is the failure mechanism when a junction diode fails due to excessive current?

A. Excessive inverse voltage

B. **Excessive junction temperature**

C. Insufficient forward voltage

D. Charge carrier depletion

This answer should be fairly common sense – put enough energy into anything and it will burn up. (See: Vader, Darth: Death Star.) It happens that in a junction diode, the junction is the part that heats the most so it overheats the soonest – then **excessive junction temperature** kills the diode. Chances are good if you've ever found a burned-up diode in a piece of equipment, it was melted right in the center. Now you know why.

E6B08 (A) Which of the following describes a type of Schottky barrier diode?

A. **Metal-semiconductor junction**

B. Electrolytic rectifier

C. PIN junction

D. Thermionic emission diode

In this question they are looking for some phrase that describes some sort of Schottky diode. Two of the items in the wrong answers aren't semiconductor devices of any sort. The only phrase that qualifies is **metal-semiconductor junction**.

There was such a thing as an "electrolytic rectifier." It wasn't a semiconductor device, though, and wasn't even solid-state. If you can imagine such a thing, it was a liquid state device.

A PIN junction diode is something completely different from a Schottky.

"Thermionic emission diode" is a 20 dollar name for a vacuum tube diode – definitely not a semiconductor based device.

E6B09 (C) What is a common use for point contact diodes?

A. As a constant current source

B. As a constant voltage source

C. **As an RF detector**

D. As a high voltage rectifier

Point contact diodes are still around, though much smaller and more sophisticated than the unit you saw in Figure 30.6. Figure 31.1 shows a present-day one, packaged in glass. The piece of geranium is on the right in that picture, and the bent wire is the "point."

A common use for point contact diodes is the very same use we had for them back in 1906; they're used as **RF detectors**.

E6B03 (B) What type of bias is required for an LED to emit light?

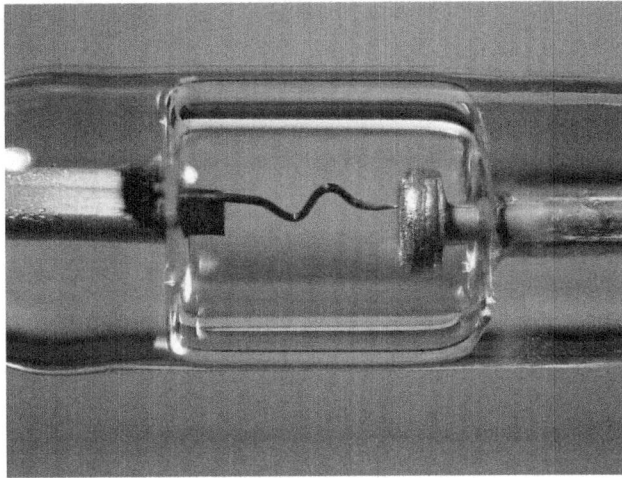

Figure 31.1: Interior of Modern Point Contact Diode. Photo by Gary Crawford, used by permission.

    A. Reverse bias
    B. **Forward bias**
    C. Zero bias
    D. Inductive bias

Like most diodes, LED's only conduct electricity when they are hooked up in **forward bias**, and they must conduct to emit light.

Diode Schematic Symbols

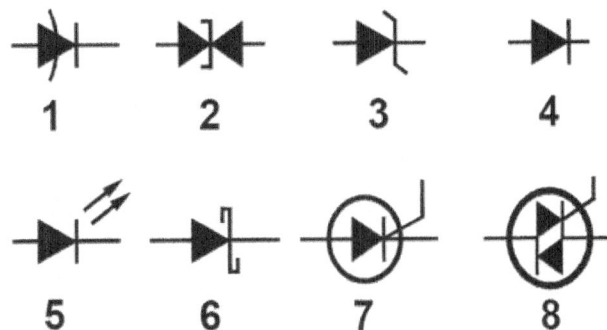

Figure 31.2: Figure E6-2 in the question pool

E6B10 (B) In Figure E6-2, what is the schematic symbol for a light-emitting diode?
    A. 1
    B. **5**
    C. 6
    D. 7

    We need to know one and only one diode symbol for questions in this section, and that's the LED symbol. It's a standard diode symbol with two arrows pointing away from it symbolizing the light rays shooting out of it, symbol number **5** in the illustration.

Figure 31.3: Light Emitting Diode

Key Concepts in This Chapter

- The most useful characteristic of a Zener diode is that it has a constant voltage drop under conditions of varying current.

- An important characteristic of a Schottky diode when used as a power supply rectifier is that it has less forward voltage drop.

- A common use of a Schottky diode is as a VHF/UHF mixer or detector.

- A semiconductor device designed for use as a voltage-controlled capacitor is a varactor diode.

- PIN diodes are useful as RF switches or attenuators because they have low junction capacitance.

- The attenuation of RF signals by PIN diodes is controlled by forward DC bias current.

- The failure mechanism when a junction diode fails due to excessive current is excessive junction temperature.

- The phrase "metal-semiconductor junction" describes a Schottky barrier diode.

- A common use for point contact diodes is as an RF detector.

- LED's require forward bias to emit light.

- In Figure E6-2 the schematic symbol for a light-emitting diode is symbol number 5.

Chapter 32

Digital IC's

This chapter covers the main section of the exam that deals with digital components, and really, it is just a handful of questions that mostly deal in broad generalities.

I can understand the challenge the exam committee faces in this area. Digital technology is advancing at such a rapid pace, anything more specific is almost certainly doomed to be obsolete almost as fast as the exam questions could be published. Each innovation brings multiple new terms into use, and just keeping up with the vocabulary can be dizzying.

What I'm saying is that by the time you read this, some of these terms may already be antiques – but the questions and "correct answers" on the exam won't change until July of 2024.

E6C09 (B) What is a Programmable Logic Device (PLD)?

A. A logic circuit that can be modified during use

B. A programmable collection of logic gates and circuits in a single integrated circuit

C. Programmable equipment used for testing digital logic integrated circuits

D. A type of transistor whose gain can be changed by digital logic circuits.

When you start up your computer, you probably see a few messages from the BIOS. Then the operating system starts up and eventually, you have a fully functioning computer. Computers face a sort of chicken or egg problem at start-up – simply put, before they can load the operating system they need to load the operating system. The solution is to have a relatively small Basic Input/Output System – the BIOS -- stored on a chip that starts up with the computer and loads the main operating system.

That BIOS on your computer lives on a chip that acts as a Programmable Logic Device, or PLD. Programmable logic devices consist of **a programmable collection of logic gates and circuits in a single integrated circuit**. The answer about a "logic circuit that can be modified during use" is *almost* right, but programming a PLD doesn't modify its circuits any more than opening up your internet browser modifies the circuits of your computer.

PLD's come in several different forms, but the key characteristic is that they are all programmable – you can, for instance, update the BIOS in your computer. PLD's are also non-volatile, meaning their memory doesn't vanish when the power gets turned off.

The strong suit of PLD's is handling specific, typically repetitive, computations. For instance, we might use a PLD to store and output a wide array of waveforms to create a digital oscillator for a transmitter. In the broad scheme of things, it's a rather "dumb" task, so by handing it off to a specialized unit we can free up computation cycles in the main CPU.

E6C02 (B) What happens when the level of a comparator's input signal crosses the threshold?

A. The IC input can be damaged

B. The comparator changes its output state

C. The comparator enters latch-up

D. The feedback loop becomes unstable

Broadly speaking, a comparator is a device that compares two inputs and generates some output based on those inputs.

In digital circuitry, a comparator compares two voltages and outputs a digital signal (0 or 1) indicating which one is larger. If input A is greater than input B, the comparator outputs a 1. If input A is less than input B, the comparator outputs a 0. You'll see a real-world comparator application when we cover analog-to-digital conversion in Chapter 43.

That point where the two inputs are equal is called the threshold. If the inputs cross the threshold, **the comparator changes its output state**, i.e., it changes from 0 to 1 or vice versa.

E6C01 (A) What is the function of hysteresis in a comparator?

A. **To prevent input noise from causing unstable output signals**

B. To allow the comparator to be used with AC input signals

C. To cause the output to change states continually

D. To increase the sensitivity

Hysteresis in any system, electronic or otherwise, is lag. Your home's thermostat has hysteresis built into it. If you set it to 72°, it is probably set up so that the heater runs until the temperature is actually around 74 degrees, then it shuts down until the temperature drops to about 70°. Otherwise, it would constantly switch the heater on and off.

In the same way, we build hysteresis into comparators, **to prevent input noise from causing unstable output signals**. They don't respond instantaneously to the inputs crossing the threshold, there's a little bit of delay – "Is this a real change, or is this just random noise?"

E6C03 (A) What is tri-state logic?

A. **Logic devices with 0, 1, and high-impedance output states**

B. Logic devices that utilize ternary math

C. Low power logic devices designed to operate at 3 volts

D. Proprietary logic devices manufactured by Tri-State Devices

Tri-state logic is not some new breakthrough computing technology using 0's, 1's, and 2's. It still uses just 0's and 1's.

What tri-state logic is, is a technology that saves a LOT of printed circuit board traces.

The "next destination" for digital outputs of all sorts is some sort of "bus." In electronics a bus is a communication system that transports data from one place to another. When you plug a USB flash drive into your computer, you're connecting to the USB bus, for example. If we were still in the days of wiring up chassis boxes, a bus would usually be where a whole bunch of wires connect together.

Typical digital buses have an inherent limitation; they can only "listen" to one input at a time. Well, technically, they *could* listen to all the inputs, but the result would be nonsense. Worse, signals from other devices on the line could affect all the devices connected to the line. Chaos.

Let's take the example of a computer's RAM. Each memory address needs a unique connection to the memory bus of the CPU. My computer has 16 billion memory addresses, so all I need is sixteen billion individual circuit traces --- hmmm --- I don't think this is going to work out!

That's where tri-state logic comes in. Tri-state logic devices have three possible output states. They're **logic devices with 0, 1 and high impedance output states**. The logic device, normally a low impedance device, can be electrically switched to a high impedance device. Since it is connected to a low impedance device at the other end, this effectively disconnects the unit. Functionally, it's like Figure 32.1.

When the Control input is neutral, the device outputs whatever the input is, either 1 or 0. When the Control input is energized, the output is neither 1 nor 0; it switches to high impedance and the device is effectively offline. It can't affect anything downstream, and nothing downstream can affect it.

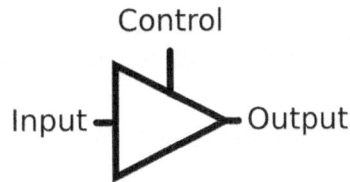

Figure 32.1: Tri-state Logic

E6C05 (D) What is an advantage of CMOS logic devices over TTL devices?

    A. Differential output capability

    B. Lower distortion

    C. Immune to damage from static discharge

    D. **Lower power consumption**

CMOS stands for Complementary Metal-Oxide Semiconductor. MOSFETs are CMOS devices; there's a thin layer of metal-oxide insulation between the gate and the channel of CMOS devices.

TTL devices are a different technology with many applications similar to CMOS. TTL stands for Transistor-Transistor Logic. TTL is somewhat of an antique now, and you'd be hard pressed to purchase a TTL device as a discrete component, but it's still present internally in CPU's.

CMOS technology has largely supplanted TTL, mostly because CMOS logic devices offer significantly **lower power consumption.**

E6C06 (C) Why do CMOS digital integrated circuits have high immunity to noise on the input signal or power supply?

    A. Large bypass capacitance is inherent

    B. The input switching threshold is about two times the power supply voltage

    C. **The input switching threshold is about one-half the power supply voltage**

    D. Bandwidth is very limited

There are actually a lot of reasons for CMOS's high immunity to noise, but the only correct one listed here is that **the input switching threshold is about one-half the power supply voltage.** For a typical CMOS device, the power supply is 5v, so the threshold is about 2.5v. A voltage below 2.5v switches the logic device's logic to one state, above 2.5v switches it to another. This means a random bit of noise has to be over 2.5v to cause a logic error. Compare that to TTL technology, with a threshold of just 1.5v. Doesn't sound like much difference, but that makes the CMOS about 50% more noise resistant.

E6C04 (C) Which of the following is an advantage of BiCMOS logic?

    A. Its simplicity results in expensive devices than standard CMOS

    B. It is immune to electrostatic damage

    C. **It has the high input impedance of CMOS and the low output impedance of bipolar transistors**

    D. All these choices are correct

BiCMOS is just what it sounds like; a system using both bipolar and CMOS transistors. Bipolar + CMOS = BiCMOS!

The idea of the desirability of mismatched impedances has come up before, back on page 187, and the advantage here stems from the same concept of intentionally mismatching impedances to limit power transfer. When logic is talking to logic, we don't want any unexpected interactions between components, and those impedance mismatches help reduce those interactions. So, since BiCMOS logic **has the high input impedance of CMOS and the low output impedance of bipolar transistors**, circuits can be designed to take advantage of that mismatch.

E6C07 (B) What best describes a pull-up or pull-down resistor?

A. A resistor in a keying circuit used to reduce key clicks

B. A resistor connected to the positive or negative supply line used to establish a voltage when an input or output is an open circuit

C. A resistor that ensures that an oscillator frequency does not drift

D. A resistor connected to an op-amp output that prevents signals from exceeding power supply voltage

A pull-up or pull-down resistor is **a resistor connected to the positive or negative supply line used to establish a voltage when an input or output is an open circuit.**

Here's an example that should make it more clear. Pull-up resistors, and less commonly, pull-down resistors, are often used when the input to a digital device is a button or switch. So let's use the reset button on your computer. Somewhere deep inside your computer is an input that is waiting for a signal from that switch. Until it gets the signal from the switch, its output is 0, and we'll say that in this case, a high voltage input makes that logic device send 0. The triangle with a circle on the end represents a digital logic device called a "NOT operation" – if you send it a 1, it sends out a 0, and vice versa.

Somewhere on the motherboard, something checks that output every now and then (like every millisecond or so) and says, "Nope. Still 0. Don't reset the computer." Then, the switch gets pressed, and now the logic device is sending a 1 and, boom, reset.

It is probably hooked up something like Figure 32.2.

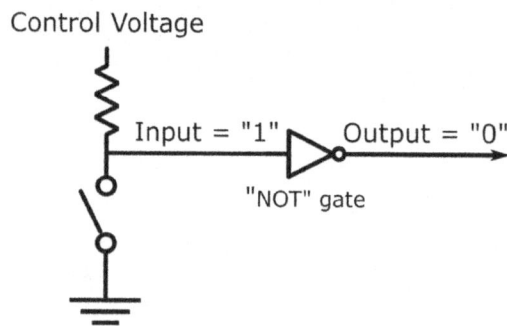

Figure 32.2: Pull-up Resistor

So long as the switch isn't closed, the control voltage is high to the input, and the device outputs 0. When the switch gets closed, the voltage to the input drops and its output goes to 1. (Notice that without the resistor, we'd have a short circuit to ground.)

The control voltage through the resistor was "pulling up the input." If we hooked up the resistor and a normally-open switch like in Figure 32.3, we'd have a pull-down resistor – when the switch is open, the voltage to the input gets pulled down.

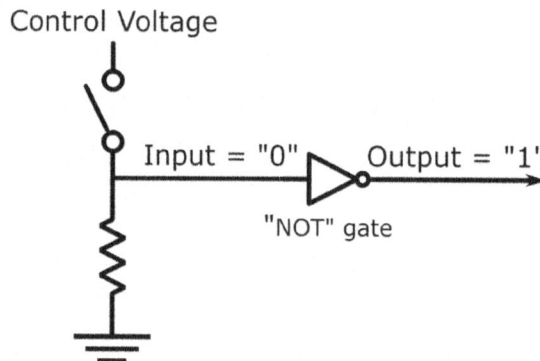

Figure 32.3: Pull-down Resistor

Why would we bother with adding that pull-down resistor? Isn't the voltage at the input going to be zero so long as that switch is open? On paper, yes. In reality, not necessarily. Logic

204

gates actually have three states, of which only two are useful. In the example above with the NOT gate, we'll say +5 V puts the gate in the "0" state, and 0 V puts it in the "1" state. So far, so good. But what if that input is ungrounded and there's some stray electromagnetic energy floating around, most likely from nearby components? What if that stray energy causes the input to be energized at, say, +2.5 V? That might put the gate into an "indeterminate" state. We've committed a dreadful digital sin by leaving the input "floating" and we can't predict whether it will output a 1 or a 0. As my friend Dennis explained to me, we have created a circuit that is saying, "I don't know what I'm going to do, but when I do it, you probably won't like it."

Key Concepts in This Chapter

- A Programmable Logic Device is a programmable collection of logic gates and circuits in a single integrated circuit.
- When the level of a comparator's input signal crosses the threshold, the comparator changes its output state.
- The function of hysteresis in a comparator is to prevent input noise from causing unstable output signals.
- Tri-state logic uses logic devices with 0, 1, and high impedance output states.
- An advantage of CMOS logic devices over TTL devices is lower power consumption.
- CMOS digital integrated circuits have high immunity to noise on the input signal or power supply because the input switching threshold is about one-half the power supply voltage.
- An advantage of BiCMOS logic is it has the high input impedance of CMOS and the low output impedance of bipolar transistors.
- A pull-up or pull-down resistor is best described as a resistor connected to the positive or negative supply line used to establish a voltage when an input or output is an open circuit.

Chapter 33

Inductors and Piezoelectric Devices

Inductors

E6D04 (B) Which materials are commonly used as a core in an inductor?

    A. Polystyrene and polyethylene

    B. **Ferrite and brass**

    C. Teflon and Delrin

    D. Cobalt and aluminum

None of the materials listed, except **ferrite and brass**, would make a useful core for an inductor.

Ferrite has the effect of increasing the inductance of a coil when it is used as a core, compared to air or even to iron.

Brass is commonly used in variable inductors.

We often find variable inductors in resonant circuits that require adjustment; literally "fine tuning." We can vary the inductance of an inductor several ways, but in terms of mechanics, the easiest way is to move some core material in or out of the coil. Usually this is accomplished with a screw mechanism that moves a "slug" – a cylinder of core material – in and out of the center of the coil.

Figure 33.1: Ferrite Slug Variable Inductor

Since ferrite has a very high inductance index, it's used in variable inductors that need a broad range of adjustment.

Brass has a lower inductance index than air! By inserting a brass slug in an inductor, we lower the inductance. Brass is used in variable inductors that don't need as much range but do need precision.

E6D11 (B) Which type of slug material decreases inductance when inserted into a coil?
 A. Ceramic
 B. **Brass**
 C. Ferrite
 D. Powdered iron

A material that decreases inductance when inserted into a coil is **brass**. All the other materials listed are either neutral (some ceramics) or increase inductance.

E6D10 (A) What is a primary advantage of using a toroidal core instead of a solenoidal core in an inductor?
 A. **Toroidal cores confine most of the magnetic field within the core material**
 B. Toroidal cores make it easier to couple the magnetic energy into other components
 C. Toroidal cores exhibit greater hysteresis
 D. Toroidal cores have lower Q characteristics

Toroidal means "doughnut shaped", but saying "toroidal" sounds much more impressive.

 Toroidal inductors are very common, and there's a good reason for that; **toroidal cores confine most of the magnetic field within the core material**. That way, the inductor isn't radiating great gobs of magnetic field out into the other components creating noise and interference. Toroidal inductors keep their magnetic business to themselves.

 A "Solenoidal core" is a cylindrical core, such as you'd find in – you guessed it! – a solenoid.

E6D05 (C) What is one reason for using ferrite cores rather than powdered-iron in an inductor?
 A. Ferrite toroids generally have lower initial permeability
 B. Ferrite toroids generally have better temperature stability
 C. **Ferrite toroids generally require fewer turns to produce a given inductance value**
 D. Ferrite toroids are easier to use with surface mount technology

Technically, ferrite is a ceramic, though not much like the ceramic in your coffee cup. It's made of iron oxide – rust -- and various other metals. Those other metals vary depending on the application for the material. Ferrite has very high *permeability,* giving it a very high inductive index – putting a ferrite core in a coil boosts the inductance of the coil a lot.

 Powdered-iron cores are iron mixed with epoxy, and powdered iron's inductive index is lower than ferrite's.

Figure 33.2: Toroidal Inductor

 Because of ferrite's high inductive index, **ferrite toroids generally require fewer turns to produce a given inductance value**. Fewer turns means a smaller and lighter coil.

E6D08 (B) What is one reason for using powdered-iron cores rather than ferrite cores in an inductor?
 A. Powdered-iron cores generally have greater initial permeability
 B. **Powdered-iron cores generally maintain their characteristics at higher currents**
 C. Powdered-iron cores generally require fewer turns to produce a given inductance
 D. Powdered-iron cores use smaller diameter wire for the same inductance

Inductors are ferociously complicated things when one gets down into the nitty-gritty details of their performance. For instance, as the magnetic field goes in and out of the core, electric currents swirl around inside the core. They're called eddy currents. They sap energy from the circuit, and generate their own magnetic fields, which also interact with the core and the coils, altering the value of the inductor. This can get very significant at high current levels.

Powdered iron cores are just powdered iron mixed with epoxy and molded, usually into a toroid shape. Because it's mostly just a wad of epoxy, the core isn't a very good conductor, and that turns out to be an advantage in high current situations because internal eddy currents are reduced. **Powdered-iron cores generally maintain their characteristics at higher currents**.

E6D06 (D) What core material property determines the inductance of an inductor?
    A. Thermal impedance
    B. Resistance
    C. Reactivity
    D. **Permeability**

Permeability is the measure of how intensely a magnetic field "permeates" a material. More precisely, it's how much magnetic field the material can store. Ferrite, as noted, has very high permeability. Air has almost zero permeability. Different core materials give us different inductance values because of their **permeability**.

E6D12 (C) What is inductor saturation?
    A. The inductor windings are over-coupled
    B. The inductor's voltage rating is exceeded causing a flashover
    C. **The ability of the inductor's core to store magnetic energy has been exceeded**
    D. Adjacent inductors become over-coupled

When we run out of permeability, we hit saturation. The material is as full of magnetism as it can get. **The ability of the inductor's core to store magnetic energy has been exceeded.** You might wonder if this is a problem . . .

E6D01 (A) Why should core saturation of an impedance matching transformer be avoided?
    A. **Harmonics and distortion could result**
    B. Magnetic flux would increase with frequency
    C. RF susceptance would increase
    D. Temporary changes of the core permeability could result

Once the core is saturated, the inductor starts creating distortion – it's like any other overdriven component – and distortion creates unwanted harmonics. The correct answer is **harmonics and distortion could result**.

E6D09 (C) What devices are commonly used as VHF and UHF parasitic suppressors at the input and output terminals of a transistor HF amplifier?
    A. Electrolytic capacitors
    B. Butterworth filters
    C. **Ferrite beads**
    D. Steel-core toroids

It just wouldn't be a ham exam if we didn't mention **ferrite beads** somewhere along the line, right? At the Extra level we learn they have a really fancy name – Technicians and Generals have to just call them ferrite beads, but Extras get to call them "parasitic suppressors."

Be sure to work that into all your conversations. Your friends will be very impressed.

E6D13 (A) What is the primary cause of inductor self-resonance?
    A. **Inter-turn capacitance**
    B. The skin effect
    C. Inductive kickback
    D. Non-linear core hysteresis

When we look at the turns of a coil we see a conductor, then some insulation or air, then a conductor and – oh, my, this coil is also a capacitor.

This can be so significant it even has a name; **Inter-turn capacitance**. True, it's not much of a capacitor, but any time we combine an inductor with a capacitor we get a resonant circuit, and depending on the coil and the operating frequency, it is possible to end up with a self-contained resonant circuit. Great, if that was the intended design, not so great if it wasn't.

E6D07 (A) What is current in the primary winding of a transformer called if no load is attached to the secondary?

    A. **Magnetizing current**

    B. Direct current

    C. Excitation current

    D. Stabilizing current

A simple transformer consists of a primary winding and a secondary winding. Alternating current in the primary winding creates alternating magnetic lines of force that cross the secondary winding and induce a current in it – assuming the secondary's current has somewhere to go, which is called the load.

If there's no load attached to the secondary, the primary current is just magnetizing (and demagnetizing) whatever magnetizable material it encounters. So if there's no load, we call the current in the primary the **magnetizing current**. It's normally a smaller amount of current than is present once we attach a load to the secondary, because once current starts to flow in the secondary it depletes that magnetic field.

If you've ever wondered if that little wall wart power supply for your cell phone uses more power when the phone is plugged into it, yes, it does, and this is why.

Piezoelectric Devices

E6D02 (A) What is the equivalent circuit of a quartz crystal?

    A. **Motional capacitance, motional inductance, and loss resistance in series, all in parallel with a shunt capacitor representing electrode and stray capacitance**

    B. Motional capacitance, motional inductance, loss resistance, and a capacitor representing electrode and stray capacitance all in parallel

    C. Motional capacitance, motional inductance, loss resistance, and a capacitor representing electrode and stray capacitance all in series

    D. Motional inductance and loss resistance in series, paralleled with motional capacitance and a capacitor representing electrode and stray capacitance

Before we jump into the meat of this question, let's cover what an equivalent circuit is.

An equivalent circuit is a tool for understanding and/or communicating the workings of a particular component, circuit or device. In the previous section we covered inter-turn capacitance in a coil which can cause a coil to become self-resonant. We could draw an equivalent circuit of a simple coil by drawing a coil in parallel with a capacitor. The capacitor represents the inter-turn capacitance of the coil, not a separate component.

This question is asking about the equivalent circuit of a quartz crystal.

In electronics, "quartz crystal" means an electronic component constructed out of a crystal of quartz (or sometimes special ceramic material) connected to two conductor plates. Like Figure 33.3

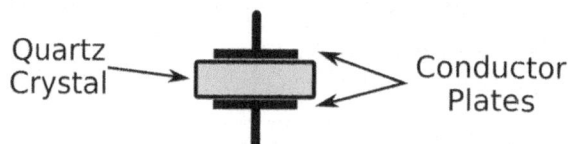

Figure 33.3: Crystal

Quartz crystal has piezoelectric properties. That means, if you had a piece of quartz and hooked a battery to either side of it, it would move – just a tiny bit. And if you had an incredibly sensitive voltmeter hooked up to that quartz and squeezed it, you'd see a tiny voltage being generated.

If we create a component like the one in Figure 33.3 and feed it an alternating current, the quartz will vibrate – and depending mainly on its size, it will resonate at some particular frequency. It becomes an oscillator, and a very precise one at that.

This question asks us what the equivalent circuit, using "normal" components would be. The answer is a bit complicated!

That crystal acts a lot like a capacitor – it resists passing low frequencies. So we know we'd have a capacitor in our equivalent circuit. Not only that, when we feed it an alternating current, it's a capacitor that's changing value constantly.

The crystal also acts as though it has internal inductance – it's not a coil, but it acts like one. It resists passing high frequencies. So we'll have a coil in our equivalent circuit as well. And it's a resistor, so we'll have a resistor. So far, we have a capacitor, a coil, and a resistor in series – all the elements of an LRC series resonant circuit.

But wait, there's more! Consider that those two electrodes are separated by, for our purposes, a not-very-good conductor. Aha, we know what that is! It's a capacitor. There's some stray capacitance from the component leads, the electrodes on either side of the crystal, and maybe even the container. We represent all that with a capacitor in parallel. We end up with the circuit shown in Figure 33.4.

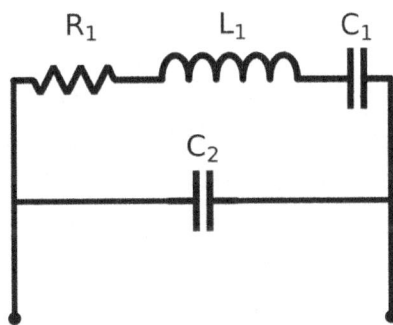

Figure 33.4: Equivalent Circuit of a Quartz Crystal

That's the equivalent circuit for a quartz crystal, and it's a series-parallel resonant circuit. L_1 and C_1, all qualities created by the motion of the crystal, and the exam refers to them as "motional inductance" and "motional capacitance." The resistance, R_1, simply comes from the loss through the crystal.

When a capacitor is hooked in parallel across a series circuit, it's often called a shunt capacitor, and that's the term the exam uses for the capacitance represented by C_2.

Put that all together, and the correct answer is; **Motional capacitance, motional inductance, and loss resistance in series, all in parallel with a shunt capacitor representing electrode and stray capacitance.**

All well and good, except all the possible answers sound a lot alike. So, remember that *the motional components are in series and the shunt is the only thing in parallel*. The correct answer is the one with those qualities listed.

E6D03 (A) Which of the following is an aspect of the piezoelectric effect?

  A. **Mechanical deformation of material by the application of a voltage**

  B. Mechanical deformation of material by the application of a magnetic field

  C. Generation of electrical energy in the presence of light

  D. Increased conductivity in the presence of light

The piezoelectric effect – pronounced pea-AY-zoh – is what we call the ability of certain materials to change size in response to an electric charge. It's the **mechanical deformation of material by the application of a voltage.**

Lots of materials exhibit the effect, including many gems, and even cane sugar crystals, but the material we use the most for electronics is very pure quartz machined to very precise sizes.

Key Concepts in This Chapter

- Materials commonly used as a core in an inductor include ferrite and brass.

- Brass decreases inductance when inserted into a coil.

- A primary advantage of using a toroidal core instead of a solenoidal core in an inductor is that toroidal cores confine most of the magnetic field within the core material.

- One reason for using ferrite cores rather than powdered-iron in an inductor is ferrite cores generally require fewer turns to produce a given inductance value.

- One reason for using powdered-iron cores rather than ferrite cores in an inductor is powdered-iron cores generally maintain their characteristics at higher currents.

- The core material property that determines the inductance of an inductor is permeability.

- The definition of inductor saturation is the point at which the ability of the inductor's core to store magnetic energy has been exceeded. Core saturation leads to harmonics and distortion.

- Ferrite beads are commonly used as VHF and UHF parasitic suppressors at the input and output terminals of a transistor HF amplifier.

- The primary cause of inductor self-resonance is inter-turn capacitance.

- If no load is attached to the secondary winding of a transformer, the current in the primary winding is called the magnetizing current.

- The equivalent circuit of a quartz crystal is motional capacitance, motional inductance, and loss resistance in series, all in parallel with a shunt capacitor representing electrode and stray capacitance.

- An aspect of the piezoelectric effect is mechanical deformation of material by the application of a voltage.

Chapter 34

Analog IC's and Component Packaging

Analog IC's

E6E03 (D) Which of the following materials is likely to provide the highest frequency of operation when used in MMICs?
    A. Silicon
    B. Silicon nitride
    C. Silicon dioxide
    D. **Gallium nitride**

MMICs are Monolithic Microwave Integrated Circuits, and once you've said that a few times you'll join the rest of the world in calling them "mimics." They're a family of semiconductor devices designed to operate in the microwave range.

Gallium nitride is a hot topic in semiconductor engineering circles. It is already widely used in microwave applications because, like its cousin gallium arsenide that we covered in Chapter 29, it can operate at much higher frequencies than most silicon compounds. It also has great promise in high power-handling capacity and its size advantage alone over silicon devices is rather amazing. Figure 34.1 shows a relative size comparison for two devices operating at the same voltage.

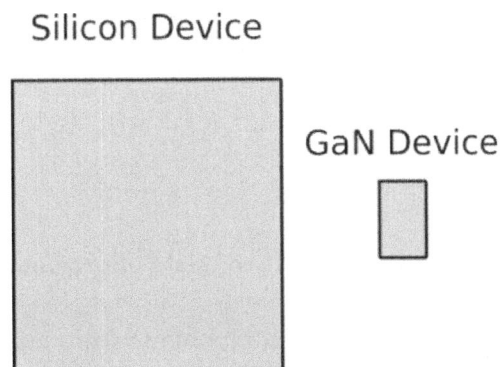

Silicon Device

GaN Device

Figure 34.1: Size Comparison of Silicon vs. Gallium Nitride Devices

All this adds up to making **gallium nitride** very desirable for MMICs. For memory purposes, remember that **S**ilicon is **S**low, but oh, **G**ee **G**allium is fast! (Sure, you groan now, but you'll be thanking me on exam day. You're welcome.)

By the way, the exam committee uses the term "microwave" a few times in the question pool, and you might wonder what, precisely, those microwave frequencies are. There is no official ITU frequency range for that term, but common radio-frequency engineering usage is microwaves start at about 1 GHz and go up to about 100 GHz.

E6E04 (A) Which is the most common input and output impedance of circuits that use MMICs?

    A. **50 ohms**

    B. 300 ohms

    C. 450 ohms

    D. 10 ohms

The most common input and output impedance of circuits that use MMICs is the same as our favorite ham transmitter output impedance, **50 ohms**.

E6E06 (D) What characteristics of the MMIC make it a popular choice for VHF through microwave circuits?

    A. The ability to retrieve information from a single signal even in the presence of other strong signals

    B. Plate current that is controlled by a control grid

    C. Nearly infinite gain, very high input impedance, and very low output impedance

    D. **Controlled gain, low noise figure, and constant input and output impedance over the specified frequency range**

The characteristics of the MMIC that make it a popular choice for VHF through microwave circuits are **controlled gain, low noise figure, and constant input and output impedance over the specified frequency range.**

That's quite a mouthful. Let's do this one by process of elimination, using a little basic semiconductor knowledge and common sense.

The first wrong answer is "the ability to retrieve information from a single signal even in the presence of other signals." That's not really a characteristic of a MMIC, it's a characteristic of a quality receiver. While MMICs might be part of that receiver, they don't possess that ability on their own.

The next is "plate current that is controlled by a control grid." Ha! They're describing a vacuum tube. That can't be right!

How about "nearly infinite gain, very high input impedance, and very low output impedance" – well, we could stop at "infinite gain" since there's no such practical thing, but we also know from the previous question that the input and output impedances are typically the same, at 50 ohms.

That leaves us with the correct answer, **controlled gain, low noise figure, and constant input and output impedance over the specified frequency range**.

E6E07 (D) What type of transmission line is used for connections to MMICs?

    A. Miniature coax

    B. Circular waveguide

    C. Parallel wire

    D. **Microstrip**

We covered microstrip back in Chapter 27. It consists of precision printed circuit conductors above a ground plane.

It's possible to construct several types of components from microstrip; the antenna in your mobile phone is almost certainly a patch of microstrip. Figure 34.2 shows a photomicrograph of a bandpass filter that consists of nothing but accurately placed microstrip.

E6E08 (A) How is power supplied to the most common type of MMIC?

    A. **Through a resistor and/or RF choke connected to the amplifier output lead**

    B. MMICs require no operating bias

    C. Through a capacitor and RF choke connected to the amplifier input lead

    D. Directly to the bias voltage (VCC IN) lead

MMICs are typically powered **through a resistor and/or RF choke connected to the amplifier output lead**. The RF choke prevents the signal from escaping its intended path

214

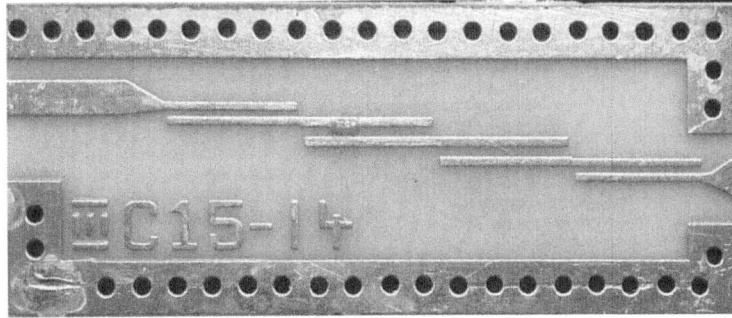
Figure 34.2: Microstrip Bandpass Filter

and ending up in the power supply, and there's probably a resistor in the circuit as well.

We can use logic to figure this one out if we know a key fact, which is that MMICs very seldom *have* a "VCC IN" lead. (VCC, more properly written V_{CC}, is the collector voltage.) The answer that includes that phrase is wrong.

"MMICs require no operating bias" would mean MMICs are quite remarkable indeed, because an amplifier that required no operating bias would be somehow creating energy from nothing.

We'd never supply DC power through a capacitor, capacitors don't pass DC.

Electronic Component Packaging

E6E02 (A) Which of the following device packages is a through-hole type?

A. **DIP**

B. PLCC

C. Ball grid array

D. SOT

Electronic components come in a variety of shapes, known as "packages" or "form factors," to accomodate various design considerations and manufacturing systems.

Components that mount on printed circuit boards come in two broad families of packaging; through-hole types and surface-mounted types.

Through-hole components are mounted on the side of the circuit board that is opposite to the traces – the strips of metal that serve as the "wiring" of the printed circuit board. Their leads go through holes in the circuit board then are soldered to a connection point. Through-hole devices are reasonably easy for humans to solder.

Surface-mount components are placed on the same side of the circuit board as the traces where they connect to the appropriate connection point. Surface-mount components are easy for machines to solder, not so easy for humans.

Figure 34.3 shows a couple of 555 timer chips.

The chip on the upper left is a through-hole component. Those metal "legs", or leads, poke through a hole in the circuit board. (Or they stick "through a hole" in a socket on the circuit board. That's not true for the other components listed in the wrong answers.) The chip on the lower right is a surface mount component – rather than leads it has tabs that get soldered. Even though they clearly must have "leads" of some sort to connect to the circuit board, surface-mount components are said to be leadless because the leads are so minimal.

The chip on the upper left is a classic **DIP**, for "Dual Inline Package" configuration; it has dual rows of inline leads. In the answers to the question, **DIP** is the only type of through-hole component listed.

Figure 34.3: 555 Timer Chips. Through-hole mount (L) and Surface Mount (R)

E6E10 (D) What advantage does surface-mount technology offer at RF compared to using through-hole components?

A. Smaller circuit area

B. Shorter circuit-board traces

C. Components have less parasitic inductance and capacitance

D. **All these choices are correct**

Take a look at those 555 timers in Figure 34.3 and you'll see the dramatic size difference in those otherwise identical components. Because the surface-mount chips are so much smaller, they offer the advantages of **smaller circuit area and shorter circuit-board traces.** Because the leads are so much shorter, **the components have less parasitic inductance and capacitance.**

E6E12 (C) Why are DIP through-hole package ICs not typically used at UHF and higher frequencies?

A. Too many pins

B. Epoxy coating is conductive above 300 MHz

C. **Excessive lead length**

D. Unsuitable for combining analog and digital signals

You learned back in Chapter 27 that in microwave circuits it is important to avoid **excessive lead length** to avoid unwanted inductive reactance; through-hole package IC's have much longer leads than surface-mount packages.

E6E09 (D) Which of the following component package types would be most suitable for use at frequencies above the HF range?

A. TO-220

B. Axial lead

C. Radial lead

D. **Surface mount**

For frequencies above HF, our best choice is **surface mount** component packages. None of the other types of components listed in the possible answers are **surface mount** components.

E6E11 (D) What is a characteristic of DIP packaging used for integrated circuits?

A. Package mounts in a direct inverted position

B. Low leakage doubly insulated package

C. Two chips in each package (Dual In Package)

D. **A total of two rows of connecting pins placed on opposite sides of the package (Dual In-line Package)**

If you've played around with electronic stuff for a while, you've no doubt encountered a DIP switch – one of the annoying things shown in Figure 34.4.

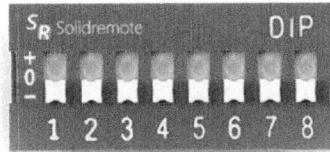

Figure 34.4: DIP Switch

The back side of a DIP switch – or any other Dual In-line Package component, such as the through-hole 555 timer in Figure 34.3 – has **a total of two rows of connecting pins placed on opposite sides of the package.** It's a **Dual In-line Package.**

Key Concepts in This Chapter

- As a semiconductor material, gallium nitride provides a higher frequency of operation than any silicon compound.
- The most common input and output impedance of circuits that use MMIC's is 50 ohms.
- The characteristics of the MMIC that make it a popular choice for VHF through microwave circuits are controlled gain, low noise figure, and constant input and output impedance over the specified frequency range.
- The type of transmission line used for MMICs is called microstrip.
- For the most common type of MMIC, power is supplied through a resistor and/or RF choke connected to the amplifier output lead.
- Because of the smaller size of the components, surface-mount technology offers the advantages of smaller circuit area and shorter circuit-board traces. The components also have less parasitic inductance and capacitance.
- DIP through-hole package IC's are not typically used at UHF and higher frequencies because of excessive lead length.
- The most suitable component package type for use at frequencies above the HF range is surface mount.
- One type of through-hole type device package is a DIP, for dual-inline package. DIP packages used for integrated circuits have a total of two rows of connecting pins placed on opposite sides of the package.

Chapter 35

Optical Components

Photoconductivity

E6F02 (A) What happens to the conductivity of a photoconductive material when light shines on it?

    A. **It increases**
    B. It decreases
    C. It stays the same
    D. It becomes unstable

When light hits a photoconductive material, the material's conductivity **increases.**

Photoconductive devices are made of semiconductor material. Now that you've studied semiconductors, you might already suspect the mechanism at work here. Remember, semiconductors conduct when they have lots of free electrons floating around and lots of holes for those electrons to flow to. Photons come bombing in to the semiconductor's atoms and knock loose electrons, creating free electrons and holes, so it's just as though you hit the semiconductor with some voltage, and it conducts more easily.

Photoconductors are useful for all sorts of things, from the "magic eye" that keeps a garage door from closing when something is blocking it to highly sophisticated scientific measuring instruments. In amateur radio we use them in optoisolators and optical shaft encoders.

E6F06 (A) Which of these materials is most commonly used to create photoconductive devices?

    A. **A crystalline semiconductor**
    B. An ordinary metal
    C. A heavy metal
    D. A liquid semiconductor

The material listed that is affected the most by photoconductivity is **a crystalline semiconductor.**

Ordinary metal isn't photoconductive, nor is a heavy metal.

Liquid semiconductors do exist, but they're so exotic you're never going to see one – they require furnaces that can melt semiconductor material. Even if you see one, don't expect to use it as a photoconductor – thus far, they're pretty weak in that area.

E6F03 (D) What is the most common configuration of an optoisolator or optocoupler?

    A. A lens and a photomultiplier
    B. A frequency modulated helium-neon laser
    C. An amplitude modulated helium-neon laser
    D. **An LED and a phototransistor**

An optoisolator and an optocoupler sound like they'd be opposites, but they are the same thing. Common ones consist of **an LED and a phototransistor** inside an opaque package. You can actually create one yourself by pointing a couple of LED's at each other and wrapping them with electrical tape.

They're used to electrically isolate one part of a device from another, and more often in digital applications than analogue, though not always. Let's say we have some sort of digital circuit that, for whatever reason, is inherently full of high frequency electrical noise. We don't want that getting into the next stage of whatever we are building, so we use an optoisolator between the stages and in effect it filters out the noise, leaving us with a nice clean signal at the next stage.

We don't use huge numbers of optoisolators for our gear, but one related hobby does – the DIY musician community positively loves optoisolators for their homebrew effects boxes, which are those little boxes you see guitarists stomping on to change their sound. Optoisolators let them easily isolate their "stomp box" (that's what they call them!) from the electrical nightmare that is the average stage's ground system.

E6F08 (C) Why are optoisolators often used in conjunction with solid-state circuits when switching 120 VAC?

   A. Optoisolators provide a low impedance link between a control circuit and a power circuit
   B. Optoisolators provide impedance matching between the control circuit and power circuit
   C. **Optoisolators provide a very high degree of electrical isolation between a control circuit and the circuit being switched**
   D. Optoisolators eliminate the effects of reflected light in the control circuit

Optoisolators do just what the name says they do. **Optoisolators provide a very high degree of electrical isolation between a control circuit and the circuit being switched**.

This is a good thing – accidentally hooking an average transistor directly into 120 VAC has unpleasant results, so isolating solid-state circuits from wall voltage is important.

E6F05 (A) Which describes an optical shaft encoder?

   A. **A device that detects rotation of a control by interrupting a light source with a patterned wheel**
   B. A device that measures the strength of a beam of light using analog to digital conversion
   C. A digital encryption device often used to encrypt spacecraft control signals
   D. A device for generating RTTY signals by means of a rotating light source

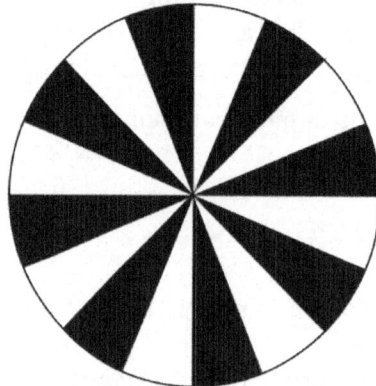

Figure 35.1: Typical Optical Encoder Disc.

Here's an application of photoconductivity.

An optical shaft encoder is a **device which detects rotation of a control by interrupting a light source with a patterned wheel.**

It uses a light source – most likely an LED – and a photoconductive semiconductor, much like an optoisolator. Between the light source and the semiconductor is a rotating transparent disc with opaque marking, mounted on a shaft. Perhaps something like Figure 35.1.

Other models use an opaque disc with holes in it, but they work the same way.

As the shaft rotates the disc – most likely because your hand is spinning the knob attached to the shaft – the disc interrupts the light to the photoconductive semiconductor. Count the pulses, and the machine knows the input that has been made.

This illustration shows a very simple design – they can get much cleverer and more complex with multiple patterns printed on the disc being read by multiple led/semiconductor pairs to build, for instance, a single knob that can control multiple functions on a radio.

Photovoltaic Effect

E6F04 (B) What is the photovoltaic effect?
    A. The conversion of voltage to current when exposed to light
    B. **The conversion of light to electrical energy**
    C. The conversion of electrical energy to mechanical energy
    D. The tendency of a battery to discharge when exposed to light

Photoconductivity and the photovoltaic effect are related but quite different. Both are caused by photons jostling electrons loose from their orbits. But while photoconductivity causes a semiconductor to conduct more easily, the photovoltaic effect in, for instance, a solar cell actually turns the cell into a charged unit like a battery.

The conversion of light to electrical energy is called the photovoltaic effect.

Figure 35.2: Photovoltaic (Solar) Cell

E6F09 (D) What is the efficiency of a photovoltaic cell?
    A. The output RF power divided by the input DC power
    B. Cost per kilowatt-hour generated.
    C. The open-circuit voltage divided by the short-circuit current under full illumination
    D. **The relative fraction of light that is converted to current**

The common name for a photovoltaic cell is solar cell, and the efficiency of a photovoltaic cell is defined much like the efficiency of an amplifier. With an amplifier, it's "what percentage of the electricity that went in the power supply came out of the output as signal?" For a photovoltaic cell, it's "what percentage of the light that went in came out of the output as electricity?" As the test puts it, it's **the relative fraction of light that is converted to current**.

For the sorts of relatively inexpensive cells that we're likely to be using, peak efficiency is presently in the 12% to 25% area. More exotic cells have hit as high at 47%, as of 2020.

E6F10 (B) What is the most common type of photovoltaic cell used for electrical power generation?
    A. Selenium
    B. **Silicon**
    C. Cadmium Sulfide
    D. Copper oxide

The most common type of photovoltaic cell used for power generation is made of the same stuff most diodes, transistors and IC's are made of; **silicon**.

E6F11 (B) What is the approximate open-circuit voltage produced by a fully-illuminated silicon photovoltaic cell?
    A. 0.1 V
    B. **0.5 V**
    C. 1.5 V
    D. 12 V

Solar panels are made up of individual photovoltaic cells. Each cell creates about **0.5 V** when fully illuminated. It doesn't matter what size it is when it comes to the voltage, but a larger cell can deliver more current at that half a volt.

E6F01 (C) What absorbs the energy from light falling on a photovoltaic cell?
   A. Protons
   B. Photons
   C. **Electrons**
   D. Holes

Let's keep this simple, without launching into quantum theory and all that. When photons come bombing into the silicon wafer, they hit **electrons**. The electrons suddenly have too much energy to stay in orbit around their nucleus and they're floating around without a home. That means the silicon is electrically charged and the light energy has been converted to electrical energy.

Key Concepts in This Chapter

- When light shines on photoconductive material its conductivity increases.

- Crystalline semiconductors are affected by photoconductivity.

- The most common configuration of an optoisolator or optocoupler is an LED and a phototransistor.

- Optoisolators are often used in conjunction with solid-state circuits when switching 120 VAC because optoisolators provide a very high degree of electrical isolation between a control circuit and the circuit being switched.

- An optical shaft encoder is a device which detects rotation of a control by interrupting a light source with a patterned wheel.

- The photovoltaic effect is the conversion of light to electrical energy.

- The efficiency of a photovoltaic cell is the relative fraction of light that is converted to current.

- The most common type of photovoltaic cell used for electrical power generation is silicon.

- The approximate open-circuit voltage produced by a fully-illuminated silicon photovoltaic cell is 0.5 V.

- Electrons absorb the energy from light falling on a photovoltaic cell.

When you're done with the Practice Exam for this chapter, find out just how much progress you're making by taking Progress Check #7.

Chapter 36

Digital Circuits

Digital Logic and Components

Back in 1847 an English mathematician and educator, George Boole, was thinking about logic. Specifically, he was developing a mathematics of logic. Of course, math is, almost by definition, logical, but this was math about logic.

Figure 36.1: George Boole

To vastly oversimplify his very large contribution to the field, he formalized a mathematical way of representing things like, "This thing is true, but this other thing is false." In the beginning, he was using T's and F's, but somewhere along the way he changed to 1's for true and 0's for false. In Boolean math, a true thing plus another true thing equals true, while a true thing plus a false thing equals false, so 1+1=1 and 1+0=0.

Logicians looked at this, stroked their beards, nodded sagely, and ...well, that was about it in terms of impact on society.

Fast forward almost 100 years and we find one Claude Shannon, soon to be known as the Father of Information Theory, doing his graduate work at MIT. Shannon, who knew Boole's work, was trying to design logic circuits and realized that relays could represent those 1's and 0's – and digital logic was born. (And if that wasn't enough, Shannon also built what may have been the original Useless Machine; that desk toy with a switch on a little box. If you switch it on, a mechanical arm comes out of the box and switches the switch off. Shannon was a bit of a character.)

In this section, you'll learn about several types of logic elements, including gates and flip-flops.

We're not going to concern ourselves with how these devices function internally. We'll leave that to the people with the Electrical Engineering degrees. We'll just consider them magic boxes.

What's important to know for the exam is how each magic box responds to various inputs.

Multivibrators

The first class of digital circuits we'll look at is multivibrators. They don't really vibrate in the way, say, a Harley-Davidson hardtail motorcycle vibrates, but they "vibrate" from one state to another, and they can do it more than once, so they're multivibrators.

Multivibrators are classified by whether they stay in a particular state once they are put there by a signal. Bi-stable multivibrators are stable in both the 0 and 1 position – tell them to switch to state 1, they switch there and stay there. Monostable multivibrators are stable in only one of the states. Switch them to state 1, they stay there for some set amount of time, then switch back to 0 – they're only stable in one state, unstable in the other. Finally, there are astable multivibrators. They aren't stable at all – they continuously switch from one state to another with no input whatsoever.

If all this sounds like terribly foreign words only an engineer could love, consider it's a lot shorter to say "monostable multivibrator" than it is to say, "a circuit that, when triggered, changes state for a certain duration then reverts back to the original state."

E7A01 (C) Which circuit is bi-stable?
    A. An "AND" gate
    B. An "OR" gate
    C. **A flip-flop**
    D. A bipolar amplifier

Flip-flops have all sorts of uses, but the one we're most likely to encounter is in a frequency divider.

Here's how a flip-flop works. Think of it as a toggle switch. Each time it gets a particular signal at the input, it toggles from one side to the other. We'll say one side is named Zero and one side is named One. We'll also say that this particular flip-flop toggles whenever it sees a 1 at the input.

So now we have a machine that functions as shown in Table 36.1.

Frequency Table	
Flip-Flop	
Input	**Output**
1,0,1,0	1, ,0, ,

Table 36.1: Flip-Flop

At the input we see a 1, then a 0, another 1, and another zero. But the flip-flop only responds to 1's. (It's "bi-stable" – both the 0 and the 1 state are stable until it sees the correct triggering signal.) So its switch is set to 1 ....then there's a pause ....then 0 ....then another pause ....

Notice, the flip flop is changing state at one-half the original frequency. Obviously, this isn't true if the input looks like 1,0,0,0,0,0,1,0,0,1,0,1, etc. But what if our input is a sine wave? We could tell the flip-flop to only change state at the peak positive voltage of the sine wave.

Figure 36.2 shows a sine wave and the output from a flip-flop that would result from the sine wave being its input.

We have a flip-flop that triggers whenever the voltage goes above 5V positive. On the first positive peak, it sets itself to the One side. On the second positive peak, it flip-flops over to the Zero side. Nothing happens on the negative peaks.

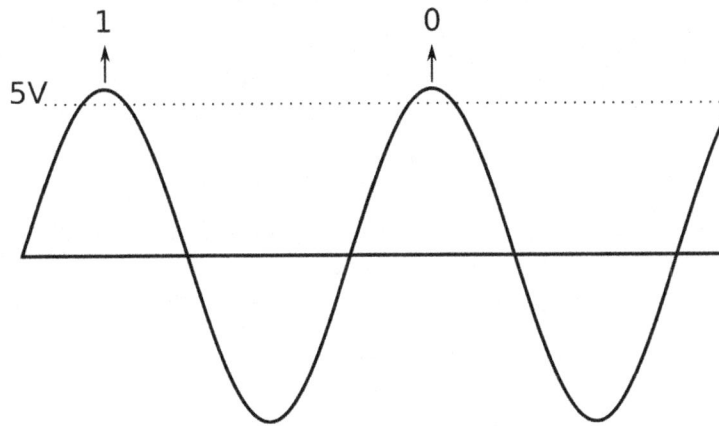

Figure 36.2: Sine-Wave Showing Flip-Flop Output

Frequency Table	
Decade Counter	
Input	**Output**
	Continued on next page.
Frequency Table	
1,1,1,1,1,1,1,1,1,1	...1

Table 36.2: Decade Counter

How is this useful? You've probably noticed, the higher the frequency a piece of equipment has to handle, the more complicated and expensive things get in terms of design and manufacture. If we can make a part of a circuit operate properly at a fraction of the frequencies it needs to measure, that's a winner! Imagine, for instance, a frequency counter for high frequencies. Do we really need to count each and every cycle, or could we divide it down by 2 and still have a reasonably accurate instrument? Yes, we could. (In practice, we could divide it by much more than 2 and *still* have an accurate frequency counter.)

In this case, we just need a flip-flop that can handle the input frequency. Then we can sample the output of the flip-flop at half the input frequency. Remember, it's bi-stable – so it holds onto that "1" or "0" until the next 1 comes along at the input. Then we count the 1's (and keep track of the time) and multiply by two and we've measured the frequency on the cheap. That's a frequency divider, and they're useful in all sorts of devices. We could even stack up two of them and divide the input frequency by 4.

E7A03 (B) Which of the following can divide the frequency of a pulse train by 2?

   A. An XOR gate
   B. **A flip-flop**
   C. An OR gate
   D. A multiplexer

As noted in the previous question, **a flip-flop** can divide a frequency by 2.

E7A04 (B) How many flip-flops are required to divide a signal frequency by 4?

   A. 1
   B. **2**
   C. 4
   D. 8

It takes **2** flip-flops to divide a signal frequency by 4. The first flip-flop divides the frequency by 2. The second divides the output of the first by 2 so **2** flip-flops to divide by 4.

E7A02 (A) What is the function of a decade counter?

A. **It produces one output pulse for every 10 input pulses**

B. It decodes a decimal number for display on a seven-segment LED display

C. It produces 10 output pulses for every input pulse

D. It decodes a binary number for display on a seven-segment LED display

A decade counter counts – don't get ahead of me! – decades! It counts groups of 10, then signals, "Yep, that's one group of ten!" **It produces one output pulse for every ten input pulses**.

Among many other uses, decade counters can be used to divide an input frequency by 10.

E7A05 (D) Which of the following is a circuit that continuously alternates between two states without an external clock?

A. Monostable multivibrator

B. J-K flip-flop

C. T flip-flop

D. **Astable multivibrator**

We said that flip-flops are bi-stable – the switch stays wherever it is until the flip-flop receives the next trigger pulse.

This question is about a component that continuously alternates between two states without an "external clock." (The "clock" is what we called a "trigger" when we covered flip-flops. The clock generates a series of pulses that tell all the other components to go to the next step.)

This is a component that isn't stable at all – it's constantly changing state all on its own. And that's why it's called an **astable multivibrator**. It doesn't get a clock pulse because, typically, it *is* the clock.

Frequency Table	
Astable Multivibrator	
Input	**Output**
(No input)	1,0,1,0,1,0,1,0,1,0...

Table 36.3: Astable Multivibrator

You can see the astable multivibrator has no input – just output, an endless string of 1, 0, 1, 0, 1, 0 ....

E7A06 (A) What is a characteristic of a monostable multivibrator?

A. **It switches momentarily to the opposite binary state and then returns to its original state after a set time**

B. It produces a continuous square wave oscillating between 1 and 0

C. It stores one bit of data in either a 0 or 1 state

D. It maintains a constant output voltage, regardless of variations in the input voltage

We've had bi-stable and astable circuit components. What's a monostable component?

Frequency Table	
Monostable Multivibrator	
Input	**Output**
1	1 (Pause) 0

Table 36.4: Monostable Multivibrator

It's one that **switches momentarily to the opposite binary state and then returns to its original state after a set time**. (See Table 36.4.)

When would we want something that does this? Any time we want to generate a longer timing pulse every time the monostable multivibrator sees some relatively brief signal of some sort.

All these devices that contain "stable" in the name can be made from the good old 555 timer, an IC that is used in countless applications. It's all in how the 555 is hooked up.

Logic Gates

The next three questions ask about various sorts of logic gates.

Digital computers are composed of logic gates. Your desktop computer has millions upon millions of logic gates in the CPU and other components. They're the hardware equivalent of software's "if/then statements." "If such and such is true, then do this."

Because all digital logic operates on 1's and 0's, logic gates operate on two voltages, known as "logic 0" and "logic 1." The gate looks at the inputs, sees if they are at logic 0 or logic 1, then creates an output that is either logic 0 or logic 1. The name of the gate tells us what the states at the inputs have to be to generate a logic 1 output. If the logic gate is a two-input AND gate, for instance, if input A is at logic 1 AND input B is at logic 1, then the output X is logic 1. Otherwise, it is logic 0. (By convention, gate inputs and outputs are usually labeled with letters rather than numbers, using letters from the first half of the alphabet for inputs and from the last half for outputs.)

Logic gates come in seven delicious digital flavors. We'll cover three of them for the extra exam.

E7A10 (C) What is a truth table?

    A. A table of logic symbols that indicate the high logic states of an op-amp

    B. A diagram showing logic states when the digital device output is true

    C. **A list of inputs and corresponding outputs for a digital device**

    D. A table of logic symbols that indicate the logic states of an op-amp

"Truth table" is a term from logic, and particularly Boolean logic. Boolean logic is that logic that drives all our digital devices.

In Boolean logic, 1 is "true" and 0 is "false." So we can think of logic gates as truth detectors. The AND gate asks, "Is it true that input A is at logic 1 AND input B is at logic 1?" If it is, it signals, "Yep, true!" with an output of 1.

A truth table describes **a list of inputs and corresponding outputs for a digital device**.

Each type of gate has its own truth table, and you can also construct truth tables for combinations of devices, so the rows and columns multiply quite rapidly as one adds components.

E7A07 (D) What logical operation does a NAND gate perform?

    A. It produces logic 0 at its output only when all inputs are logic 0

    B. It produces logic 1 at its output only when all inputs are logic 1

    C. It produces logic 0 at its output if some but not all inputs are logic 1

    D. **It produces logic 0 at its output only when all inputs are logic 1**

NAND means "not AND."

"Not AND" has a special meaning in digital logic. In normal, non-binary logic, "not" could mean almost anything, right? If we say the number of apples Cletus has is "not seven", Cletus might have five, six, eight, or even zero apples. Cletus might have all the apples in the state of Washington! But since there are only 1's and 0's in binary logic, "not" always means "the opposite of." "Not 1" equals 0, and "not 0" equals 1.

So long as one or both inputs are at logic 0, the output of a NAND gate is 1. If both are 0, it outputs 1. If A is 1 and B is 0, it still outputs 1.

NAND Gate		
INPUT A	**INPUT B**	**OUTPUT**
0	0	1
0	1	1
1	0	1
1	1	0

Table 36.5: NAND Gate Truth Table

Of course, if A is 0 and B is 1, it STILL outputs 1, but if both inputs are at logic 1, the NAND gate outputs logic 0.

You can see the truth table for a NAND gate in Table 36.5.

A NAND gate **produces logic "0" at its output only when all inputs are logic "1."** Put another way around, if any input is at logic "0", the output is at logic "1."

E7A08 (A) What logical operation does an OR gate perform?

   A. **It produces logic 1 at its output if any or all inputs are logic 1**
   B. It produces logic 0 at its output if all inputs are logic 1
   C. It only produces logic 0 at its output when all inputs are logic 1
   D. It produces logic 1 at its output if all inputs are logic 0

An OR gate **produces logic 1 at its output if any or all inputs are logic 1.**

If both inputs are 0, the OR gate outputs 0 but if either A *or* B *or* A *and* B are at logic 1, then the OR gate outputs a 1.

OR Gate		
INPUT A	**INPUT B**	**OUTPUT**
0	0	0
0	1	1
1	0	1
1	1	1

Table 36.6: OR Gate Truth Table

E7A09 (C) What logical operation is performed by an exclusive NOR gate?

   A. It produces logic 0 at its output only if all inputs are logic 0
   B. It produces logic 1 at its output only if all inputs are logic 1
   C. **It produces logic 0 at its output if only one input is logic 1**
   D. It produces logic 1 at its output if only one input is logic 1

An exclusive NOR gate, also known as an XNOR gate, outputs a 1 if both inputs are 0 or both inputs are 1.

To understand the XNOR gate, it might be useful to first understand its cousin, the NOR gate. A NOR gate, as opposed to an XNOR gate, is the logical mirror of an OR gate. It produces logic "0" at its output if any input is logic "1."

So for a NOR gate, life is like this:

If there's one input at logic 1 – that's a big nope for the NOR gate! It outputs 0.

If a NOR gate sees logic 1 at both inputs, it also outputs 0. You could almost call a NOR gate a "NEITHER" gate!

NOR Gate		
INPUT A	**INPUT B**	**OUTPUT**
0	0	1
0	1	0
1	0	0
1	1	0

Table 36.7: NOR Gate Truth Table

The only time a NOR gate will output 1 is if both inputs are at 0.

The XNOR gate still outputs logic 1 if both inputs are logic 0, but it also outputs logic 1 if *both* inputs are logic 1.

XNOR Gate		
INPUT A	**INPUT B**	**OUTPUT**
0	0	1
0	1	0
1	0	0
1	1	1

Table 36.8: XNOR Gate Truth Table

Put more simply, the NOR gate outputs logic 0 if *any* input is at logic 1; the XNOR gate **produces logic 0 at its output if *only one* input is logic 1** The way I think of it is as a BOTH gate; it produces logic 1 if both inputs are 0 or if both inputs are 1.

E7A11 (D) What type of logic defines "1" as a high voltage?

A. Reverse Logic

B. Assertive Logic

C. Negative logic

D. **Positive Logic**

We can design a digital circuit to use a relatively high voltage as either 0 or 1, with the relatively low voltage representing the opposite.

If we choose to use high voltage to represent 1, that's called **positive logic**. You can probably guess what it's called if we use high voltage to represent 0, right?

Digital Component Schematic Symbols

The schematic symbols for logic gates show the inputs and the output and indicate the type of gate by the shape between the inputs and output.

You'll need to know three symbols for logic gates that may appear on the exam. You'll need to identify them from "Figure E6-3", shown as Figure 36.6. The symbols to remember are numbers 2, 4, and 5.

Figure 36.3 shows an AND gate and a NAND or "not and".

Each symbol shows two inputs coming in from the left (in this case) and an output going out the right. Notice the vertical line where the inputs come in – it's straight. Then notice the output of the AND gate comes straight out the right, while the NOT AND gate's output passes through a small circle.

AND   NOT AND

Figure 36.3: AND Gate (L) and NAND Gate (R)

The little circle on the output of the NAND gate is what makes the AND gate symbol into a "not and" or NAND symbol, and that holds true through the other symbols as well.

Figure 36.4 shows an OR gate and a NOR gate and you can see the pattern remains consistent.

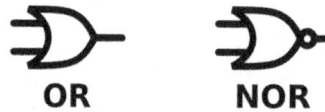

OR   NOR

Figure 36.4: OR Gate (L) and NOR Gate (R)

The OR and NOR gates are distinguished by the curved line on the input side. I think of it as part of an "O", for OR.

Once again, adding a little circle to the output side adds an N to the front and makes the symbol a NOR gate:

The only symbol on the exam that doesn't precisely follow this pattern is the one in Figure 36.5.

Figure 36.5: NOT Operation

That one is a "NOT operation", not really a gate *per se*. It still gets an "N" at the front thanks to that little circle on the right, though.

Figure E6-3

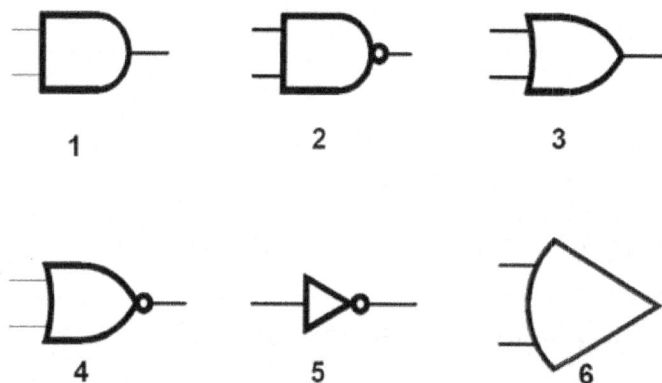

Figure 36.6: Figure E6-3 in the Question Pool

E6C08 (B) In Figure E6-3, what is the schematic symbol for a NAND gate?

    A. 1

    B. **2**

    C. 3

    D. 4

The symbol for a NAND gate is at position number **2** on figure E6-3.

230

E6C10 (D) In Figure E6-3, what is the schematic symbol for a NOR gate?

A. 1
B. 2
C. 3
D. **4**

Here we're looking for a NOR gate, so we know it will have that little circle on the right, for the N, and that the line across the inputs will be curved like the O in OR. It's symbol number **4** on figure E6-3.

E6C11 (C) In Figure E6-3, what is the schematic symbol for the NOT operation (inverter)?

A. 2
B. 4
C. **5**
D. 6

Here's that oddball symbol for a NOT operation, also known as an inverter. A NOT operation is the teenager of logic – whatever comes in, it says the opposite! If the input is 0, it outputs a 1. If the input is 1, it outputs a 0.

Since it starts with an N, we know we're looking for that little circle on the output side. And since it doesn't operate like the AND, NAND, OR, or NOR gates, with two inputs, we can look for something that just has one input and an output. It's at position number **5** on figure E6-3.

You might be wondering about symbol number 6 on E6-3. Unless it is so obscure that I just couldn't find an example anywhere, it's bogus. I think it's an "alternative operational amplifier" symbol with the output snipped off.

Key Concepts in This Chapter

- A flip-flop is a bi-stable circuit and can be used to divide a frequency in two.

- To divide a frequency by four, 2 flip-flops are required.

- The function of a decade counter digital IC is to produce one output pulse for every ten input pulses.

- A circuit that continuously alternates between two states without an external clock is an astable multivibrator.

- A monostable multivibrator switches momentarily to the opposite binary state and then returns to its original state after a set time.

- A truth table is a list of inputs and corresponding outputs for a digital device.

- A NAND gate produces logic "0" at its output only when all inputs are logic "1."

- An OR gate produces logic "1" at its output if any or all inputs are logic "1."

- An exclusive NOR gate produces logic "0" at its output if only 1 input is logic "1."

231

- The type of logic that defines "1" as a high voltage is positive logic.

- In Figure E6-5 the schematic symbol for a NAND gate is symbol number 2.

- The schematic symbol for a NOR gate is figure 4.

- The schematic symbol for the NOT operation or inverter is symbol number 5.

Chapter 37

Amplifier Classes

Let's go back and review what we covered on this subject at the General Class level. Amplifiers are characterized by their "class." Not a measure of quality, nor of the amplifier's social standing, class simply refers to the type of amplification circuit the amplifier uses. That circuitry determines the efficiency of the amplifier as well as its linearity. Linearity is a measure of how closely the output signal matches the input signal. Linearity may or may not be a critical measurement, depending on what mode we'll be transmitting.

Class A amplifiers use a single transistor (or tube) in the final stage to amplify the entire waveform – the positive side and the negative side. They are quite linear; what goes in is what comes out, only louder. However, that linearity comes at a cost. That single transistor is conducting current constantly, and that, in turn, makes the Class A amplifier woefully inefficient. 30% efficiency is about par for the course for a Class A amplifier.

Consider the implications for an imaginary Class A 1500-watt HF power amplifier. To get 1500 watts out, you would need to put in 5000 watts. Even if we imagine you have a 100% efficient power supply converting your wall power to DC for the transmitter, that's still 45 amps of 110-volt power! That's way, way beyond the capacity of normal house wiring. It's not going to work. (You can see why those high wattage amplifiers usually come out of the box set up to run on 220 volts. That cuts the amperage needed in half and gets it back down into the range of normal house wiring. But, 22.5 amps is still a lot!)

The exam doesn't ask you about Class B amplifiers, but let's cover them briefly just for perspective. Class B amplifiers use two transistors, one amplifying the positive side of the signal, one amplifying the negative side. Each half operates through 180° of the signal cycle. Each transistor gets a little rest while the other works, and because of this we gain a hefty bit of efficiency, up to about 50%. What we lose is linearity – there's some distortion created just as the signal crosses the zero line when one transistor switches off before the other switches on. In fact, it's called zero crossing distortion.

The Class AB amplifier uses two transistors as well, but set up so that their operation overlaps just a bit past the zero crossing, eliminating that zero crossing distortion. Put more technically, each side of an AB amplifier operates for more than 180° but less than 360° of the signal cycle. We lose a few percentage points of efficiency, gain a lot in linearity. Most high-end audio amplifiers these days are AB class.

The Class C amplifier takes us back to one transistor in the final stage but now that transistor is amplifying only about one-half of the positive cycle of the input waveform. What happens to the rest of it? It gets generated by a resonant circuit across the output. In chapter 39, we'll talk about "ringing" in a filter as something undesirable; in a Class C amplifier, we take advantage of ringing. The transistor puts out a short burst of signal, then the resonant circuit sort of fills in the rest.

Efficiency? You bet! In theory, up to 90%, but in practice more like 80%. That electricity hog of a final transistor is only sucking power a quarter of the time. As you might guess, though, we just about throw linearity out the window for any but the simplest signals. Terrible for SSB, many digital signals, or just about anything except CW or FM. Wonderful if we just want to crank up the power on a very simple RF carrier.

Class D amplifiers can be amazingly efficient, as high as 95%. They are also known as "switching amplifiers" and operate in a way quite similar to switching power supplies.In a Class D amplifier, the input signal goes into what's called a comparator. As you might guess from the name, the comparator compares the input signal to another input, which is a triangle wave. It calculates the difference and that difference generates an output that is a series of pulses of varying width. It's very similar to the scheme that creates pulse width modulation. In the case of the Class D amplifier, that switching signal then drives the final amplification section. The final transistor isn't used as a traditional amplifier, it's used as a switch. It recreates the original signal as a series of very high frequency pulses. While it is not a digital process, the process is analogous to digital sampling. What this means is the transistor is always either fully on or fully off.

What comes out of the final is a perfectly awful hash of the original signal and the switching signal, but the pulses are at a much higher frequency than the original signal. We run the output through a low pass filter, and that shaves off the pulses leaving a reasonable facsimile of the original signal.

On paper, Class D amplifiers could be 100% efficient with 0% distortion. Of course, in practice it doesn't work out that way, but efficiency can be pushed into the 90% area while keeping distortion acceptably low – indeed, a well-designed Class D can equal the performance of a Class AB amplifier.

Amplifier Classes			
Type	Efficiency	Linearity	Notes
A	Low	High	Single tube or transistor
B	Moderate	High	Zero crossing distortion
AB	Moderate	High	No zero crossing distortion
C	High	Low	Only usable on CW and angle modulation modes
D	Very High	High	Can equal performance of AB class

Table 37.1: Amplifier Classes & Characteristics

E7B01 (A) For what portion of the signal cycle does each active element in a push-pull Class AB amplifier conduct?

A. More than 180 degrees but less than 360 degrees

B. Exactly 180 degrees

C. The entire cycle

D. Less than 180 degrees

The Class AB amplifier is the one that uses just a bit of overlap between the amplification of the positive part of the sine wave and the negative part. Put in the language of the exam, each active element operates in **more than 180 degrees but less than 360 degrees** of the signal cycle.

E7B02 (A) What is a Class D amplifier?

A. A type of amplifier that uses switching technology to achieve high efficiency

B. A low power amplifier that uses a differential amplifier for improved linearity

C. An amplifier that uses drift-mode FETs for high efficiency

D. A frequency doubling amplifier

Class D amplifiers **use switching technology to achieve high efficiency.**

E7B03 (A) Which of the following components form the output of a class D amplifier circuit?

    A. **A low-pass filter to remove switching signal components**

    B. A high-pass filter to compensate for low gain at low frequencies

    C. A matched load resistor to prevent damage by switching transients

    D. A temperature compensating load resistor to improve linearity

Class D amplifiers use **a low-pass filter** on the output **to remove switching signal components.**

E7B14 (B) Why are switching amplifiers more efficient than linear amplifiers?

    A. Switching amplifiers operate at higher voltages

    B. **The power transistor is at saturation or cutoff most of the time**

    C. Linear amplifiers have high gain resulting in higher harmonic content

    D. Switching amplifiers use push-pull circuits

When a transistor is at saturation, it presents almost no resistance to current flow. Very little power is dissipated in that state, just as very little power is dissipated in a straight wire. (There's no E to multiply with I to make W.)

When a transistor is at cutoff, it's almost like an open switch, so it dissipates very little power in that state as well. (There's no I to multiply with E.) It's only between cutoff and saturation that the transistor presents varying values of resistance, and therefore dissipates power.

The power transistor – that's the final amplifier transistor – in a switching amplifier doesn't operate as a linear amplifier, but as a switch. It's **at saturation or cutoff most of the time**, resulting in low power dissipation.

E7B06 (B) Which of the following amplifier types reduces even-order harmonics?

    A. Push-push

    B. **Push-pull**

    C. Class C

    D. Class AB

A **push-pull** is a *type* of amplifier – not a *class* of amplifier – that reduces even order harmonics.

A push-pull amplifier uses two transistors or vacuum tubes arranged so that one "pushes" the signal through the load while the other "pulls" it through. This is accomplished by splitting the signal in two and putting one half 180° out of phase.

Push-pull amplifiers can be created as Class A's, Class B's, Class AB's, or Class C's.

The wrong answers to this question include two *classes* of amplifier, class AB and class C. That's not what they're looking for. And there are push-push amplifiers, but they actually *emphasize* even order harmonics.

E7B07 (D) Which of the following is a likely result when a Class C amplifier is used to amplify a single-sideband phone signal?

    A. Reduced intermodulation products

    B. Increased overall intelligibility

    C. Signal inversion

    D. **Signal distortion and excessive bandwidth**

If you take away one practical lesson from this chapter, let it be this one. Class C amplifiers are efficient, but they produce lots of distortion. If you use one to amplify your single-sideband phone signal your result will be **signal distortion and excessive bandwidth**. (The bandwidth spreads because of all the harmonic distortion, which creates "extra" frequencies. Not like Amateur Extra frequencies – those are good! Harmonic distortion creates additional unwanted frequencies.) Class C amplifiers are fine for CW and any of the angle modulations, such as FM, but dreadful for any amplitude modulation.

Key Concepts in This Chapter

- A class AB amplifier operates through more than 180 degrees but less than 360 degrees of the signal cycle.

- A class D amplifier is a type of amplifier that uses switching technology to achieve high efficiency.

- The component that forms the output of a Class D amplifier circuit is a low-pass filter to remove switching signal components.

- Switching amplifiers are more efficient than linear amplifiers because the power transistor is at saturation or cutoff most of the time, resulting in low power dissipation.

- The amplifier type that reduces or eliminates even order harmonics is a push-pull amplifier.

Chapter 38

Amplifier Technology & Schematic Diagrams

A Few Notes on Reading Schematics

Since the exam and this course feature quite a few schematics, let's talk a little about how to read them, because there are some questions about one coming up right around the corner and there will be more.

First, as I've mentioned elsewhere, Ben Franklin came up with the notion of positive and negative charges, and he called the positive charge positive because he decided – and no one knows why – that electricity flowed from positive to negative. He couldn't have known one way or the other, so he had a 50/50 chance and lost the coin toss. That was around 1750. We didn't know about electrons and their negative charge until 1897. In the ensuing 140 some odd years, a lot of schematic diagrams had been made, and they all assumed that electricity flowed from positive to negative. It really doesn't matter a bit, in terms of the device working, after all. So rather than create mass confusion between "old schematics" and "new schematics" people just kept using the old model and it is still with us today. (It was even longer until we really understood that positive charges flowed as well ... so maybe Ben was only 50% wrong.) If you're studying this text or another text and the writer says, "Current flows from the positive leg of the power supply through resistor R1 to the negative leg", they haven't lost their mind, they're speaking fluent schematic, or more properly, "conventional flow" as opposed to "electron flow."

There's a "rule" that, when possible, schematics should be drawn, generally, to read from left to right, with input on the left and output on the right. "Rule" really isn't the right word, though; it's more like a vague suggestion that this might be a good idea if it isn't too much trouble. Still, it's worth looking for input on the left and output on the right.

If there's a power supply connection it is usually designated by a "+" or "−." If there are multiple power supply connections for different voltages, the voltages should be shown.

Component values are rarely shown in either the exam schematics or our teaching schematics in the course, unless they are relevant.

You should know that schematics use some shortcuts to give us fewer little lines to trace all over a page. One very common one is the use of a ground symbol.

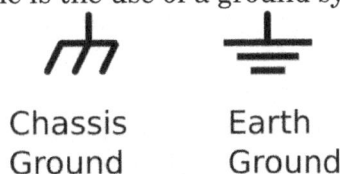

Chassis
Ground

Earth
Ground

Very officially speaking, the symbol on the left is "chassis ground" and the symbol on the right is "earth ground," meaning it indicates a wire is run to an actual ground stake driven into the dirt. In practice, you'll see them used interchangeably. We'll stick with chassis grounds in this course, since the exam does the same.

We use that symbol to simplify reading a schematic. This:

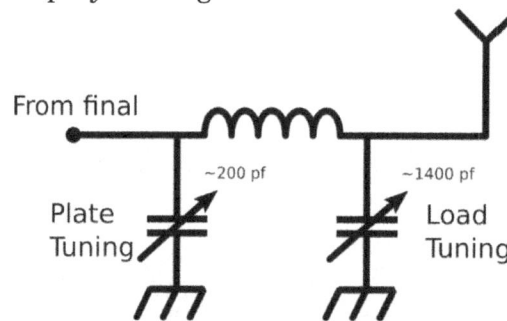

Is electrically the same as this:

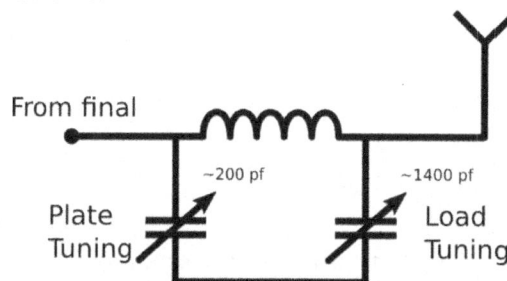

Finally, none of the schematics on the exam, nor any of the teaching schematics are complete devices. They're all just circuits within the device. For instance, in the schematic shown above, there "should" be a ground for "the other side" of the antenna – or at least an "other side." Neither is shown. Neither is the ground side of the imaginary final stage that exists somewhere off to the left in that diagram.

Amplifier Technology

E7B05 (C) What can be done to prevent unwanted oscillations in an RF power amplifier?
    A. Tune the stage for maximum SWR
    B. Tune both the input and output for maximum power
    C. **Install parasitic suppressors and/or neutralize the stage**
    D. Use a phase inverter in the output filter

We've all experienced the "scrEEE!!" of a public address system feeding back. Somebody has the volume turned up too high, then someone turns on the microphone and we get that awful howl. We've connected the output of the amplifier to its own input and round and round the sound goes getting amplified to the limits of the amp and, in the process, wiping out everyone's ears and rendering the system temporarily useless.

Consider this: For a particular system in a particular room, that feedback is always the same frequency. It isn't random, it's a product of that big resonant circuit formed by the microphone, the room, the amplifier and the speakers. That concept is important.

The same process can and will occur inside an RF power amplifier when the output gets mixed with the input – and especially at high power levels, it's very hard to keep that output out of the input. Every amplifier has a characteristic resonant frequency created by its own internal capacitances and inductances – and that's the frequency that's going to get amplified over and over making our amplifier as useless as that squealing PA system in the high school cafeteria. In the worst case, the amp will overload and pop its fuses or fry its output tubes.

Happily, when you designed your RF amplifier, you took all this into account and made that resonant frequency somewhere well above the frequency we intend to amplify, and/or included other feedback controls, or we'd be out of luck.

One method amplifier designers use to control feedback is called "**parasitic suppression**." It's just an RF filter on the output of the final amplifier stage. In older designs, the **parasitic suppressor** was often a coil and a resistor in parallel, but more modern equipment has such

238

high gain, more effective designs are demanded – the idea is still the same, though. Filter out the RF at the resonant frequency, and it can't get back into the input and get re-amplified.

Another technique is to **neutralize the stage**. This is especially used in vacuum tube amplifiers. It's simply a matter of deliberately adding some of the output back into the input, but adding it 180 degrees out-of-phase. As feedback starts, the level of the output goes up, and so does the level of the out-of-phase signal being fed to the input, so the feedback cancels itself out. Adjusted properly, the out-of-phase signal cancels out just enough of the input to prevent oscillation.

E7B08 (C) How can an RF power amplifier be neutralized?

    A. By increasing the driving power

    B. By reducing the driving power

    C. **By feeding a 180-degree out-of-phase portion of the output back to the input**

    D. By feeding an in-phase component of the output back to the input

An RF power amplifier – or any other amplifier, for that matter – can be neutralized **by feeding a 180 degree out-of-phase portion of the output back to the input**.

E7B09 (D) Which of the following describes how the loading and tuning capacitors are to be adjusted when tuning a vacuum tube RF power amplifier that employs a Pi-network output circuit?

    A. The loading capacitor is set to maximum capacitance and the tuning capacitor is adjusted for minimum allowable plate current

    B. The tuning capacitor is set to maximum capacitance and the loading capacitor is adjusted for minimum plate permissible current

    C. The loading capacitor is adjusted to minimum plate current while alternately adjusting the tuning capacitor for maximum allowable plate current

    D. **The tuning capacitor is adjusted for minimum plate current, and the loading capacitor is adjusted for maximum permissible plate current**

Vacuum tubes, by their very nature, are high impedance devices – typically in the thousands of ohms. This presents us with a technical challenge, since our antennas are, normally, low impedance devices, say, 50 ohms. That's quite the impedance mismatch, and not much power is going to get to our antenna without some sort of matching network. That's the function of a Pi-network output circuit.

The Pi-network gets its name from its schematic diagram which resembles the Greek letter π – it has nothing to do with the mathematical value of pi.

Figure 38.1 shows a Pi-network output circuit.

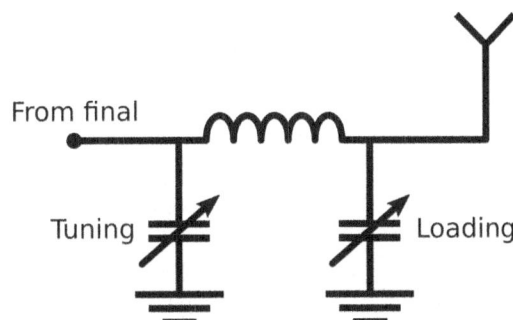

Figure 38.1: Pi-network

Sure enough, if you use your imagination, it resembles that Greek letter.

The "final" or "plate" is the anode of the final amplifier tube. In effect, the tuning capacitor on the left matches that part of the network to the impedance of that final stage of the amp. It's in parallel with the inductance – the coil at the top – making a parallel resonant circuit. When we tune that part of the pi-network to resonance with the Tuning capacitor, the output current of the plate will be at minimum.

Then we tune that loading capacitor to match the impedance of the antenna and the plate current starts heading back up. When it peaks at the maximum permissible amount, our network is all matched up and we have "tuned up the transmitter." (In practice, it's usually a back and forth process – a little more load, a little less plate, oops, little less load, little more plate ...but you get the idea.)

To tune most vacuum tube RF amplifiers, **the tuning capacitor is adjusted for minimum plate current, and the loading capacitor is adjusted for maximum permissible plate current.**

E7B16 (A) What is the effect of intermodulation products in a linear power amplifier?

    A. **Transmission of spurious signals**

    B. Creation of parasitic oscillations

    C. Low efficiency

    D. All these choices are correct

We could define intermodulation distortion as any unwanted amplitude modulation of the desired signal. It is created by non-linearity. The most common cause is overmodulation of some component or stage of a device. That overmodulation turns the tops of our nice, round sine waves into something resembling nasty, harmonic laden square waves – and square waves with a lot of energy to boot. Now we have all sorts of frequencies interacting with each other creating even more chaos. If those harmonics all show up at the output, we'll be transmitting lots of frequencies we had no intention of transmitting, and that's the textbook definition of **transmission of spurious signals**.

E7B17 (A) Why are odd-order rather than even-order intermodulation distortion products of concern in linear power amplifiers?

    A. **Because they are relatively close in frequency to the desired signal**

    B. Because they are relatively far in frequency from the desired signal

    C. Because they invert the sidebands causing distortion

    D. Because they maintain the sidebands, thus causing multiple duplicate signals

We covered this same ground back in Chapter 20 except it was about intermodulation products in receivers. Now they're asking about RF amplifiers, but the logic is the same.

Odd-order harmonics are the original frequency multiplied by an odd integer.

When a frequency near some desirable frequency combines with the second harmonic of that desirable frequency, the sum comes out to something in the third-order harmonic range, so we call it an "odd-order intermodulation distortion product." Put mathematically, $2f_1 + f_2 \approx 3f_1$. The difference, though, is usually the real concern; $2f_1 - f_2 \approx f_1$. We just shorthand all that by saying we're concerned about, "third order distortion products" or "odd order distortion products." So our concern about odd-order intermodulation distortion products is that they are relatively close in frequency to the desired signal.

E7B18 (C) What is a characteristic of a grounded-grid amplifier?

    A. High power gain

    B. High filament voltage

    C. **Low input impedance**

    D. Low bandwidth

We know this is a vacuum tube amp because only vacuum tubes have grids.

A grounded-grid tube amplifier is "self-neutralizing." One source of unwanted feedback in a triode tube amplifier is the capacitance formed between the grid and the plate. In a grounded grid amplifier, the grid, which in normal operation would be controlling the signal passing through the tube, is connected to ground. This takes its contribution to the tube's self-capacitance out of the system. The signal is applied to the cathode, which is the low input impedance the correct answer is talking about.

E7B04 (A) Where on the load line of a Class A common emitter amplifier would bias normally be set?

A. **Approximately halfway between saturation and cutoff**
B. Where the load line intersects the voltage axis
C. At a point where the bias resistor equals the load resistor
D. At a point where the load line intersects the zero bias current curve

This is the one and only time the term "load line" appears anywhere in the Extra exam questions. Don't worry about that term. We'll boil it down to simple language. Study further at your leisure if you want to become a designer of circuits.

While, in execution, this topic can get quite technical, it boils down to a simple concept. Every transistor has a cutoff point and a saturation point. If bias is below the cutoff voltage, the transistor doesn't conduct. If bias is above the saturation point, the transistor is completely on and won't amplify any more.

The question tells us this is a Class A amplifier, so we want this transistor to amplify the entire cycle. To make that happen, we want our bias signal to stay below saturation and above cutoff, so we set our bias **approximately halfway between saturation and cutoff**. See Figure 38.2.

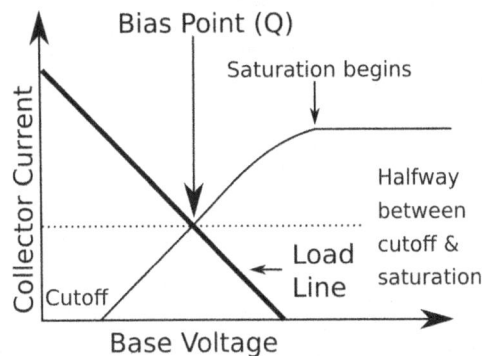

Figure 38.2: Saturation, Cutoff, and the Load Line

Amplifier Schematic Diagrams

The next three questions deal with Figure E7-1, which is a schematic of part of an amplifier.

On the Technician and General exams, they just asked you to identify components on schematic diagrams. At the Extra class level, they ask about the purposes of some of those things in the circuit.

You might want to do an instant review of the anatomy of a bipolar junction transistor schematic symbol, to better understand some of this section. Figure 38.3 shows an NPN and a PNP bipolar junction transistor.

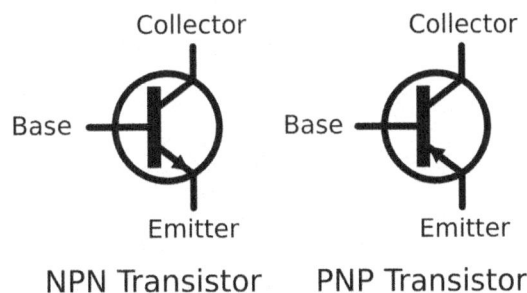

NPN Transistor   PNP Transistor

Figure 38.3: NPN and PNP Transistor Schematic Symbols

E7B10 (B) In Figure E7-1, what is the purpose of R1 and R2?

    A. Load resistors

    B. **Voltage divider bias**

    C. Self bias

    D. Feedback

Figure E7-1

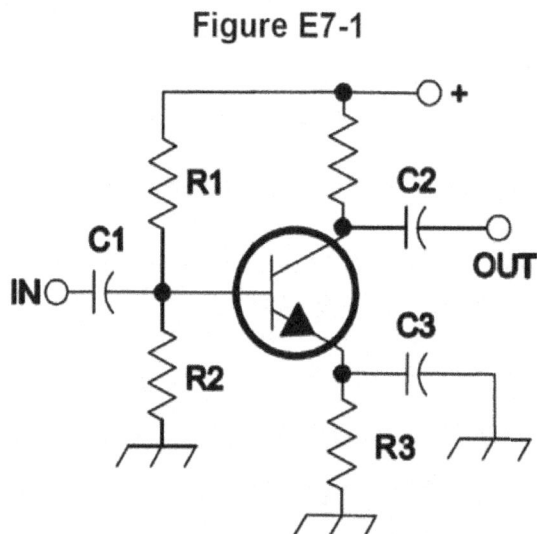

Figure 38.4: Figure E7-1 in the Question Pool

We can see R1 and R2 are connected to the positive side of the power supply and form a circuit that eventually goes to ground – in other words, they connect to the negative terminal of the power supply. We can also see they're connected to the base of that transistor, so we can guess they have something to do with bias – I don't see any other source of bias voltage!

R1 and R2 form what's called a *voltage divider*. Let's plug in some values and you'll see how it works.

Let's say the power supply is supplying +12v, and we want +0.3v of bias voltage at the base.

0.3v is $\frac{1}{40}$ of 12v. What we want to do is divide that 12v of voltage between R1 and R2 so we get 0.3v at that dot between them.

I pick a total amount of resistance for the two; I'll go with 1,000,000 ohms. Why? I don't need a bunch of current, just voltage. Now I divide the resistance like this:

$$R1 = 1,000,000 \times \frac{39}{40} = 975,000\Omega$$

$$R2 = 1,000,000 \times \frac{1}{40} = 25,000\Omega$$

Ohm's Law tells me that the current through R1 and R2 will be

$$I = \frac{E}{R} = \frac{12}{1,000,000} = 12 \times 10^{-6} = 12\ microamps$$

Now let's see what the voltage drop across R1 is.

$$E = I \times R = (12 \times 10^{-6}\ amps) \times 975,000\Omega = 11.7v$$

After R1, we have just what we want, 0.3v. So why have R2? If R2 wasn't there, we wouldn't have any voltage at all – it would have all been dropped by R1.

That's how we get **voltage divider bias.**

E7B11 (D) In Figure E7-1, what is the purpose of R3?

    A. Fixed bias

    B. Emitter bypass

    C. Output load resistor

    D. **Self bias**

Transistors in general have a bit of a self-destructive tendency. As the transistor passes current, its temperature rises. As the temperature rises, more current flows through the collector, which raises the temperature, which lets more current flow and – well, you can see it won't be long before our transistor has turned into a sad little puddle of melted silicon. This phenomenon is called thermal runaway. That's the problem R3 is set up to address. It is connected to the emitter of that NPN transistor. (We know it is the emitter because the emitter is always the part of the bipolar junction transistor schematic symbol with the arrow, and we know it is an NPN because the arrow is Not Pointing iN.)

The emitter of an operating NPN transistor is always negatively charged relative to the base. So as current flows through the transistor, negative voltage is applied through R3, then through R2, where it makes the positive base bias voltage just a little lower, reducing the current through the transistor. By carefully selecting the values of those resistors, we can make the transistor "keep control of itself" by using **self bias** as a feedback mechanism – sort of a built-in thermostat. If you're thinking this concept sounds familiar, it's the same general principle as *neutralization*, discussed on page 239.

E7B15 (C) What is one way to prevent thermal runaway in a bipolar transistor amplifier?

    A. Neutralization

    B. Select transistors with high beta

    C. **Use a resistor in series with the emitter**

    D. All these choices are correct

R3 in Figure E7-1 is that **resistor in series with the emitter** providing self-bias to prevent thermal runaway.

E7B12 (C) What type of amplifier circuit is shown in Figure E7-1?

    A. Common base

    B. Common collector

    C. **Common emitter**

    D. Emitter follower

There are three distinct ways to hook up a bipolar junction transistor to create a single-stage amplifier. These ways – engineers call them "topologies" – are common base, common collector, and **common emitter**.

By the time all the biasing, coupling, and filtering elements get added into a schematic, it isn't always easy to tell at a glance which topology is used in a particular design, but it's a fairly safe bet that the big clue is which lead of the transistor *isn't* connected to the input or output. That leaves that lead connected to ground and, in fact, another name for a **common emitter** amplifier is "grounded emitter."

In Figure E7-1, we see the input connected to the base and the output connected to the collector. That means the amplifier circuit in Figure E7-1 is a **common emitter**.

E7B13 (D) Which of the following describes an emitter follower (or common collector) amplifier?

    A. A two-transistor amplifier with the emitters sharing a common bias resistor

    B. A differential amplifier with both inputs fed to the emitter of the input transistor

    C. An OR circuit with only one emitter used for output

    D. **An amplifier with a low impedance output that follows the base input voltage**

Each type of transistor amplifier has its own set of inherent characteristics and is useful for different applications.

Transistor Amplifier Types						
Type	Input	Output	Input Impedance	Output Impedance	Voltage Gain	Current Gain
Common Emitter	Base	Collector	Moderate to High	Moderate	High	High
Common Base	Emitter	Collector	Low	Moderate to High	High	≈ 0
Common Collector	Base	Emitter	Moderate to High	Low	≈ 0	High

The *common collector* amplifier has the output connected to the emitter. The emitter's voltage "**follows**" the base voltage; in other words, the base and emitter voltages will always be very close in value. That's why it is also known as an *emitter follower* amplifier – a dismally misleading name because it sounds like the amplifier is somehow following the emitter, and that's not at all what's going on. The common collector amplifier won't provide any voltage gain at all, but will provide lots of current gain. Since the output's voltage is relatively low and its current high, it is a **low impedance output.** That's relevant to us because it makes it easier to match the amplifier's output to the relatively low impedances of most amateur radio gear, particularly our feed lines and antennas.

Key Concepts in This Chapter

- To prevent unwanted oscillations in an RF power amplifier, install parasitic suppressors and/or neutralize the stage.

- An RF power amplifier can be neutralized by feeding a 180 degree out-of-phase portion of the output back to the input.

- When tuning a vacuum tube RF power amplifier that employs a Pi-network output circuit, the tuning capacitor is adjusted for minimum plate current, and the loading capacitor is adjusted for maximum permissible plate current.

- The effect intermodulation products produce in a linear power amplifier is transmission of spurious signals.

- Odd-order rather than even-order intermodulation distortion products are of concern in linear power amplifiers because they are relatively close in frequency to the desired signal.

- One characteristic of a grounded-grid amplifier is low input impedance.

- The bias of a Class A common emitter amplifier would normally be set approximately half-way between saturation and cutoff on the load line.

- In Figure E7-1, the purpose of R1 and R2 is voltage divider bias.

- In Figure E7-1, the purpose of R3 is self bias.

- One way to prevent thermal runaway in a bipolar transistor amplifier is to use a resistor in series with the emitter.

- The type of amplifier circuit shown in Figure E7-1 is a common emitter.

- An emitter follower amplifier, also known as a common collector amplifier, has a low impedance output that follows the base input voltage.

Chapter 39

Filters and Matching Networks

This chapter covers a variety of filters and matching networks. (You'll probably notice that many matching networks *are* filters; they're resonant circuits.)

I'll give you a schematic for each of them, but those are just learning aids, not for your deep analysis, nor for memorization. None of the schematics will be on the exam.

The actual exam questions focus mostly on either the general design of the filter or on its output characteristics, and those are important to know.

Matching Networks

E7C04 (C) How does an impedance-matching circuit transform a complex impedance to a resistive impedance?

    A. It introduces negative resistance to cancel the resistive part of impedance

    B. It introduces transconductance to cancel the reactive part of impedance

    C. **It cancels the reactive part of the impedance and changes the resistive part to a desired value**

    D. Reactive currents are dissipated in matched resistances

Recall those phasor diagrams, those impedance triangles.

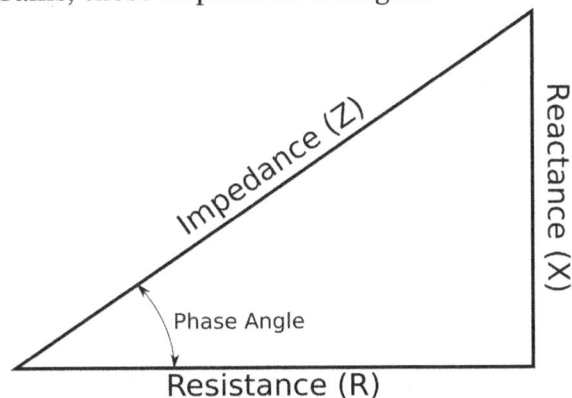

When an impedance contains reactance, whether inductive or capacitive, it's referred to as a *complex impedance*. When we build an impedance matching circuit we want to cancel out the reactance component and end up with a pure resistance that is the same number of ohms as the device we're trying to match. If we're trying to match a 50-ohm antenna, we want to end up with 50 ohms of resistance, and no reactance.

We do that by adding the opposite type of reactance. In the above triangle, we have an "excess" of inductive reactance, because the reactance point is above zero. Maybe we have an antenna that's just a bit too long for the frequency we're trying to use.

To cancel out that inductive reactance, we'd add enough capacitive reactance to balance it out, so that hypotenuse is now right on the X axis of the graph. We don't want a triangle; we just want a straight line. Then we might add a resistor – in series if our resulting impedance

245

is too low, in parallel if it is too high. Those two steps **cancel the reactive part of the impedance and change the resistive part to a desired value** are how we transform a complex impedance to a resistive impedance.

E7C01 (D) How are the capacitors and inductors of a low-pass filter Pi-network arranged between the network's input and output?

   A. Two inductors are in series between the input and output, and a capacitor is connected between the two inductors and ground

   B. Two capacitors are in series between the input and output, and an inductor is connected between the two capacitors and ground

   C. An inductor is connected between the input and ground, another inductor is connected between the output and ground, and a capacitor is connected between the input and output

   D. **A capacitor is connected between the input and ground, another capacitor is connected between the output and ground, and an inductor is connected between input and output**

Figure 39.1 shows our Pi-network again.

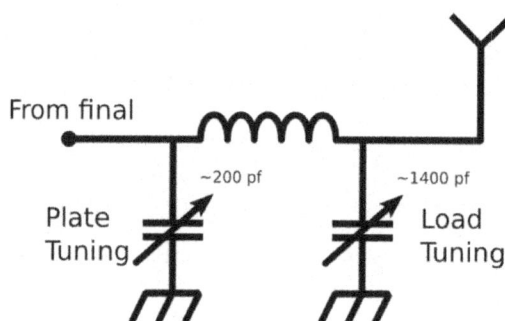

Figure 39.1: Simple Pi-network

A capacitor is connected between the input and ground, another capacitor is connected between the output and ground, and an inductor is connected between input and output. (A classic case of "a picture is worth 27 words.")

E7C12 (A) What is one advantage of a Pi-matching network over an L-matching network consisting of a single inductor and a single capacitor?

   A. **The Q of Pi-networks can be controlled**

   B. L-networks cannot perform impedance transformation

   C. Pi-networks are more stable

   D. Pi-networks provide balanced input and output

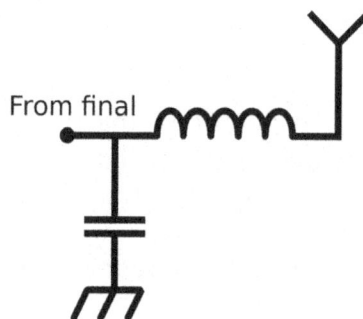

Figure 39.2: Typical L-Network

As shown in Figure 39.2, an L-matching network typically uses fixed value components and only works over a narrow range of frequencies and impedances. The choice of values of those components is usually a rather narrow range dictated by what components are actually

246

available, the frequency range to be passed, and the impedances to be matched. That leaves the choices about the Q of that network even more limited, and there's typically no adjustability to it at all.

The Pi-network works over a wider range of impedances and frequencies, and has, typically, a couple of variable capacitors, so **the Q of Pi-networks can be controlled**.

E7C07 (B) Which describes a Pi-L-network used for matching a vacuum tube final amplifier to a 50-ohm unbalanced output?

A. A Phase Inverter Load network

B. **A Pi-network with an additional series inductor on the output**

C. A network with only three discrete parts

D. A matching network in which all components are isolated from ground

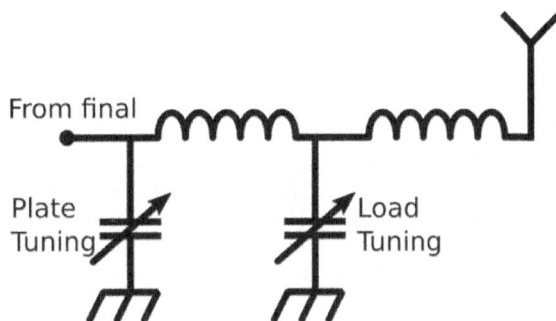

Figure 39.3: Pi-L Network

If it looks like a Pi-network but there's an "extra" coil just before the output, that's a Pi-L network. It's just **a Pi-network with an additional series inductor on the output**.

E7C03 (A) What advantage does a series-L Pi-L-network have over a regular Pi-network for impedance matching between the final amplifier of a vacuum-tube transmitter and an antenna?

A. **Greater harmonic suppression**

B. Higher efficiency

C. Does not require a capacitor

D. Greater transformation range

If we think in terms of equivalent circuits for Figure 39.3, a Pi-L network equals a Pi-network plus an L-network, as shown in Figure 39.4. In other words, we've added another filter to our filter.

Figure 39.4: Pi-L-network Equivalent Circuit

Of course, in the real world, we don't need that extra capacitor, and we don't need the extra coil in the middle. We just add the additional coil before the output to create the circuit shown in Figure 39.3.

Adding that extra L-network greatly increases the low-pass filtering that naturally occurs to some extent in a Pi-network, and means **greater harmonic suppression**.

Filters

E7C02 (C) Which of the following is a property of a T-network with series capacitors and a parallel shunt inductor?

A. It is a low-pass filter
B. It is a band-pass filter
C. **It is a high-pass filter**
D. It is a notch filter

A T-network is any three impedances, whether resistors, capacitors, or inductors, connected so two are in series and one is a parallel shunt to ground, with the shunt connected between the two series impedances.

In this case, we have two capacitors in series, with a shunt to ground through an inductor. The circuit is shown in Figure 39.5.

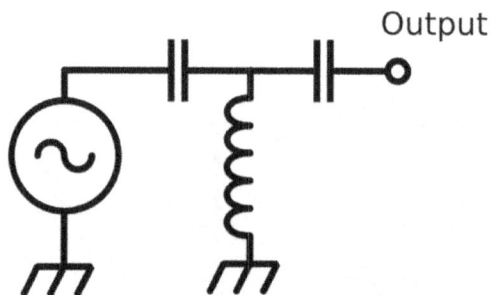

Figure 39.5: T-Network

With these components in this arrangement, we end up with a **high-pass filter** – it passes everything above a certain frequency. The capacitors will pass high frequencies while the inductor sends the low frequencies to ground, sucking them right out of the signal. If we put inductors where the capacitors are and a capacitor where the inductor is, the filter would work in just the opposite way and we'd have a low-pass filter.

E7C05 (D) Which filter type is described as having ripple in the passband and a sharp cutoff?

A. A Butterworth filter
B. An active LC filter
C. A passive op-amp filter
D. **A Chebyshev filter**

There are several types of **Chebyshev filters**, but they all have ripple in the passband and a sharp cutoff. (In this case, "ripple" is unwanted peaks and valleys in the frequency response.) In other words, they're a type of "brick wall" filter – for when we want absolutely everything below a certain frequency and almost nothing above that frequency. A schematic of the most common type looks like a Pi-network on steroids:

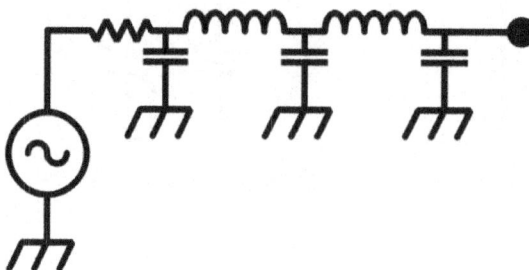

Figure 39.6: Chebyshev Filter

Effectively it's a filter stacked on a filter stacked on another filter. Figure 39.7 shows a typical frequency response of one type of Chebyshev filter.

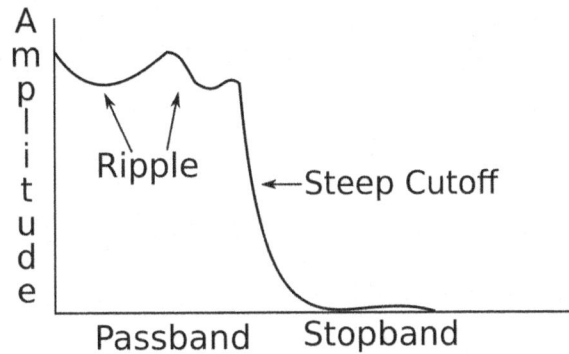

Figure 39.7: Chebyshev Frequency Response

You can see it has a very steep cutoff, but the side effect of creating that steep cutoff is those ripples in the passband.

Pafnuty Chebyshev was not a Russian electronics engineer, he was a Russian mathematician. He didn't invent the filters, but it took his innovations in mathematics to predict the very complex behavior of these filters. Be glad we don't cover *that* for the Extra exam. All we need to know is that Chebyshev filters have ripple in the passband and a sharp cutoff.

E7C06 (C) What are the distinguishing features of an elliptical filter?
   A. Gradual passband rolloff with minimal stop band ripple
   B. Extremely flat response over its pass band with gradually rounded stop band corners
   C. Extremely sharp cutoff with one or more notches in the stop band
   D. Gradual passband rolloff with extreme stop band ripple

There's a whole family of elliptical filters, just as there are Chebyshev filters. The name comes from the calculus that predicts their behavior which includes elliptical functions.

Figure 39.8 shows one sort of elliptical filter.

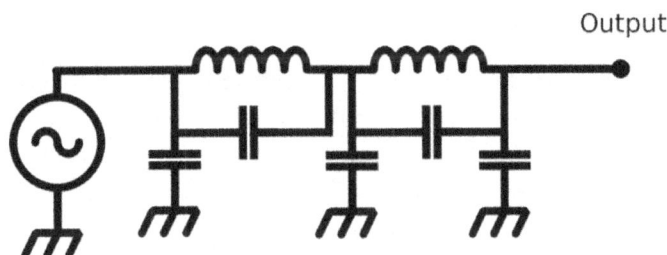

Figure 39.8: Elliptical Filter

If we need an even more **extremely sharp cutoff** than the Chebyshev filter provides, and we can live **with one or more notches in the stop band**, this might be just the ticket. Figure 39.9 shows the response of one such filter.

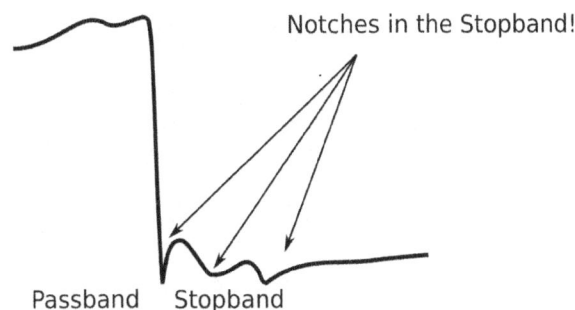

Figure 39.9: Elliptical Filter Frequency Response

Note that the exam only mentions the notch in the stopband, but the elliptical filter is known for notches in both the stop and pass bands.

For the exam, remember: Chebyshevs have ripple in the passband, ellipticals have notches in the stopband.

E7C09 (D) What is a crystal lattice filter?

A. A power supply filter made with interlaced quartz crystals

B. An audio filter made with four quartz crystals that resonate at 1 kHz intervals

C. A filter using lattice-shaped quartz crystals for high-Q performance

D. **A filter with narrow bandwidth and steep skirts made using quartz crystals**

Quartz crystals can have extraordinarily high Q's – above 100,000 – so by using them we can create **a filter with narrow bandwidth and steep skirts**.

Crystal lattice filters are often used in IF sections, the heart of a receiver's performance. Figure 39.10 shows a simplified schematic of one design of crystal lattice filter.

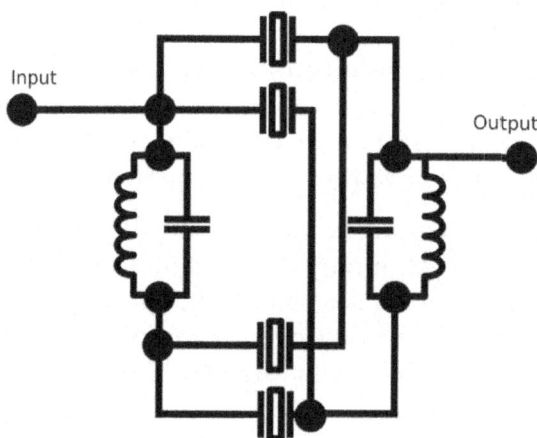

Figure 39.10: Crystal Lattice Filter

(Yes, that's simplified!)

Those symbols in Figure 39.10 that look like capacitors with a box in the middle are the crystals. Remember each of those is a resonant circuit, so between the parallel LC circuits on the sides and the various crystal connections, we end up with a lot of filtering – just what we want in our IF section, **a filter with narrow bandwidth and steep skirts**.

E7C08 (A) Which of the following factors has the greatest effect on the bandwidth and response shape of a crystal ladder filter?

A. **The relative frequencies of the individual crystals**

B. The DC voltage applied to the quartz crystal

C. The gain of the RF stage preceding the filter

D. The amplitude of the signals passing through the filter

It's not important for passing the exam, but a crystal *ladder* filter is different from the previous question's crystal *lattice* filter. A crystal ladder filter looks something like Figure 39.11.

Figure 39.11: Crystal Ladder Filter

If you rotate the page 90°, you can see why it's called a crystal "ladder" filter.

You'll recall from the chapter on resonant circuits that a crystal is its own self-contained resonant circuit, so we can create filters with crystals. Those filters are typically very accurate with regard to frequency, and have very high Q's – so, narrow and steep frequency response.

As you might guess, the factor that has the greatest effect in helping determine the bandwidth and response shape of a crystal ladder filter is – something about the crystals! It happens to be **the relative frequencies of the individual crystals.** Specifically, the crystals need to be closely matched in frequency and other characteristics for best results.

E7C10 (B) Which of the following filters would be the best choice for use in a 2-meter repeater duplexer?

    A. A crystal filter

    B. **A cavity filter**

    C. A DSP filter

    D. An L-C filter

A **cavity filter** hardly looks like an electronic device at all – it looks more like some strange sort of plumbing!

These are essential for repeater stations. They're very high Q bandpass filters; they have very narrow passbands. There's an input and output that take coaxial cable connectors. Inside, the coaxial cables feed coupling loops – just loops of wire. Between the coupling loops there's a "resonator" or inner conductor, and the outer case is conductive as well.

The electrical length of that inner conductor can be adjusted by moving it up and down with the tuning knob.

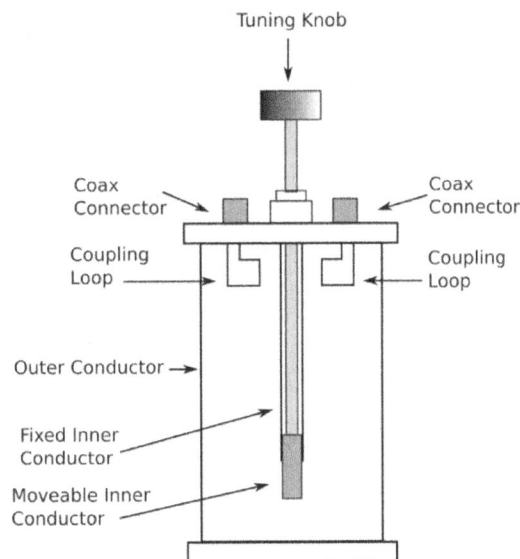

Figure 39.12: Cavity Filter

You could describe this as a very tightly tunable antenna in a resonant can. The input coupling loop radiates the input signal, causing the conductors to resonate only at the precise frequency they're set for. They won't pass anything but that frequency. It's all based on the size of the cavity – hence the name – and the frequency is fine-tuned by the length of that tunable conductor in the center.

These can also be built to handle lots and lots of power.

Pairs of cavity filters are components of *duplexers* that allow a repeater's receiver to keep receiving even though the transmitter is transmitting at the same time. One cavity filter protects the receiver from the transmit signal, and one filters the transmit signal to prevent any spurious radiation and keep it narrow so it gets filtered out by the receiver's cavity filter.

Key Concepts in This Chapter

- An impedance-matching circuit transforms a complex impedance to a resistive impedance by cancelling the reactive part of the impedance and changing the resistive part to a desired value.

- The capacitors and inductors of a low-pass filter Pi-network are arranged with a capacitor connected between the input and ground, another capacitor connected between the output and ground, and an inductor connected between the input and output.

- One advantage of a Pi-matching network over an L-matching network consisting of a single inductor and a single capacitor is that the Q of Pi-networks can be controlled.

- A Pi-L network can be described as a Pi-network with an additional series inductor on the output.

- An advantage a Pi-L network has over a regular Pi-network for impedance matching between the final amplifier of a vacuum-tube transmitter and an antenna is greater harmonic suppression.

- A T-network with series capacitors and a parallel shunt inductor is a high-pass filter.

- A filter type with ripple in the passband and a sharp cutoff is a Chebyshev filter.

- The distinguishing features of an elliptical filter are extremely sharp cutoff with one or more notches in the stop band.

- A crystal lattice filter is a filter with narrow bandwidth and steep skirts made using quartz crystals.

- The factor that has the greatest effect on the bandwidth and response shape of a crystal ladder filter is the relative frequencies of the individual crystals.

- The best filter choice for use in a 2-meter repeater duplexer is a cavity filter. .

Chapter 40

Power Supplies and Voltage Regulators

A few electric and electronic devices don't much care what voltage they get, or just operate straight off of batteries which are, at least in the short term, very stable. Most require some degree of precision in their voltage control. When we're choosing a voltage regulator, we have to ask, "how precise does it need to be" and, as you might guess, "what trade-offs am I willing to make?"

A voltage regulator can be an electro-mechanical device. Older cars used electro-mechanical voltage regulators to control the output of the generator or alternator, so that it would produce a reasonably constant voltage no matter the speed of the engine. Those old contraptions had several relays in them. Different amounts of voltage were required to close each relay – they had different strength springs holding them open. As the generator voltage changed, relays would open and close to bring resistors in or out of the circuit and thus keep the output close to the desired voltage. They weren't very precise, and were notoriously failure prone, but they got the job done when they were working.

Today, we use transistors and diodes instead of relays in most applications, but the principles of operation are much the same. Something compares the voltage to a standard, then does something about it. In the case of that antique electro-mechanical regulator, the standard is the tension of the springs in the relays. In a simple regulated power supply, it might be the breakdown voltage of a Zener diode – but there's still a standard, what's called a reference voltage.

I should mention in passing that every voltage regulator and power supply in this chapter is a DC power supply. AC voltage regulators and power supplies are very different creatures, but not part of the Extra exam nor are they at all common in amateur radio.

Linear Regulators

E7D01 (D) How does a linear electronic voltage regulator work?

    A. It has a ramp voltage as its output

    B. It eliminates the need for a pass transistor

    C. The control element duty cycle is proportional to the line or load conditions

    D. **The conduction of a control element is varied to maintain a constant output voltage**

A linear electronic voltage regulator uses some component that is operating in its linear range. For instance, it might use a transistor operating between cutoff and saturation, similar to

253

an amplifier. This is as opposed to a switching voltage regulator, which might use that same transistor, but use it only at cutoff or saturation, using the transistor as a switch.

Whether we use a transistor, a vacuum tube, or even a motor moving a contact on a mechanical potentiometer, in a linear regulator **the conduction of a control element is varied to maintain a constant output voltage.**

E7D03 (A) What device is typically used as a stable voltage reference in a linear voltage regulator?

    A. **A Zener diode**
    B. A tunnel diode
    C. An SCR
    D. A varactor diode

This should feel at least a little familiar, since we've covered this before in Chapter 30. Remember this circuit?

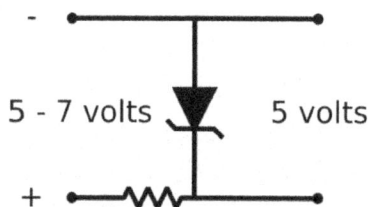

That's a very simple voltage regulator, using a Zener diode for the reference voltage. The Zener is hooked up in reverse bias. As the voltage rises above the desired 5 volts, the Zener breaks down and conducts, dropping the voltage back down to 5 volts. When the voltage across the output drops to whatever that resistor causes it to be, the Zener goes out of breakdown, and back and forth it goes, averaging out to the 5 volts we want.

E7D04 (B) Which of the following types of linear voltage regulator usually make the most efficient use of the primary power source?

    A. A series current source
    B. **A series regulator**
    C. A shunt regulator
    D. A shunt current source

A **series regulator** makes the most efficient use of the primary power source. You'll see why when we look at its opposite number, the shunt regulator.

If a regulator is a series regulator, the device directly controlling the voltage is in series with the input and the load, as in the schematic shown in Figure 40.1.

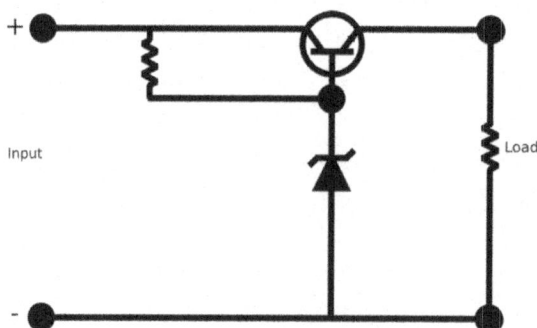

Figure 40.1: Series Regulator

In that circuit, the transistor is used as a variable resistor. When a transistor is used this way, we call it a "pass transistor" because it's passing current. The base voltage of the transistor is regulated by the Zener diode, so that the resistance between the collector and emitter increases if the input voltage goes up and decreases if it goes down.

E7D11 (D) What is the function of the pass transistor in a linear voltage regulator circuit?

    A. Permits a wide range of output voltage settings

    B. Provides a stable input impedance over a wide range of source voltage

    C. Maintains nearly constant output impedance over a wide range of load current

    D. **Maintains nearly constant output voltage over a wide range of load current**

The transistor in a linear voltage regulator circuit is the "adjustment valve," **maintaining a nearly constant output voltage over a wide range of load current.**

E7D13 (C) What is the equation for calculating power dissipated by a series linear voltage regulator?

    A. Input voltage multiplied by input current

    B. Input voltage divided by output current

    C. **Voltage difference from input to output multiplied by output current**

    D. Output voltage multiplied by output current

The equation for calculating power dissipation by a series connected linear voltage regulator is **voltage difference from input to output multiplied by output current.**

$$(E_{input} - E_{output}) \times I_{output}$$

We end up with $E \times I$ which is the formula for watts. All we're doing here is answering the question, "How many watts disappeared in the voltage regulator?" So, let's say we put 10 volts in, got 8 volts out, so we lost 2 volts along the way, and the output current is 2 amps. 2 times 2 is 4 – yeah, I know, I got it the first time, too! – so we lost 4 watts inside the circuit.

E7D05 (D) Which of the following types of linear voltage regulator places a constant load on the unregulated voltage source?

    A. A constant current source

    B. A series regulator

    C. A shunt current source

    D. **A shunt regulator**

If a linear voltage regulator isn't a series regulator, it's a shunt regulator.

    The series regulator has a weakness. All the current used by the load must go through that transistor, so for your average "buy it from Amazon.com" transistor, the upper limit is going to be about 15 amps. After that, your transistor goes up in smoke. There are also limits to how much the input supply voltage can vary before the transistor runs out of ability to correct it.

    That's why we also have shunt regulators. Rather than controlling the supply current through the transistor, the transistor acts in parallel across the load to become a variable resistor to ground. Excess voltage is shunted off to ground, which means the load can draw lots more amps.

    If you think about it, that first, simple Zener diode-based regulator we looked at was a shunt regulator.

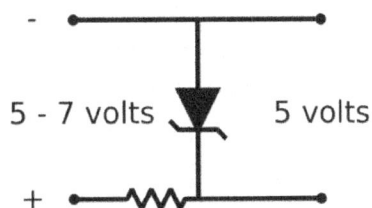

    Once the voltage gets excessive, it gets shunted off to the negative rail of the power supply, and the voltage is regulated.

    Figure 40.2 shows a simplified schematic of a slightly more elaborate one that will yield more precise voltage control.

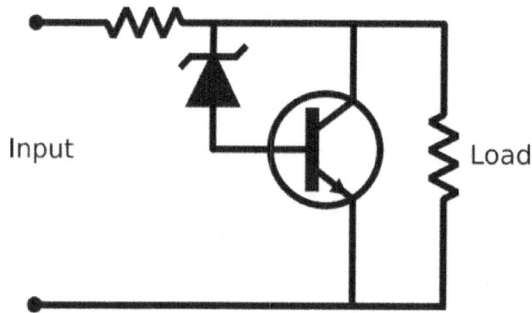

Figure 40.2: Shunt Regulator

Now the transistor is connected across the supply lines, parallel to the load. As the voltage rises above the reference voltage (the breakdown voltage of the Zener diode) the base of the transistor is energized and it acts as a variable resistor, shunting some of the excess voltage to the opposite rail. The higher the input voltage goes, the more the transistor conducts. When the voltage lowers, the transistor resists more, or even goes into cutoff and shunts nothing to ground. That way the input voltage source sees a consistent load.

This regulator can supply lots of current, because most of the current is going through the load. Problem solved, right? We'll just make all the regulators in the world shunt regulators. It's even exactly the same number of parts, so the cost should be about the same.

Well ... no. For one thing, notice that in the series regulator in Figure 40.1, there's a resistor in the line to the transistor base and the Zener diode. Not much of the current from the input is going through there – really, just enough to tickle the base of the transistor into action, a fraction of the total current. That's why a series regulator makes the most efficient use of the primary power source.

In the shunt regulator (Figure 40.2), every single amp of current for the load must go through that resistor near the input.

That's going to dissipate a lot of our power as heat, not as useful watts going to the load.

For that and some other more exotic reasons, the shunt regulator is not the end all and be all of voltage regulation, either.

We have three questions about Figure E7-2.

E7-2 is a linear voltage regulator, in this case a series regulator. Ignore the circle drawn around the Zener diode at D1; that's still a Zener diode. Imagine all those ground symbols gone and replaced by a single line across the bottom and you'll see, it's the same diagram you saw on page 254 with a few capacitors thrown into the mix.

E7D08 (C) What type of circuit is shown in Figure E7-2?

    A. Switching voltage regulator

    B. Grounded emitter amplifier

    C. **Linear voltage regulator**

    D. Monostable multivibrator

E7-2 is a diagram of a **linear voltage regulator**.

Figure E7-2

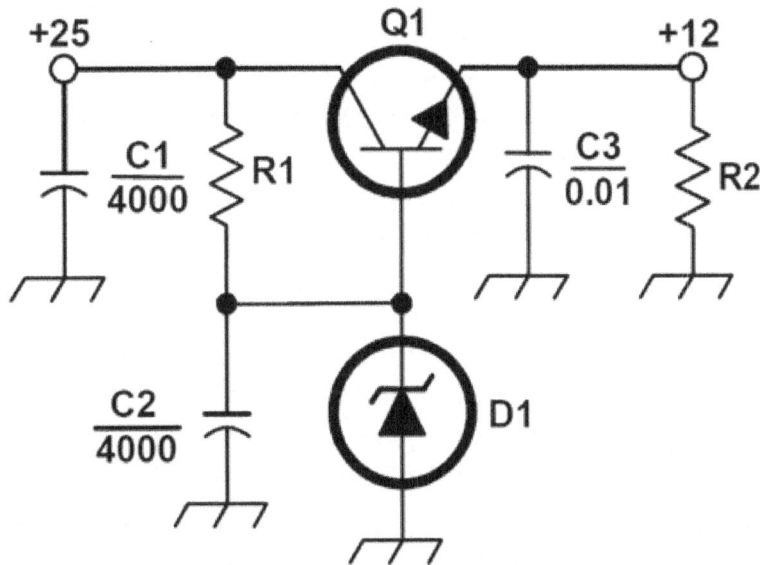

Figure 40.3: Figure E7-2 in the Question Pool

E7D06 (C) What is the purpose of Q1 in the circuit shown in Figure E7-2?

    A. It provides negative feedback to improve regulation

    B. It provides a constant load for the voltage source

    C. **It controls the current supplied to the load.**

    D. It provides D1 with current

The purpose of Q1 – Transistor 1 – is **to control the current supplied to the load**.

Let's look at the wrong answers just to reduce any possible confusion. Does it provide negative feedback? No. It's not hooked up to anything that could respond to negative feedback.

Does it provide a constant load for the voltage source? No, this is a series regulator; it's the shunt regulator that provides a constant load for the voltage source.

Does it provide the Zener diode with current? No – there's no current coming off the base of the transistor to flow through the Zener diode.

Since the other three answers are definitely out, we'll go with **it controls the current supplied to the load.**

E7D07 (A) What is the purpose of C2 in the circuit shown in Figure E7-2?

    A. **It bypasses rectifier output ripple around D1**

    B. It is a brute force filter for the output

    C. To self-resonate at the hum frequency

    D. To provide fixed DC bias for Q1

Capacitors block DC and pass AC; the ripple that is part of the output of the rectifier isn't exactly AC, but it is a changing voltage, and that will also pass through a capacitor.

C2 **bypasses rectifier output ripple around D1,** Diode 1. If the ripple didn't get bypassed around the diode, it would energize that transistor base, where it would "modulate the signal", which is the output of the regulator, and we'd have unwanted ripples in our DC.

E7D02 (C) What is a characteristic of a switching electronic voltage regulator?

    A. The resistance of a control element is varied in direct proportion to the line voltage or load current

B. It is generally less efficient than a linear regulator

C. **The controlled device's duty cycle is changed to produce a constant average output voltage**

D. It gives a ramp voltage at its output

Imagine you want to dim the light in your dining room to half of full brightness. One way you could do it would be with a big variable resistor that reduces the voltage supplied to the light. That would be like a linear voltage regulator.

Another way you could do it, if you were very talented in a very strange way, would be to switch the light off and on really fast so it was only on half of the time. That would be like a switching voltage regulator.

In a switching electronic voltage regulator, whatever device we are using as a switch gets switched off and on in varying proportions to create the voltage we desire. Voltage too low? Leave the switch on a little longer, bring the average up. Too high? Turn it off a little longer. We're *changing the duty cycle*. In a switching power supply, **the controlled device's duty cycle is changed to produce a constant average output voltage**.

E7D10 (C) What is the primary reason that a high-frequency switching type high-voltage power supply can be both less expensive and lighter in weight than a conventional power supply?

A. The inverter design does not require any output filtering

B. It uses a diode bridge rectifier for increased output

C. **The high frequency inverter design uses much smaller transformers and filter components for an equivalent power output**

D. It uses a large power factor compensation capacitor to recover power from the unused portion of the AC cycle

Just as with switching type amplifiers, the fact that the transistor is operated only in short bursts, and is always either completely on or completely off makes this type of power supply extremely efficient. You've probably noticed that VHF and UHF radios are smaller and lighter than their HF companions – in large part because higher frequencies not only can use smaller components, but often must do so. That's also why the switching type power supply is smaller and lighter; **the high frequency inverter design uses much smaller transformers and filter components for an equivalent power output.**

E7D12 (C) What is the dropout voltage of an analog voltage regulator?

A. Minimum input voltage for rated power dissipation

B. Maximum output voltage drops when the input voltage is varied over its specified range

C. **Minimum input-to-output voltage required to maintain regulation**

D. Maximum that the output voltage may decrease at rated load

There comes a time in every regulator's life that there's just not enough input voltage being supplied to supply the output voltage that's needed. That's called the drop-out voltage. The drop-out voltage is the **minimum input-to-output voltage required to maintain regulation.**

E7D14 (D) What is the purpose of connecting equal-value resistors across power supply filter capacitors connected in series?

A. Equalize the voltage across each capacitor

B. Discharge the capacitors when the voltage is removed

C. Provide a minimum load on the supply

D. **All these choices are correct**

Those resistors across the power supply filter capacitors are called *bleeder resistors*. Filter

258

capacitors tend to be big capacitors, holding a lot of charge. The *primary* purpose of a bleeder resistor is to slowly discharge those capacitors when the unit is turned off, for safety reasons.

Another reason for the bleeder resistor is to **equalize the voltage across each capacitor**, which improves voltage regulation.

Yet another reason is to create a constant **minimum load** on the supply, which helps prevent voltage fluctuations in light load situations.

For this question **all these choices are correct**.

E7D15 (D) What is the purpose of a step-start circuit in a high-voltage power supply?
  A. To provide a dual-voltage output for reduced power applications
  B. To compensate for variations of the incoming line voltage
  C. To allow for remote control of the power supply
  D. **To allow the filter capacitors to charge gradually**

In a high voltage power supply, we want to sort of "bring the lights up gradually." Those big output capacitors are going to suck up a lot of electrons, and they can be damaged by the current if they're charged too quickly.

That's why high-voltage power supplies often incorporate a step-start circuit that **allows the filter capacitors to charge gradually.**

Solar Charge Controllers

E7D09 (C) What is the main reason to use a charge controller with a solar power system?
  A. Prevention of battery undercharge
  B. Control of electrolyte levels during battery discharge
  C. **Prevention of battery damage due to overcharge**
  D. Matching of day and night charge rates

The main reason – indeed, the only reason I can think of – to use a charge controller with a solar power system is the **prevention of battery damage due to overcharge**. Solar cells don't know your battery is fully charged – they'll happily keep charging it until terrible things happen inside it, like the release of all the H and O in the H_2SO_4 – the sulfuric acid – leaving you with nothing but sulfides in your now recently deceased battery.

Key Concepts in This Chapter

- A linear electronic voltage regulator works by varying the conduction of a control element to maintain a constant output voltage.

- The device typically used as a stable reference voltage in a linear voltage regulator is a Zener diode.

- The type of linear voltage regulators that make the most efficient use of the primary power source are usually series regulators.

- The transistor in a linear voltage regulator circuit maintains nearly constant output voltage over a wide range of load current.

- The equation for calculating power dissipation by a series connected linear voltage regulator is voltage difference from input to output multiplied by output current.

- The type of linear voltage regulator that places a constant load on the unregulated voltage source is a shunt regulator.

- The device shown in Figure E7–2 is a linear voltage regulator.

- The purpose of Q1 in the circuit shown in Figure E7-2 is to control the current supplied to the load.

- The purpose of C2 in the circuit shown in Figure E7-2 is to bypass ripple from the rectifier around D1.

- One characteristic of a switching electronic voltage regulator is the controlled device's duty cycle is changed to produce a constant average output voltage.

- The primary reason that a high-frequency switching type high voltage power supply can be both less expensive and lighter in weight than a conventional power supply is that the high frequency inverter design uses much smaller transformers and filter components for an equivalent power output.

- The drop-out voltage of an analog voltage regulator is the minimum input-to-output voltage required to maintain regulation.

- One purpose for connecting equal-value resistors across power supply filter capacitors connected in series is to equalize the voltage across each capacitor. Other reasons include to discharge the capacitors when the voltage is removed, and to provide a minimum load on the supply.

- The purpose of a "step-start" circuit in a high voltage power supply is to allow the filter capacitors to charge gradually.

- The main reason to use a charge controller with a solar power system is to prevent battery damage due to overcharge.

When you complete the fasttrackham.com Practice Exam for this chapter you'll be ready to take Progress Check #8.

Progress Check #8

Chapter 41

Modulation & Demodulation, Part I

Modulation

Baseband

E7E07 (B) What is meant by the term baseband in radio communications?

    A. The lowest frequency band that the transmitter or receiver covers

    B. **The frequency range occupied by a message signal prior to modulation**

    C. The unmodulated bandwidth of the transmitted signal

    D. The basic oscillator frequency in an FM transmitter that is multiplied to increase the deviation and carrier frequency

The baseband is where all modulation starts. The "baseband" is **the frequency range occupied by a message signal prior to modulation**. It's what is going to be somehow combined with our carrier to create the signal. For phone, the baseband is what's coming out of our microphone. For digital signals, it's probably what's coming out of the sound card of our computer.

"Baseband" will come up again when we discuss some more exotic forms of modulation; just keep in mind, it just means "the frequency range occupied by a message signal prior to modulation."

Creating FM & Phase Modulation

E7E01 (B) Which of the following can be used to generate FM phone emissions?

    A. A balanced modulator on the audio amplifier

    B. **A reactance modulator on the oscillator**

    C. A reactance modulator on the final amplifier

    D. A balanced modulator on the oscillator

Now that we've taken our deep dive into reactance back in Chapter 22, the phrase "**reactance modulator**" should make some sense to you, and from our discussions on diodes you can probably make a good guess at the sort of component we would use to *vary* the *reactance* in an oscillator.

The question asks us about FM, frequency modulation. That means we need to vary the frequency of the carrier. The carrier frequency gets created by **the oscillator** – not by the audio amplifier, nor by the final amplifier, which are a couple of the possibilities in the wrong answers.

There are any number of ways to build a reactance modulator, but they all end up being this basic idea. We start with a nice, stable (and simplified) oscillator:

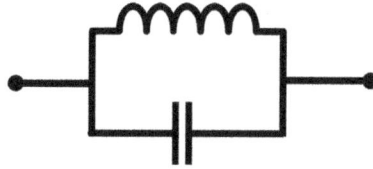

Then we make the reactance variable:

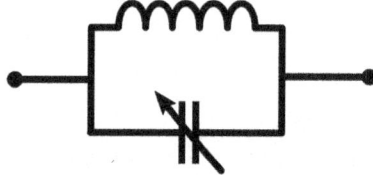

We can make the capacitance variable, as in the example above, or we can make the inductance variable; either way, we'll be changing the resonant frequency of that resonant circuit.

Of course, there's more to the circuit than that, since we have to make that reactance vary in response to the signal we want to transmit, but that's the idea at the core of it.

Two of the wrong answers suggest we would use a balanced modulator. That's not going to give us FM; it would give us SSB.

E7E02 (D) What is the function of a reactance modulator?
    A. To produce PM signals by using an electrically variable resistance
    B. To produce AM signals by using an electrically variable inductance or capacitance
    C. To produce AM signals by using an electrically variable resistance
    D. To produce PM or FM signals by using an electrically variable inductance or capacitance

The function of a reactance modulator is **to produce PM (phase modulated) or FM signals by using an electrically variable inductance or capacitance**.

Pre-Emphasis & De-Emphasis

E7E05 (D) What circuit is added to an FM transmitter to boost the higher audio frequencies?
    A. A de-emphasis network
    B. A heterodyne suppressor
    C. A heterodyne enhancer
    D. A pre-emphasis network

Some FM transmitters have a circuit to boost the higher audio frequencies called a **pre-emphasis network**. We need these because of the inherent character of the demodulators on the receiving end. They're noisy, and they get noisier the higher the transmitted audio frequency. Not a little noisier either – the noise rises at 6 dB per octave. Since the frequencies of speech occupy about five octaves, we're talking about around 30 dB of noise at the top end.

To knock down that noise, we can put a -6 dB per octave filter in the receiver – but then we'd have very muffled sounding speech coming out of the receiver.

To compensate, then, we "pre-emphasize" the audio at the transmitting end by 6 dB per octave, then "de-emphasize" on the receiving end, and everything comes out the way it should. We crank the treble way up when we send the signal, then crank it way down when we receive it.

E7E06 (A) Why is de-emphasis commonly used in FM communications receivers?
    A. For compatibility with transmitters using phase modulation
    B. To reduce impulse noise reception
    C. For higher efficiency
    D. To remove third-order distortion products

As you might recall from your General studies, there are two systems to create frequency modulation, but only one is practical for our purposes. In shorthand form, it's phase modulation.

Phase modulators don't need a separate pre-emphasis network. In a sort of mirror image of the 6 dB per octave noise gain in FM *demodulators*, phase modulators inherently add a 6 dB per octave rise in amplitude of the transmitted signal. **For compatibility with transmitters using phase modulation**, we must use de-emphasis in the receiver.

Without de-emphasis on the receiving end, our signal would sound tinny *and* noisy.

Creating SSB

E7E04 (A) What is one way a single-sideband phone signal can be generated?

A. **By using a balanced modulator followed by a filter**
B. By using a reactance modulator followed by a mixer
C. By using a loop modulator followed by a mixer
D. By driving a product detector with a DSB signal

One way a single-sideband phone signal can be generated is **by using a balanced modulator followed by a filter**.

In a single-sideband transmitter, we create the RF carrier with an oscillator, then feed the RF carrier into a balanced modulator, which modulates the carrier with the signal we want to send via SSB.

Why a balanced modulator? By its very nature, a balanced modulator creates both the upper and lower sidebands and suppresses the carrier – the carrier never comes out of the balanced modulator, so, no need to filter it out.

The balanced modulator creates the sidebands, then a filter strips out either the upper or lower sideband, leaving us with a single sideband.

Demodulation

Mixers

E7E08 (C) What are the principal frequencies that appear at the output of a mixer circuit?

A. Two and four times the original frequency
B. The square root of the product of input frequencies
C. **The two input frequencies along with their sum and difference frequencies**
D. 1.414 and 0.707 times the input frequency

Whenever two frequencies mix, we end up with **the two input frequencies along with their sum and difference frequencies**. That's the principal behind, among other things, intermodulation distortion and a superheterodyne receiver's IF stage.

E7E09 (A) What occurs when an excessive amount of signal energy reaches a mixer circuit?

A. **Spurious mixer products are generated**
B. Mixer blanking occurs
C. Automatic limiting occurs
D. A beat frequency is generated

Like any other electronic circuit, a mixer circuit can be overdriven to the point that what were nice pure sine waves become nasty, harmonic-rich square waves. Once all those harmonics start mixing, we'll get all sorts of unwanted additional frequencies. If it happens in our transmitter, we call the result "spurious emissions." In the case of a mixer circuit, we call those "**spurious mixer products.**" Look for the key word, **spurious**.

Next, we launch into various sorts of demodulators. The demodulator is the receiver's counterpart to the transmitter's modulator. The job of a demodulator is to extract the baseband – the information that is carried by the modulated carrier.

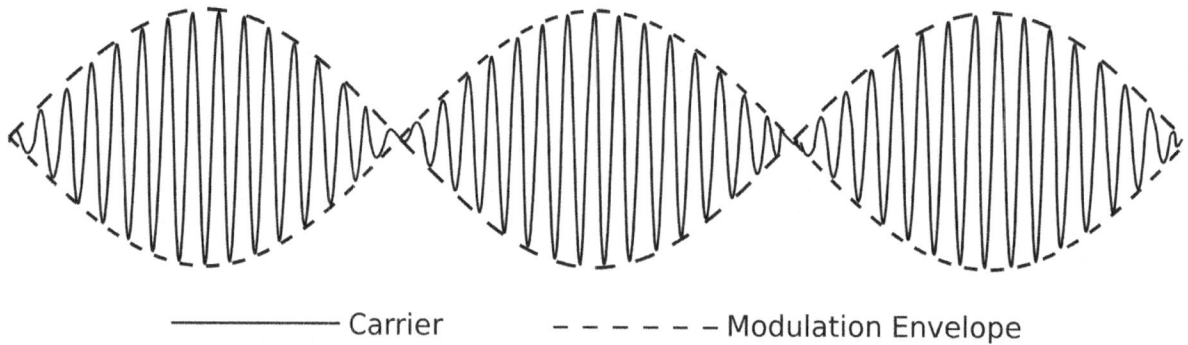

——————— Carrier   – – – – – – Modulation Envelope

Figure 41.1: Amplitude Modulated Carrier

Demodulating AM

E7E10 (A) How does a diode envelope detector function?

A. **By rectification and filtering of RF signals**
B. By breakdown of the Zener voltage
C. By mixing signals with noise in the transition region of the diode
D. By sensing the change of reactance in the diode with respect to frequency

We can make a detector for AM signals using nothing more than a 5 cent diode. In fact, if we add an antenna, a set of high impedance headphones, and a ground we can make a whole AM radio receiver.

Remember that an AM signal comes to us as a modulated carrier. It's a radio frequency wave with a modulation envelope imposed on it. The modulated carrier looks something like Figure 41.1.

That modulation envelope is, you'll notice, symmetrical. For each positive peak there's an equal negative peak. To hear the transmitted audio we need to get rid of half of that signal – because positive plus equal negative adds up to zero. That's called "rectification" and that's what the diode does, doing its one-way valve thing. Then, once the diode rectifies the signal, it's useful to filter out the radio frequency and just leave the audio.

A diode envelope detector works **by rectification and filtering of RF signals**.

Figure 41.2 shows the simplest possible device to do exactly that.

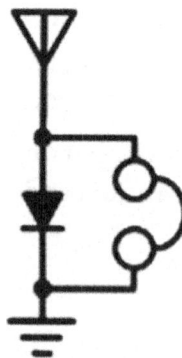

Figure 41.2: Primitive Radio

The diode rectifies the signal and those headphones certainly aren't capable of reproducing radio frequencies, so they're a perfectly fine filter. The headphones plus the diode are the "diode envelope detector" in this fine, if drastically limited, radio.

Demodulating SSB

E7E11 (C) Which type of detector is used for demodulating SSB signals?

    A. Discriminator

    B. Phase detector

    C. **Product detector**

    D. Phase comparator

We can't use our simple diode detector radio set for single-sideband and have any sort of satisfactory result. For single-sideband, we need a **product detector.**

That diode detector is an *envelope detector*. For the purpose of understanding a *product detector* you might think of the envelope detector as a "subtractive detector." It takes the original modulation envelope and subtracts the positive or negative half, leaving us with the original audio.

We won't go into the advanced mathematics of SSB, which looks like this:

$$y\left(T\right) = \left(C + m\left(T\right)\cos\left(\omega\right)\cos\left(\omega\right)\right)$$

Just notice there is a lot of multiplying going on in that equation. To get the original audio carried by the SSB signal, we need to detect those multiplications, i.e., the products. That's why an SSB detector is called a *product* detector.

Demodulating FM

E7E03 (D) What is a frequency discriminator stage in a FM receiver?

    A. An FM generator circuit

    B. A circuit for filtering two closely adjacent signals

    C. An automatic band-switching circuit

    D. **A circuit for detecting FM signals**

The detector in an FM receiver must reverse the process that created the FM carrier. It needs to translate the variations in frequency into the audio that created the signal.

One **circuit for detecting FM signals** is called the frequency discriminator. In this context, to "detect" a signal is to demodulate it. We'll discuss another way to demodulate FM in chapter 46 in the section on phase-locked loops, starting on page 301.

Key Concepts in This Chapter

- The term baseband refers to the frequency components present in the modulating signal.

- To generate FM phone emissions we can use a reactance modulator on the oscillator.

- The function of a reactance modulator is to produce PM or FM signals by using an electrically variable inductance or capacitance.

- One way a single-sideband phone signal can be generated is by using a balanced modulator followed by a filter.

- The circuit added to an FM transmitter to boost the higher audio frequencies is called a pre-emphasis network.

- De-emphasis is commonly used in FM communications receivers for compatibility with transmitters using phase modulation.

- One way of generating a single-sideband signal is by using a balanced modulator followed by a filter.

- The principal frequencies that appear at the output of a mixer circuit are the two input frequencies along with their sum and difference frequencies.

- When an excessive amount of signal energy reaches a mixer circuit spurious mixer products are generated.

- A diode envelope detector functions by rectification and filtering of RF signals.

- The type of detector used for demodulating SSB signals is a product detector.

- The frequency discriminator stage in an FM receiver is a circuit for detecting FM signals.

Chapter 42

Modulation and Demodulation, Part II

Modulation Index & Deviation Ratio

E8B01 (A) What is the modulation index of an FM signal?
  A. **The ratio of frequency deviation to modulating signal frequency**
  B. The ratio of modulating signal amplitude to frequency deviation
  C. The type of modulation used by the transmitter
  D. The bandwidth of the transmitted signal divided by the modulating signal frequency

With AM, it's easy to see the 100% modulation point with an oscilloscope. Since an unmodulated carrier just looks like a sine wave, we're at 100% when the modulation envelope scrunches that sine wave down to zero.

But what about FM? How much frequency change is 100%? 100% of what? 1 Hz of deviation? 1 kHz? 1 MHz? How about a signal that deviates from, say, the 10-meter band to the gigahertz band? There's no physical law that says we couldn't use that kind of deviation. (There's a whole lot of FCC law about that, though, not to mention that it's a wildly impractical scheme.) What's 100%? With FM – and other "angle modulated" signals, we need another measure, one that relates to the maximum frequency we want to transmit, since that frequency will give us the widest deviation.

When we speak of deviation, we're talking about how much we change the frequency of the carrier – and that's going to automatically end up being a "plus and minus" number, since the carrier will swing both higher and lower than the center frequency. However, when we do deviation calculations, we don't worry about the "plus and minus" stuff, we're only concerned with how far from center we're going to deviate. (Math folks would say we want the "absolute value" of the deviation.) So, an "8 kHz deviation" means we're going up 8 kHz and down 8 kHz and end up occupying about 16 kHz of bandwidth.

The value we use to indicate how much modulation is being applied to an FM signal is called the *modulation index* or *deviation ratio*. The modulation index is **the ratio of frequency deviation to modulating signal frequency**

The formula for modulation index is:

$$Modulation\,Index = \frac{\Delta f}{f_{max}}$$

In that formula, f_{max} is the highest frequency component in the transmitted signal (in Hz), and Δf is the carrier's maximum deviation (also in Hz) from its center frequency. (The Greek letter delta, Δ, is commonly used to designate "amount of change.") So if we want to transmit everything up to 8 kHz, and that causes our carrier to change frequency by 8 kHz higher and 8 kHz lower, that's a modulation index of 1. (8000/8000 = 1.)

Interesting to know, but trivial, right? Yes. Until someone sets a legal limit on the modulation index. Ours happens to be, by law, 1.0 on frequencies below 29.0 MHz. Then, we can work it this way:

$$\Delta f = f_{max} \times 1.0$$

...and now we know by how many Hz our carrier can deviate or the maximum frequency we can transmit. Much more useful.

E1C09 (B) [97.307] What is the highest modulation index permitted at the highest modulation frequency for angle modulation below 29.0 MHz?

A. 0.5

B. **1.0**

C. 2.0

D. 3.0

Angle modulation is the name for the class of modulation that includes Frequency Modulation and, more relevantly for the HF bands, Phase Modulation.

Below the 10-meter band, our maximum permitted modulation index for angle modulated signals is **1.0**.

Practically speaking, this limit, combined with bandwidth limitations, means no FM Phone transmissions below the upper reaches of the 10-meter band. (Remember, the bandwidth of the signal is double the deviation.)

E8B02 (D) How does the modulation index of a phase-modulated emission vary with RF carrier frequency?

A. It increases as the RF carrier frequency increases

B. It decreases as the RF carrier frequency increases

C. It varies with the square root of the RF carrier frequency

D. **It does not depend on the RF carrier frequency**

Look again at that formula for modulation index.

$$Modulation\,Index = \frac{\Delta f}{f_{max}}$$

There's nothing in that formula about the carrier's frequency, only the amount the frequency changes. The formula doesn't care if you're deviating a 70 cm carrier or a 160-meter carrier. Modulation index **does not depend on the RF carrier frequency**.

E8B03 (A) What is the modulation index of an FM-phone signal having a maximum frequency deviation of 3000 Hz either side of the carrier frequency when the modulating frequency is 1000 Hz?

A. **3**

B. 0.3

C. 3000

D. 1000

The modulation index we choose to use is more or less arbitrary – it's limited by law below 29.0 MHz to 1.0, but not in the higher frequencies. For this question, they ask about an FM-phone signal having a maximum frequency deviation of 3000 Hz either side of the carrier frequency when the modulating frequency is 1000 Hz.

$$Modulation\,Index = \frac{\Delta f}{f_{max}} = \frac{3000}{1000} = 3$$

E8B04 (B) What is the modulation index of an FM-phone signal having a maximum carrier deviation of plus or minus 6 kHz when modulated with a 2 kHz modulating frequency?

A. 6000

B. **3**

C. 2000

D. 1/3

We know the formula for modulation index is

$$Modulation\,Index = \frac{\Delta f}{f_{max}}$$

So let's plug in our numbers.

$$Modulation\,Index = \frac{6000}{2000} = 3$$

E8B09 (B) What is deviation ratio?

A. The ratio of the audio modulating frequency to the center carrier frequency

B. **The ratio of the maximum carrier frequency deviation to the highest audio modulating frequency**

C. The ratio of the carrier center frequency to the audio modulating frequency

D. The ratio of the highest audio modulating frequency to the average audio modulating frequency

That definition sounds vaguely familiar, doesn't it? It's the same as modulation index. Deviation ratio (a.k.a. modulation index) is **the ratio of the maximum carrier frequency deviation to the highest audio modulating frequency**.

E8B05 (D) What is the deviation ratio of an FM-phone signal having a maximum frequency swing of plus-or-minus 5 kHz when the maximum modulation frequency is 3 kHz?

A. 60

B. 0.167

C. 0.6

D. **1.67**

This works exactly the same way as the modulation index calculations.

$$Deviation\,ratio = \frac{\Delta f}{f_{max}} = \frac{5000}{3000} = 1.67$$

E8B06 (A) What is the deviation ratio of an FM-phone signal having a maximum frequency swing of plus or minus 7.5 kHz when the maximum modulation frequency is 3.5 kHz?

A. **2.14**

B. 0.214

C. 0.47

D. 47

Yes, it's the same problem with different numbers.

$$Deviation\,ratio = \frac{\Delta f}{f_{max}} = \frac{7500}{3500} = 2.14$$

You're never going to have a deviation ratio under 1.0 on the exam, so you can automatically toss out any answers that are under 1.0 if they're asking you about deviation ratio or modulation index.

Multiplexing

E8B08 (D) What describes Orthogonal Frequency Division Multiplexing?

    A. A frequency modulation technique that uses non-harmonically related frequencies

    B. A bandwidth compression technique using Fourier transforms

    C. A digital mode for narrow-band, slow-speed transmissions

    D. **A digital modulation technique using subcarriers at frequencies chosen to avoid intersymbol interference**

I'll bet you own an Orthogonal Frequency Division Multiplexing, or OFDM device. If you have a reasonably modern wi-fi router, it uses OFDM. If you have a DSL connection, that uses OFDM. Digital broadcast television is OFDM as is digital broadcast radio. If your mobile phone uses a 4G network, that's OFDM, too.

First, "orthogonal" is just a $10 word that means "related to a 90° angle." In the case of OFDM the 90° angle is a phase angle between a baseband and a carrier.

To multiplex is to combine two or more signals into one. FM stereo stations multiplex several signals into one carrier, for instance. In this case, we're multiplexing two or more digital data streams.

Orthogonal Frequency Division Multiplexing is a modulation scheme that essentially transmits digital data over a lot of very narrow bandwidth signals simultaneously rather than one broad bandwidth signal. Each channel is slower than the overall transmission rate, so each individual symbol is transmitted over a longer time, and the spaces between the symbols are longer as well. Properly timed, these spaces help eliminate **intersymbol interference**, and the slower symbol rate is more immune to interference.

The actual mechanism of creating the signal is to combine two (or even more) digital signals into a "baseband" signal which then modulates a carrier, creating numerous **subcarriers** – what you and I call sidebands in the HF world. All the interrelated frequencies are worked out so the subcarriers fall at frequencies that also help **avoid intersymbol interference**.

Look for the key words **subcarriers** and **intersymbol interference**.

Ham radio modes that use OFDM include MT63, Q15X25 (or NEWQPSK), Robust PACKET, and WinDRM, all high-speed digital modes.

E8B07 (A) Orthogonal Frequency Division Multiplexing is a technique used for which type of amateur communication?

    A. **High speed digital modes**

    B. Extremely low-power contacts

    C. EME

    D. OFDM signals are not allowed on amateur bands

OFDM is used for **high-speed digital modes**.

E8B10 (B) What is frequency division multiplexing?

    A. The transmitted signal jumps from band to band at a predetermined rate

    B. **Two or more information streams are merged into a baseband, which then modulates the transmitter**

    C. The transmitted signal is divided into packets of information

    D. Two or more information streams are merged into a digital combiner, which then pulse position modulates the transmitter

Frequency division multiplexing, FDM, is very similar to orthogonal frequency division multiplexing, as the names would imply. In fact, the distinction is very slight, since OFDM is a sub-category of FDM. When we speak of frequency division multiplexing, we're talking about a single wideband signal carrying multiple data streams, each in a different part of its bandwidth. OFDM accomplishes that with subcarriers, FDM does it without the subcarriers. They both start with a baseband, so watch for that key word **baseband**, the signal that modulates the carrier.

Here's a really easy way to think of Frequency Division Multiplexing. Think of the commercial AM broadcast band. That's one giant crude but effective Frequency Division Multiplexing scheme – a bunch of different signals, all spread out across a big chunk of bandwidth.

C4FM, also known as Yaesu Fusion, is an amateur mode that uses a form of Frequency Division Multiplexing.

E8B11 (B) What is digital time division multiplexing?

A. Two or more data streams are assigned to discrete sub-carriers on an FM transmitter

B. **Two or more signals are arranged to share discrete time slots of a data transmission**

C. Two or more data streams share the same channel by transmitting time of transmission as the sub-carrier

D. Two or more signals are quadrature modulated to increase bandwidth efficiency

FDM and OFDM multiplex signals in the frequency domain. Digital time division multiplexing multiplexes signals in the time domain.

Digital time division multiplexing doesn't have any fancy subcarriers or trick basebands. It's a time-sharing scheme. Let's say we have two digital files to send, File A and File B. Digital time division multiplexing sends stuff from File A for x amount of time, then File B, then File A, etc., until they're both complete.

Key Concepts in This Chapter

- The modulation index of an FM signal is the ratio of frequency deviation to modulating signal frequency.

$$Modulation\ Index\ or\ Deviation\ Ratio = \frac{\Delta f}{f_{max}}$$

- The highest modulation index permitted at the highest modulation frequency for angle modulation below 29.0 MHz is 1.0.

- The modulation index of a phase-modulated emission does not depend on the RF carrier frequency.

- The modulation index of an FM-phone signal having a maximum frequency deviation of 3000 Hz either side of the carrier frequency when the modulating frequency is 1000 Hz is 3.

- The modulation index of an FM-phone signal having a maximum carrier deviation of plus or minus 6 kHz when modulated with a 2 kHz modulating frequency is 3.

- The term deviation ratio means the ratio of the maximum carrier frequency deviation to the highest audio modulating frequency. It is the same as the modulation index.

- The deviation ratio of an FM-phone signal having a maximum frequency swing of plus-or-minus 5 kHz when the maximum modulation frequency is 3 kHz is 1.67.

- The deviation ratio of an FM-phone signal having a maximum frequency swing of plus or minus 7.5 kHz when the maximum modulation frequency is 3.5 kHz is 2.14.

- Orthogonal Frequency Division Multiplexing can be described as a digital modulation technique using subcarriers at frequencies chosen to avoid intersymbol interference.

- Orthogonal Frequency Division Multiplexing is a technique used for high-speed digital modes.

- Frequency division multiplexing is created by merging two or more information streams into a baseband, which then modulates the transmitter.

- Digital time division multiplexing can be described as two or more signals arranged to share discrete time slots of a data transmission.

Chapter 43

Software Defined Radio

Figure 43.1: IC-7300

E7F01 (C) What is meant by direct digital conversion as applied to software defined radios?

    A. Software is converted from source code to object code during operation of the receiver

    B. Incoming RF is converted to a control voltage for a voltage controlled oscillator

    C. **Incoming RF is digitized by an analog-to-digital converter without being mixed with a local oscillator signal**

    D. A switching mixer is used to generate I and Q signals directly from the RF input

Here's an easy prediction: the software defined radio is the future of ham radios, at least in the HF bands. The technology is getting incredibly inexpensive, and the capabilities are just amazing.

The underlying truth of software defined radio is that if there's a mathematical formula for something, a computer can do that something – and all our forms of modulation and demodulation can be described by precise mathematical formulas.

As you have seen, the IF section of analog receivers is both the most critical for good performance and the most vulnerable to a lot of potential problems. If we could somehow get the IF section out of the picture and replace it with something better, we open up a lot of possibilities for enhanced performance. That's almost the whole reason software defined radio is desirable.

The earlier in the reception process we can convert an incoming signal to digital, the sooner we can start digitally processing it. As we'll soon see, though, that takes some computing power that wasn't available to us common folk until relatively recently.

With the actual incoming RF digitized, and with sufficiently clever algorithms, we don't need an IF mixer, IF filter, or really much of anything else that often gives us such technical challenges in analog receivers. We can do all that in the digital domain.

In direct digital conversion (as applied to software defined radios) **incoming RF is digitized by an analog-to-digital converter without being mixed with a local oscillator signal**.

Software defined receivers can consist of a conversion box that digitizes the incoming RF and sends it to your computer, which does the rest of the digital processing. Today, for under $200[1], you can buy a receiver conversion box that covers from 1 kHz to 2 GHz. You plug your antenna into it, hook it up to your computer with a USB cable, and you're in the SDR receiver business.

Software defined transceivers are on the market as well, including the IC-7300 above as well as radios made by the Flex company, and more are on the way.

E8A08 (C) Why would a direct or flash conversion analog-to-digital converter be useful for a software defined radio?

    A. Very low power consumption decreases frequency drift

    B. Immunity to out-of-sequence coding reduces spurious responses

    C. **Very high speed allows digitizing high frequencies**

    D. All these choices are correct

A direct analog-to-digital converter in a software defined radio converts the incoming RF signal directly to digits. There's no IF section, no other "down conversion" of the frequency, it digitizes what's coming out of the RF filter right after the antenna. To do that it must be a **very high-speed** converter, and that **allows digitizing high frequencies.**

E7F05 (B) How frequently must an analog signal be sampled by an analog-to-digital converter so that the signal can be accurately reproduced?

    A. At least half the rate of the highest frequency component of the signal

    B. **At least twice the rate of the highest frequency component of the signal**

    C. At the same rate as the highest frequency component of the signal

    D. At four times the rate of the highest frequency component of the signal

It's the Nyquist-Shannon Sampling Theorem that tells us that the minimum sample rate that will allow us to reconstruct the original signal from a digital representation of that signal is **twice the rate of the highest frequency component of the signal**. For high fidelity audio that goes as high as about 22 kHz, that means the minimum sample rate we need is 44 kHz – and, it's no accident that that is almost exactly the sample rate used for Compact Discs.

Think of the implications for software defined radios. To accurately sample an RF signal up in the 2-meter band, we need around 300,000,000 samples per second. That software defined radio that goes up to 2 GHz would need to take 4,000,000,000 samples per second – way beyond the capabilities of even the fastest CPU's as of this writing, though we're getting closer. So how do they do it? Trickery – such as undersampling – and some very fast specialized semiconductors.

---

[1]I recently bought an SDR (receiver) for $39.95. It's a hacked USB television receiver that claims to cover 100 kHz to 1.7 GHz. I don't have an antenna to bring in 100 kHz (!) but the little thing works on all the ham bands. There aren't many features, and not much in the way of useful filtering, but you can listen to a lot.

E7F10 (A) What aspect of receiver analog-to-digital conversion determines the maximum receive bandwidth of a Direct Digital Conversion SDR?

    A. **Sample rate**

    B. Sample width in bits

    C. Sample clock phase noise

    D. Processor latency

Basic information theory tells us that the more information we want to transmit, the more bandwidth we need, and the more bandwidth we want to hear, the more information we'll need to process; so, the maximum receive bandwidth of a Direct Digital Conversion Software Defined Radio, or SDR, is set by the **sample rate**.

E8A09 (D) How many different input levels can be encoded by an analog-to-digital converter with 8-bit resolution?

    A. 8

    B. 8 multiplied by the gain of the input amplifier

    C. 256 divided by the gain of the input amplifier

    D. **256**

To calculate how many of anything can be represented by a given number of bits, we raise the number 2 to the power of the number of bits. (There's a key for that on your TI-30XS. It's the "[^]" key in the left-hand row under the π key.) 2^8 (2 [^] 8 on your calculator) is **256**.

E7F06 (D) What is the minimum number of bits required for an analog-to-digital converter to sample a signal with a range of 1 volt at a resolution of 1 millivolt?

    A. 4 bits

    B. 6 bits

    C. 8 bits

    D. **10 bits**

To sample a signal with a range of 1 volt at a resolution of 1 millivolt, we need to be able to represent all the numbers from 0 to 1000.

In binary, we represent everything with 1's and 0's, so we need to be able to write the number 1000 in binary. We can do some jazzy math on our calculator to figure this out – we need the $\log_2$ (the base 2 logarithm) of 1000. Unfortunately, most calculators don't do $\log_2$ with the push of a button – their logs are all base 10. But if we really, really need to calculate this, it's:

$$\frac{\log_{10}(1000)}{\log_{10}(2)} = 9.966$$

We divide the log of the number we want to represent by the log of the base we want to represent it in. Of course, we can't have 9.966 bits, we need **10 bits**.

Or, you could do what I did. "Huh …1000 is about …….lemme see …. 2^9? Nope, that's 512. Must be 2^{10}. Yeah, that's 1024. We need **10 bits** to do this." I know it isn't very elegant, but there are no bonus points for elegance.

E7F11 (B) What sets the minimum detectable signal level for a direct-sampling SDR receiver in the absence of atmospheric or thermal noise?

    A. Sample clock phase noise

    B. **Reference voltage level and sample width in bits**

    C. Data storage transfer rate

    D. Missing codes and jitter

First, believe it or not, that phrase, "in the absence of atmospheric or thermal noise" is designed to simplify this for you. It means you don't need to take noise floor into consideration, you just need to have some knowledge of how analog-to-digital converters work.

The A-to-D converter's job is to measure the incoming signal and assign each sample a value. To do that, it compares each sample to a reference voltage, which represents the highest value it is possible for the system to represent.

As you might imagine, that reference voltage is rather critical. The reason it is critical is that each possible numeric value for a sample reflects some fraction of that voltage. The range of values that a system can successfully sample depends very much on that voltage *and* on the sample width. Let's look at a couple of extreme examples.

Let's say we have a reference voltage of 3.3 volts, which is pretty typical, and a sample width of 4 bits, which is definitely not typical. A sample width of 4 bits means we can represent 16 different values. The voltage represented by each sample will be in multiples of 1/16 of 3.3, or $\approx 206\ mV$. That means the minimum detectable signal for this system is about 206 millivolts. Anything lower than that is 0.

Let's raise our sample width to 16 bits. Now each sample will be representing values of 1/65,536 of 3.3 volts, or $\approx 50\ microvolts$. This system's minimum detectable signal is about 50 microvolts; quite an improvement over that 4-bit, 206 millivolt system!

What if the reference voltage is 5 volts instead of 3.3? In the 4-bit system, that makes the minimum detectable signal even higher, at $312\ mV$. The 16-bit system deteriorates to a minimum detectable signal of 76 microvolts.

Can we make things better by lowering the reference voltage? For weak stations, yes, but that leaves us with less headroom for strong signals; in other words, less dynamic range. However, it would improve the figure for minimum detectable signal.

You can see that the two critical factors in minimum detectable signal in a direct sampling SDR receiver are **reference voltage level and sample width in bits.**

E8A02 (A) Which of the following is a type of analog-to-digital conversion?

    A. **Successive approximation**

    B. Harmonic regeneration

    C. Level shifting

    D. Phase reversal

Successive approximation is a widely used analog-to-digital conversion system.

So far as an A-to-D converter is concerned, an analog signal is simply a series of unknown voltages. The converter needs some way to measure and quantify each voltage. Successive approximation achieves that by storing that incoming voltage in a capacitor, then comparing it in a systematic way to known voltages. It is something like a game of 20 questions. Imagine the highest voltage an imaginary A-to-D converter can convert is 8.000v and the incoming voltage is 6.91v. The comparisons would go like, "Is it over 5.000?" "Yep." "Is it over 7.50?" "Nope." (We now know it's between 5 and 7.5.) "Is it over 6.25?" "Yep." And so it goes until it zeroes in on a match. The component that does the comparing is the one we covered back on page 201, the comparator.

E8A11 (A) Which of the following is a measure of the quality of an analog-to-digital converter?

    A. **Total harmonic distortion**

    B. Peak envelope power

    C. Reciprocal mixing

    D. Power factor

We're constantly looking for linearity from our electronics, and a good measure of linearity in any device – A-to-D converter or not – is **total harmonic distortion**, or THD. **Total harmonic distortion** is the ratio of the sum of the power of all harmonics of the input signal to the power of the fundamental frequency of the signal. It can be expressed as a percentage or in dB.

E7F03 (C) What type of digital signal processing filter is used to generate an SSB signal?
   A. An adaptive filter
   B. A notch filter
   C. **A Hilbert-transform filter**
   D. An elliptical filter

A Hilbert-transform filter is not like the filters we talked about in the sections on resonant circuits. This filter is a calculus-based algorithm in a computer program.

We'll talk about a couple of *transforms* as we cover digital signal processing. Transforms are calculus formulas that are typically used to either analyze complex events or create complex events.

Understanding the textbook definitions of these transforms requires at least a year of calculus, but we can understand the functions with nothing more than a little imagination. So, imagine you have two shoeboxes full of signal. One is what's left over from a high pass filtering of a signal – so it's the low part. The other is what's left after a low pass filtering of the same signal. The Hilbert transform can mathematically use either of those shoeboxes full of signal to recreate the other shoebox full of signal – which, if you think about it, is precisely what we need to do to demodulate an SSB signal.

And, all that math is reversible, so we can use it to, in essence, dismantle a signal to create a single-sideband signal as well. That's what a **Hilbert-transform filter** does, and unlike the other transform mentioned in the exam, the only use we have for a Hilbert transform is the detection or creation of an SSB signal.

Key Concepts in This Chapter

- In direct digital conversion, incoming RF is digitized by an analog-to-digital converter without being mixed with a local oscillator signal.

- A direct or flash conversion analog-to-digital converter be useful for a software defined radio because its very high speed allows digitizing high frequencies.

- To be able to accurately reproduce an analog signal, the signal must be sampled by an analog-to-digital converter at at least twice the rate of the highest frequency component of the signal.

- The aspect of receiver analog-to-digital conversion that determines the maximum receive bandwidth of a Direct Digital Conversion SDR is the sample rate.

- An analog-to-digital converter with 8 bit resolution can encode 256 levels.

- The minimum number of bits required for an analog-to-digital converter to sample a signal with a range of 1 volt at a resolution of 1 millivolt is 10 bits.

- The minimum detectable signal level for an SDR (in the absence of atmospheric or thermal noise) is set by the reference voltage level and sample width in bits.

- Successive approximation is a type of analog-to-digital conversion.

- A key measure of the quality of an analog-to-digital converter is total harmonic distortion.

- The type of digital signal processing filter used to generate an SSB signal is a Hilbert-transform filter.

Chapter 44

Digital Signal Processing

E7F02 (A) What kind of digital signal processing audio filter is used to remove unwanted noise from a received SSB signal?

    A. **An adaptive filter**
    B. A crystal-lattice filter
    C. A Hilbert-transform filter
    D. A phase-inverting filter

An adaptive filter is one that adapts to changing levels and types of noise in the incoming signal. They're not limited to SSB signals, though the question sort of implies they are. They're not even limited to radio – I use a digital adaptive filter in my sound studio to remove the last little vestiges of room noise from my audio recordings.

The adaptive filter works in the digital domain. It analyzes the incoming signal, figures out what's noise and what isn't, then deletes the noise and keeps the desired signal. At least you hope that's what it does. It very much depends on the quality of the algorithm driving the filter.

E7F07 (C) What function is performed by a Fast Fourier Transform?

    A. Converting analog signals to digital form
    B. Converting digital signals to analog form
    C. **Converting digital signals from the time domain to the frequency domain**
    D. Converting 8-bit data to 16-bit data

Here's our other transform, the Fast Fourier Transform. Joseph Fourier was a French mathematician who was not at all interested in radio, nor electricity, nor sound, nor even frequencies, but the transfer of heat. In the course of studying that, he came up with the Fourier Transform to help analyze heat transfer.

Other folks came up with ways to apply it to asteroid orbits and x-ray crystallography. Then in 1965, a couple of mathematical gents named Cooley and Tukey invented what we now know as the Fast Fourier Transform. They didn't care about radio, either. They wanted it to analyze seismographic data to detect Soviet nuclear tests.

Today it is one of the single most important pieces of mathematics in digital signal processing of all sorts. It's called the Fast Fourier Transform simply because it *is* fast – it eliminates a lot of operations from the Fourier Transform, so it can be done by a computer at lightning speed.

Here's what it does. It analyzes a signal and converts it **from the time domain to the frequency domain.** Here's what that means in plain English and pictures.

Your digital signal processor has a problem. All it "knows" at any given instant is something like this: 10110110. That's it, and it's supposed to do something useful to that. However, it doesn't really "know" *anything* except the amplitude of one sample from that. It doesn't know what frequency that is part of, nor what it is surrounded by that would give it a clue to that. That's where the Fast Fourier Transform goes to work. It takes a bunch of samples and makes "sense" of them for the digital signal processor so it knows what to do with that individual

Figure 44.1: Joseph Fourier

sample and can work its magic. Then it can do its calculations and figure out, "Aha. I need to change that 10110110 to 10110111."

Figure 44.2 shows a complex waveform – about a half-second's worth of audio.

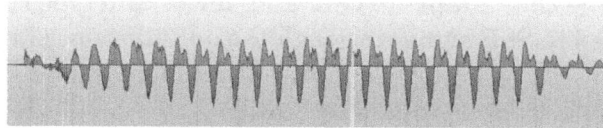

Figure 44.2: Complex Waveform Represented in Time Domain

On that graph of that waveform, the X axis, the horizontal axis, represents time. The Y axis, the vertical axis, represents amplitude. The signal is represented in the time domain. Keep in mind, that half-second contains some 22,000 samples.

Figure 44.3 shows that same complex waveform represented in the frequency domain.

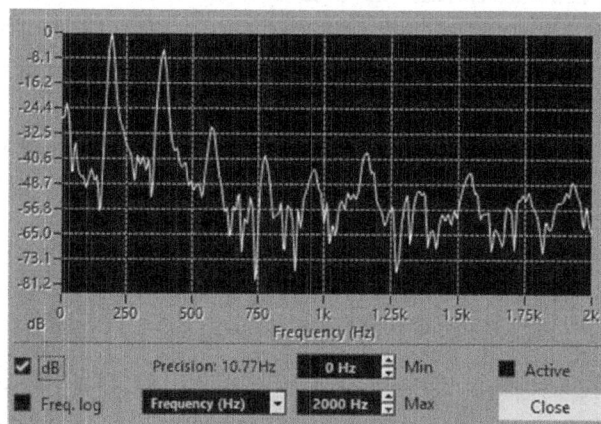

Figure 44.3: Complex Waveform Represented in Frequency Domain

In that graph, the X axis represents frequency, while the Y axis still represents amplitude. So we can see that just below 250 Hz, there's a peak at 0 dB, and another just a little lower than 0 dB at about 400 Hz, a big dip at 750 Hz, etc. That's the output of my audio software's spectrum analyzer, and it's run by a Fast Fourier Transform. (Boy, is it ever fast – I just ran a 10 minute sample through that analyzer, about 26,000,000 samples worth. It took less than a half-second to do all that analysis.)

Of course, in our digital signal processor, that Fast Fourier Transform isn't outputting a graph, it's just outputting the digits that make up that graph so that they're available for other pieces of software to process in the digital domain.

E7F08 (B) What is the function of decimation?
    A. Converting data to binary code decimal form
    B. **Reducing the effective sample rate by removing samples**
    C. Attenuating the signal
    D. Removing unnecessary significant digits

The function of decimation is **reducing the effective sample rate by removing samples**.

Why would we want to do that? We just spent all that money and effort getting the sample rate *up*!

Ah, but even though we are rich in samples, we are poor in processing power compared to the number of samples.

It's one thing to sample a signal at a high rate, it's quite another to process all those samples through the very complex algorithms of a digital signal processor. Those Fast Fourier Transforms are fast, but they're not THAT fast – and that's just one process among many. So we need to reduce the number of samples to be processed, and that's what a decimator does.

E7F09 (A) Why is an anti-aliasing digital filter required in a digital decimator?
    A. **It removes high-frequency signal components that would otherwise be reproduced as lower frequency components**
    B. It peaks the response of the decimator, improving bandwidth
    C. It removes low-frequency signal components to eliminate the need for DC restoration
    D. It notches out the sampling frequency to avoid sampling errors

Remember aliasing from the section on oscilloscopes? This is that same aliasing, caused by an insufficient sample rate.

Aliasing is signals that appear to be part of the original signal but are not.

Aliasing is created by a sampling rate too low for the sampled signal. Since the decimator's whole job is to reduce the effective sample rate, the more we decimate the more aliasing will occur. Here's how it works. We'll start with a simple sine wave, as shown in Figure 44.4.

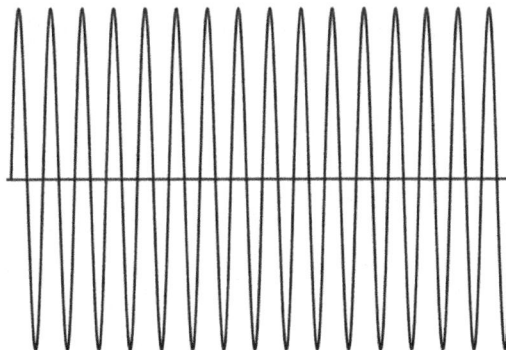

Figure 44.4: Sine Wave

We'll say our sample rate is the bare minimum, two samples per cycle. The gray and white boxes in Figure 44.5 represent the samples.

From those samples, we can faithfully recreate the original signal, at least in theory.

If we cut the sampling rate to a third of the minimum, we end up getting samples like those in Figure 44.6.

Uh oh ... now when we convert back to analog, the digital-to-analog converter doesn't see three half-cycles in each of those samples, it sees the *average* of the three half-cycles. We end up with something like Figure 44.7 ...a signal at one-third the original frequency and two-thirds the original amplitude.

The signal is traveling under an alias, masquerading as something legitimate, but it is a fake! To get rid of those aliases, an anti-aliasing filter **removes high-frequency signal components which would otherwise be reproduced as lower frequency components.**

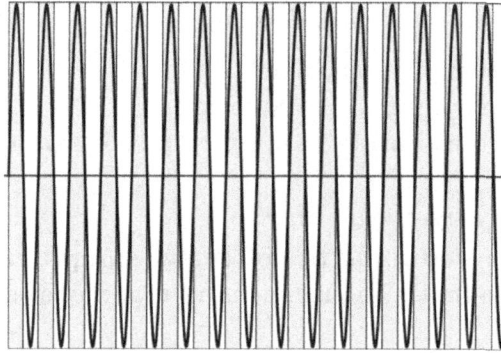

Figure 44.5: Sine Wave Sampled at Minimum Sample Rate

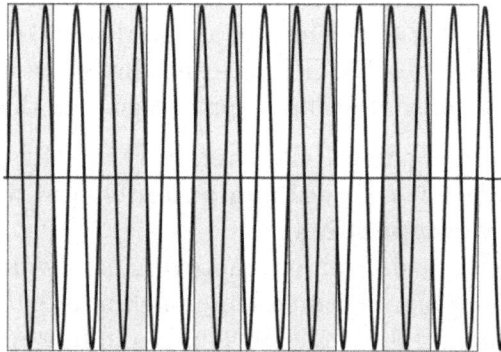

Figure 44.6: Sine Wave Sampled at $\frac{1}{3}$ Minimum Sample Rate

E8A10 (C) What is the purpose of a low-pass filter used in conjunction with a digital-to-analog converter?

 A. Lower the input bandwidth to increase the effective resolution

 B. Improve accuracy by removing out-of-sequence codes from the input

 C. **Remove harmonics from the output caused by the discrete analog levels generated**

 D. All these choices are correct

Even with a high sample rate, we still don't have a perfect representation of the original signal – our picture still has "steps" in it that weren't there before the digitization process. Those are what the answer calls "discrete analog levels," as seen in Figure 44.8.

The good news for us creators of digital magic is that those steps happen at a very high and very predictable frequency. So far as electronics can see, the steps are just high frequency signals – harmonics. So by adding a low-pass filter to our digital-to-analog converter, we filter out those high frequency "signals" and are left with the original wave.

E7F04 (D) What is a common method of generating an SSB signal using digital signal processing?

 A. Mixing products are converted to voltages and subtracted by adder circuits

 B. A frequency synthesizer removes the unwanted sidebands

 C. Varying quartz crystal characteristics emulated in digital form

 D. **Signals are combined in quadrature phase relationship**

Quadrature Modulation, which is what this question is asking about, is created by mixing two modulated RF signals with the same frequency, but 90° out of phase. That's why it's called "quadrature" – 90° is one-fourth of the way around a circle. The proportion of each signal at any given instant determines the final waveform.

Quadrature modulation is not a type of modulation, such as Amplitude Modulation or Angle Modulation. It's a modulation system, used to produce a modulated carrier of whatever type is needed.

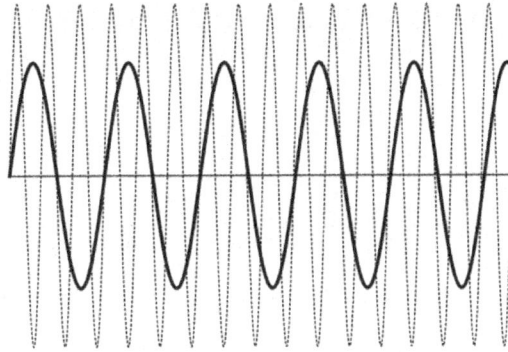

Figure 44.7: Aliasing

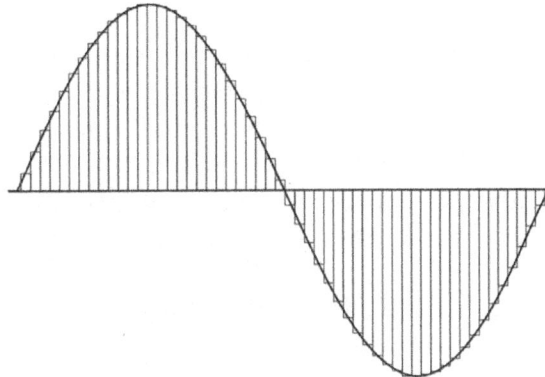

Figure 44.8: Discrete Analog Levels

The sums of those two signals create a new waveform that can be both amplitude modulated and phase modulated, so a quadrature modulator can produce any type of modulation.

E7F13 (D) What is the function of taps in a digital signal processing filter?

A. To reduce excess signal pressure levels

B. Provide access for debugging software

C. Select the point at which baseband signals are generated

D. **Provide incremental signal delays for filter algorithms**

A lot of the work of digital signal processing is accomplished through various phase manipulations, which means we need a way to manipulate signals across time. In analog systems, we can use reactances to create phase shifts, and thus filters – but we don't have any coils or capacitors in our digital domain, so we use delays.

Think of taps as "outputs that happen after set amounts of delay." See Figure 44.9.

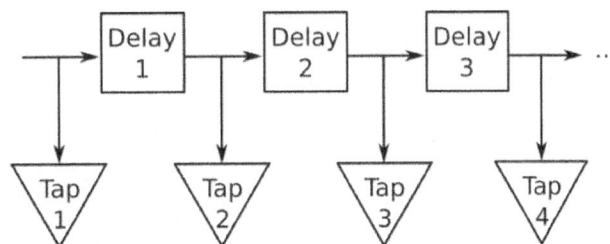

Figure 44.9: Taps

Taps **provide incremental signal delays for filter algorithms**.

E7F14 (B) Which of the following would allow a digital signal processing filter to create a sharper filter response?

    A. Higher data rate

    B. **More taps**

    C. Complex phasor representations

    D. Double-precision math routines

More taps are the equivalent of higher Q in analog resonant circuits.

E7F12 (A) Which of the following is an advantage of a Finite Impulse Response (FIR) filter vs an Infinite Impulse Response (IIR) digital filter?

    A. **FIR filters can delay all frequency components of the signal by the same amount**

    B. FIR filters are easier to implement for a given set of passband rolloff requirements

    C. FIR filters can respond faster to impulses

    D. All these choices are correct

Digital signal processors contain two types of filters, FIR filters and IIR filters. FIR stands for Finite Impulse Response, IIR for Infinite Impulse Response.

Those names don't tell you much, do they? Their actual functions and uses are all wrapped up in the calculus of digital signal processing, and the two filters are not really opposites so much as just useful for different functions. Put as simply as possible, the FIR outputs a "moving average" of its inputs, while an IIR is used for more complex outputs that can be thought of as software driven.

One characteristic of FIR is that they **delay all frequency components of the signal by the same amount**, while IIR's can delay different frequency components for different amounts of time.

E8A04 (B) What is "dither" with respect to analog-to-digital converters?

    A. An abnormal condition where the converter cannot settle on a value to represent the signal

    B. **A small amount of noise added to the input signal to allow more precise representation of a signal over time**

    C. An error caused by irregular quantization step size

    D. A method of decimation by randomly skipping samples

Dither is **a small amount of noise added to the input signal to allow more precise representation of a signal over time**. We use dither when we reduce bit depth.

It sounds quite paradoxical, doesn't it? We *add* noise to get a more precise representation of a signal? Yes, in fact, that is how it works. Let's do an extreme example.

First, be sure you're clear about the difference between sample rate and bit depth. Sample rate is "how often do we sample?" Bit depth is "how many different integers can we use to represent the sample?"

Let's say we start with a sine wave sampled with a bit depth of 16 bits. We can represent each sample's amplitude with any number from 0 to 65,536, so we can create a quite accurate representation of that waveform. You can see the result in Figure 44.10.

In Figure 44.10, we have almost no rounding errors at all – we had just the right number to represent each amplitude. We run our samples through the digital to analog converter, the original wave is accurately reproduced, and wow, high fidelity – linearity -- happens. Nifty!

Now we'll go insane and drop the bit depth to just two. Now we can only use one of five integers, 0 through 4, to represent an amplitude. If we don't have the right number, we have to round down or round up. Now our samples look like Figure 44.11.

We run those samples through the D-to-A converter and we get something like the mess in Figure 44.12 back out.

Ewwww! What you're looking at in Figure 44.12 is the result of what is called "quantization error" – an engineer's way of saying "rounding error."

From this point, there's really no getting 100% of that original signal back, but we can definitely make this ugly situation a bit better, and we do that by adding back in some noise.

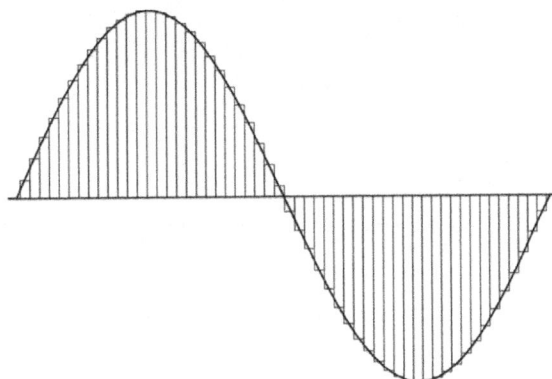

Figure 44.10: Sine Wave Sampled with Bit Depth of 16 Bits

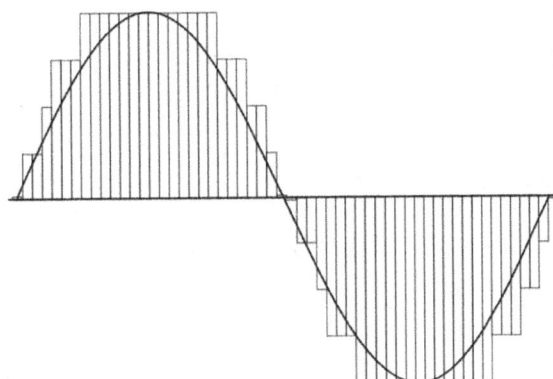

Figure 44.11: Sine Wave Sampled with Bit Depth of 4 Bits

Noise is random, so it tends to smooth out those graph lines, and we get something like Figure 44.13.

What we have really done is traded in our distortion for some noise. We haven't "created information", we've actually destroyed some information with information's opposite, which is noise.

Dithering can be applied to any digital representation of something, even of images.

If this is still a little baffling, and you'd like to explore more than you need for the exam, let me direct you to this YouTube video, put out by iZotope, a company that is, for my money, the absolute best at digital audio processing for us audio producer types.

https://www.youtube.com/watch?v=vVNzylf9sGo&feature=youtu.be

You'll even get to see dithering applied to photographs.

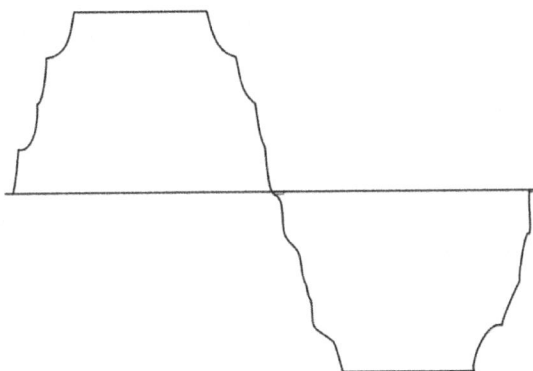

Figure 44.12: Output of Digital to Analog Conversion of Undersampled Sine Wave

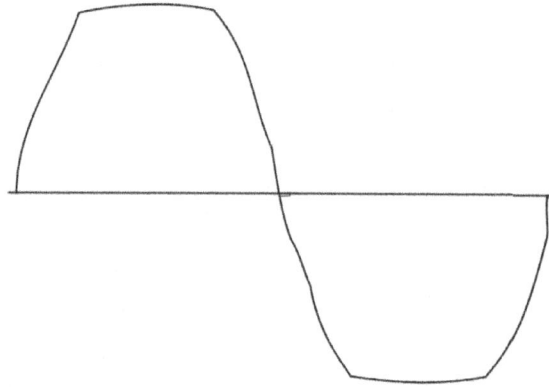

Figure 44.13: Output of Digital to Analog Conversion After Dithering

Bonus Pro Tip: Do not brag or even divulge to your friends, your spouse, or your spousal equivalent that you have been studying dithering. No good will come of it. Trust me.

Key Concepts in This Chapter

- Digital signal processing audio filters used to remove unwanted noise from a received SSB signal are adaptive filters.

- A Fast Fourier Transform can convert digital signals from the time domain to the frequency domain.

- The function of decimation with regard to digital filters is reducing the effective sample rate by removing samples.

- An anti-aliasing filter is required in a digital decimator because it removes high-frequency signal components which would otherwise be reproduced as lower frequency components.

- The purpose of a low-pass filter used in conjunction with a digital-to-analog converter is to remove harmonics from the output cause by the discrete analog levels generated.

- A common method of generating an SSB signal using digital signal processing is to combine signals with a quadrature phase relationship.

- The function of taps in a digital signal processing filter is to provide incremental signal delays for filter algorithms. More taps allow a digital signal processing filter to create a sharper filter response.

- An analog signal must be sampled by the analog-to-digital converter at twice the rate of the highest frequency component of the signal so that the signal can be accurately reproduced.

- An advantage of a Finite Impulse Response (FIR) filter vs. an Infinite Impulse Response (IIR) digital filter is that FIR filters delay all frequency components of the signal by the same amount.

- Dither is a small amount of noise added to the input signal to allow more precise representation of a signal over time.

Chapter 45

Op-Amps and Op-Amp Circuits

Op-amps are a type of tremendously versatile integrated circuit – you can buy whole books of projects featuring op-amps. Using op-amps, we can create audio amplifiers, voltage comparators, filters, rectifiers, peak detectors, analog to digital and digital to analog converters, oscillators, and more. They come in several packages or "form factors." They can be in metal cans, as shown in Figure 45.1.

Figure 45.1: Op-Amp in a Can

Or they can be in DIP or surface-mount packages.

Op-amp is short for "operational amplifier." That name actually has almost nothing to do with how we use them today. Originally, they were invented to perform the operations – addition, subtraction, etc. – in analog computers. Well, analog computers are rare items these days, but it's hard to purchase an electronic device without at least one op-amp inside.

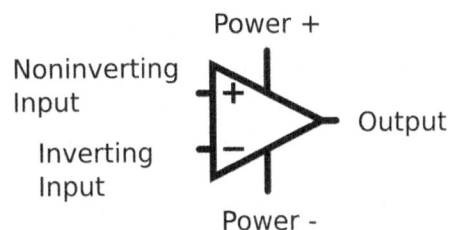

Figure 45.2: Op-amp Schematic Symbol

The op-amp is a differential amplifier. It detects and amplifies the difference between two voltages. Figure 45.2 shows a common schematic symbol of an op-amp.

Some op-amps have more inputs or outputs, but this one is by far the most common. The op-amp detects the difference between the two inputs, known as the "noninverting" and the "inverting" input, and outputs that difference. As a straight amplifier, then, one input would be, say, some audio, the other input would be grounded, and the output would be equal to the input plus the gain of the op-amp.

By the way, in rare instances – such as on this exam! – you'll see a symbol for an op-amp that looks like Figure 45.3.

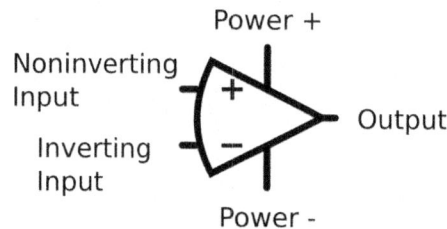

Figure 45.3: Alternative Op-amp Symbol

That's the same op-amp, just a slightly different symbol.

An "ideal" op-amp – the kind that only exists in the slightly fevered minds of engineers – has infinite gain, infinite input impedance, and zero output impedance. In reality, op-amps have high gain, very high input impedance, and very low output impedance, so in those areas they get pretty close to the ideal. Because of that nearly infinite input impedance, they can be "direct coupled" – we don't need any coupling capacitors or transformers to isolate them from the circuits that are ahead of them in the device.

That ideal op-amp also has infinite bandwidth, zero phase shift, zero noise, and some other supposed zero or infinite characteristics, but real op-amps fall far short of those ideal characteristics. Most important for the exam, they have limited bandwidth.

They're also not high-power devices. You won't be running an op-amp as the final of your 1500-watt HF amplifier.

Op-Amp Characteristics

E7G12 (A) What is an operational amplifier?

A. **A high-gain, direct-coupled differential amplifier with very high input impedance and very low output impedance**

B. A digital audio amplifier whose characteristics are determined by components external to the amplifier

C. An amplifier used to increase the average output of frequency modulated amateur signals to the legal limit

D. A RF amplifier used in the UHF and microwave regions

An integrated circuit operational amplifier – an op-amp – is **a high-gain, direct-coupled differential amplifier with very high input impedance and very low output impedance**.

You're familiar with high and low impedance. Gain is characterized as a multiple; 2× gain, or 47×, for instance. *Direct-coupled* means one stage of the amplifier is connected directly to the next stage, with no capacitor or transformer in between. What's significant about that is that a direct-coupled amplifier can amplify DC. In fact, it is sometimes called a DC amplifier.

E7G01 (A) What is the typical output impedance of an op-amp?

A. **Very low**

B. Very high

C. 100 ohms

D. 1000 ohms

The output impedance of an op-amp is **very low**.

E7G03 (D) What is the typical input impedance of an op-amp?

    A. 100 ohms

    B. 1000 ohms

    C. Very low

    D. **Very high**

Apparently, numbers don't go as high as the input impedance of an op-amp! According to the exam, the typical input impedance of an integrated circuit op-amp is **very high**. It isn't quite infinite, but it is so high that in most designs we can assume it is infinite and that there is no current flow to the input.

E7G04 (C) What is meant by the term "op-amp input offset voltage"?

    A. The output voltage of the op-amp minus its input voltage

    B. The difference between the output voltage of the op-amp and the input voltage required in the immediately following stage

    C. **The differential input voltage needed to bring the open loop output voltage to zero**

    D. The potential between the amplifier input terminals of the op-amp in an open loop condition

One key fact to know here; an op-amp is an "inverting amplifier." That means it turns the input waveform upside down, so a negative voltage at the input makes a positive voltage at the output and vice versa.

That ideal op-amp – no doubt made by the Ideal Component Company, who also make isotropic antennas – has zero input offset voltage; no voltage across the inputs is necessary to bring the output level to zero. In practice, it takes some voltage across the inputs to drive the output voltage to zero. Typically, it is in the range of two volts. It's a little bit like transistor bias, but unlike transistor bias, the need for input offset voltage creates different design problems that we have to take into consideration when designing op-amp circuits.

E7G08 (D) How does the gain of an ideal operational amplifier vary with frequency?

    A. It increases linearly with increasing frequency

    B. It decreases linearly with increasing frequency

    C. It decreases logarithmically with increasing frequency

    D. **It does not vary with frequency**

Caution! They're asking about one of those famous _Ideal_™ op-amps, not a real one. An ideal op-amp's gain **does not vary with frequency**. That ideal op-amp has infinite bandwidth, infinite gain, infinite input impedance, zero output impedance, needs no input offset voltage, and is trustworthy, loyal, helpful, friendly, courteous, kind, obedient, cheerful, thrifty, brave, clean, reverent, careful, quick and kind! That's the op-amp they're asking about in this question.

E7G06 (B) What is the gain-bandwidth of an operational amplifier?

    A. The maximum frequency for a filter circuit using that type of amplifier

    B. **The frequency at which the open-loop gain of the amplifier equals one**

    C. The gain of the amplifier at a filter's cutoff frequency

    D. The frequency at which the amplifier's offset voltage is zero

Now we're talking about a real-world op-amp, not one of those _Ideal_™ op-amps from the last question.

A real op-amp's bandwidth shrinks as its gain increases, something like you see in Figure 45.4. We could also say that Figure 45.4 shows that as frequency increases, gain decreases.

Gain-bandwidth is the answer to the question, "What's the bandwidth of this op-amp when I have just enough gain to have the output equal the input?" Mathematically, it's the gain – in

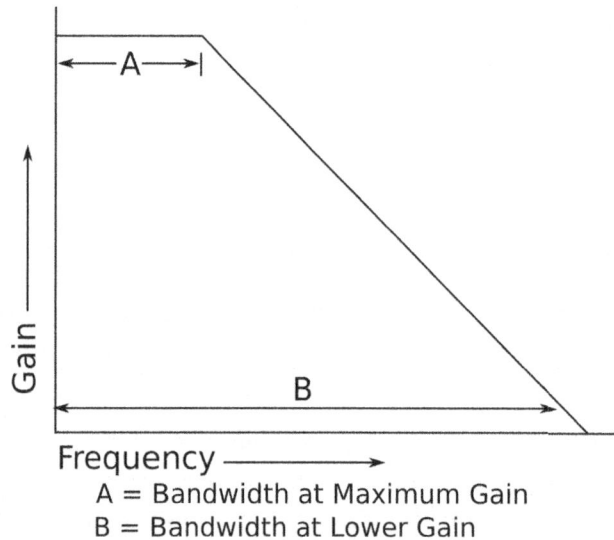

Frequency ─────→
A = Bandwidth at Maximum Gain
B = Bandwidth at Lower Gain

Figure 45.4: Gain-Bandwidth of an Op-Amp

this case a gain of 1 – multiplied by the bandwidth. Gain-bandwidth is, therefore, also known as Gain-Bandwidth Product, or GBP. GBP is a handy number to have if we're designing an op-amp circuit because it lets us easily calculate the bandwidth of an op-amp.

$$GBP = Gain \times Bandwidth$$

So ...

$$Bandwidth = \frac{GBP}{Gain}$$

E7G02 (D) What is ringing in a filter?
   A. An echo caused by a long time delay
   B. A reduction in high frequency response
   C. Partial cancellation of the signal over a range of frequencies
   D. Undesired oscillations added to the desired signal

Filters are resonant circuits and resonant circuits --- well, they resonate! They wouldn't work if they didn't resonate, but that resonance has a side effect, and that's called "ringing."

In a perfect world of ideal circuits, if we put one cycle of a sine wave into a resonant circuit, we'd get one cycle of a sine wave out, and that would be that; something like Figure 45.5.

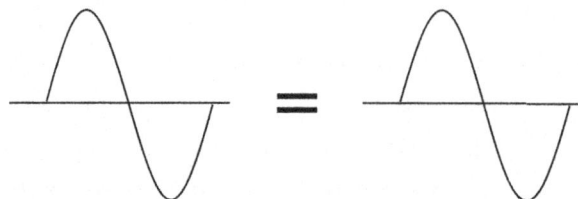

Figure 45.5: No Ringing

Alas, we live in the real world, and in the real world when we put one cycle of a sine wave into a resonant circuit and it rattles around in there for a bit, we end up with something like Figure 45.6.

The higher the Q of the circuit, the more ringing we'll get.

It's the electrical equivalent of a room with lots of reverberation. Look closely, you'll see that there is even a bit of harmonic action happening in that trace above.

290

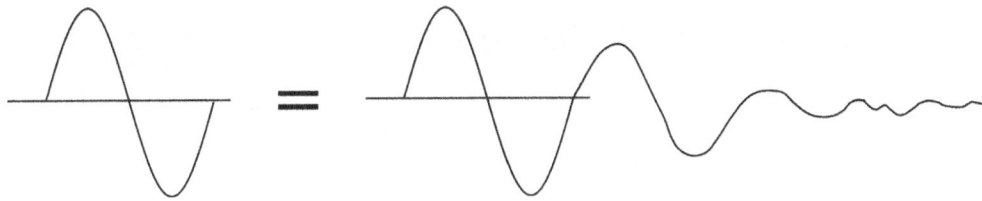

Figure 45.6: Ringing

E7G05 (A) How can unwanted ringing and audio instability be prevented in an op-amp RC audio filter circuit?

A. **Restrict both gain and Q**

B. Restrict gain but increase Q

C. Restrict Q but increase gain

D. Increase both gain and Q

They threw in that question about ringing in filters because op-amps are often used in active filter circuits. Because these circuits involve some amplification, ringing can be even more challenging to control than in passive filters.

In a passive filter, we control ringing by keeping the Q as low as possible.

In an active filter that uses op-amps, we control unwanted ringing and audio instability by **restricting both gain and Q**.

Next, we have four questions about Figure E7-3. E7-3 is just a very standard op-amp circuit, and three of the questions are really the same question with different numbers plugged in. One has a little extra twist. They're all about the gain of the op-amp.

Op-amp gain is set by two resistors, one ahead of the noninverting input and the other looping back from the output to the noninverting input. The first resistor in all these problems is called R_1, the second is R_F, the F being for "feedback." The ratio between those two resistors, R_F divided by R_1, is the gain of the op-amp.

Op-Amp Calculations

E7G07 (C) What magnitude of voltage gain can be expected from the circuit in Figure E7-3 when R1 is 10 ohms and RF is 470 ohms?

A. 0.21

B. 94

C. **47**

D. 24

We set the gain for an op-amp by changing the values of the two resistors you see in the schematic in Figure 45.7, R_1 and R_F. The ratio between the values of those two resistors will equal the gain.

$$Gain = \frac{R_F}{R_1}$$

You can keep this formula straight in your mind by thinking about the function of R_1 and remembering that an op-amp is a differential amplifier – it's going to amplify the difference between the non-inverting input and the inverting input. The higher the value of R_1, the less difference there is between those inputs, so R1 reduces gain. The higher the value of R_1 in the formula, the lower the gain.

Figure E7-3

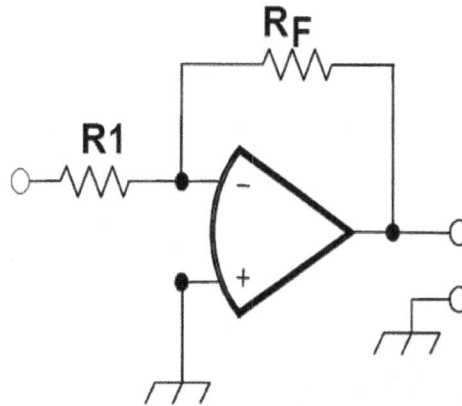

Figure 45.7: Figure E7-3 in the question pool

For the question, then, it's

$$Gain = \frac{R_F}{R_1} = \frac{470\,\Omega}{10\,\Omega} = 47 \times \ Gain$$

Notice that's the next-to-highest value in the possible answers for this op-amp gain question.

E7G10 (C) What absolute voltage gain can be expected from the circuit in Figure E7-3 when R1 is 1800 ohms and RF is 68 kilohms?

    A. 1

    B. 0.03

    C. **38**

    D. 76

$$Gain = \frac{R_F}{R_1} = \frac{68,000\,\Omega}{1800\,\Omega} = 37.77 \approx 38 \times \ Gain$$

38 is the next-to-highest value in this op-amp gain question, too.

E7G11 (B) What absolute voltage gain can be expected from the circuit in Figure E7-3 when R1 is 3300 ohms and RF is 47 kilohms?

    A. 28

    B. **14**

    C. 7

    D. 0.07

$$Gain = \frac{R_F}{R_1} = \frac{47,000\,\Omega}{3300\,\Omega} = 14.24 \approx 14 \times \ Gain$$

Whaddaya know? **14** is the next-to-highest value in this op-amp gain question, too. That won't be true for the next question, but it's not purely a gain question.

E7G09 (D) What will be the output voltage of the circuit shown in Figure E7-3 if R1 is 1000 ohms, RF is 10,000 ohms, and 0.23 volts DC is applied to the input?

    A. 0.23 volts

    B. 2.3 volts

    C. -0.23 volts

    D. **-2.3 volts**

Here's how we proceed for this question. First, we calculate the gain.

$$Gain = \frac{R_F}{R_1} = \frac{10,000\ \Omega}{1000\ \Omega} = 10 \times Gain$$

Our op-amp is set up to provide 10x gain. We're applying 0.23 volts DC to the input – that's +0.23 volts.

Key fact for this question: an op-amp is an *inverting amplifier*. It reverses the polarity of whatever hits the input – it "inverts" the input. We put +0.23 volts into a gain of 10, so when that inverts, we get **-2.3 volts** at the output. They give us the wrong choice of 2.3 volts, so be sure you get that inversion in there if you get this question.

Key Concepts in This Chapter

- An operational amplifier is a high-gain, direct-coupled differential amplifier with very high input impedance and very low output impedance.

- The typical output impedance of an integrated circuit op-amp is very low. The input impedance is typically very high.

- The term op-amp input offset voltage refers to the differential input voltage needed to bring the open loop output voltage to zero.

- The gain of an ideal operational amplifier does not vary with frequency.

- The gain-bandwidth of an operational amplifier is the frequency at which the open-loop gain of the amplifier equals one.

- The effect of ringing in a filter is undesired oscillations added to the desired signal.

- Unwanted ringing and audio instability can be prevented in an op-amp RC audio filter circuit by restricting both gain and Q.

- The formula for calculating the gain of an op-amp is the value of the resistor between the output and the input, called RF, divided by the value of the resistor between the source and the input. called R1. Gain equals RF divided by R1.

$$Gain = \frac{R_F}{R_1}$$

- In Figure E7-3, when R1 is 10 ohms and RF is 470 ohms, the gain is 47.

- In Figure E7-3, when R1 is 1800 ohms and RF is 68 kilohms, the absolute voltage gain is 38.

- In Figure E7-3, when R1 is 3300 ohms and RF is 47 kilohms, the gain is 14.

- The output voltage of the circuit shown in Figure E7-3 if R1 is 1000 ohms and RF is 10,000 ohms, and 0.23 volts DC is applied to the output is -2.3 volts. A key fact to remember for this question is that an op-amp is an inverting amplifier.

Once you complete the Practice Exam for this chapter, you'll be ready for Progress Check #9.

Progress Check #9

Chapter 46

Oscillators and Signal Sources

Analog Frequency Synthesis

That famous ideal resonant circuit, with no resistance or other loss anywhere in the circuit, would resonate forever. In the real world, to make a passive resonant circuit into an oscillator, we have to take a bit of the resonating signal and feed it back into an amplifier of some sort to overcome the losses in the circuit. That signal that gets fed back is called positive feedback. The source of that positive feedback signal determines what sort of oscillator the oscillator is.

E7H01 (D) What are three oscillator circuits used in amateur radio equipment?
A. Taft, Pierce and negative feedback
B. Pierce, Fenner and Beane
C. Taft, Hartley and Pierce
D. **Colpitts, Hartley and Pierce**

It helps to know that the committee was having a bit of fun with the wrong answers to this one. The "Taft-Peirce" company (yes, Peirce) was an early sewing machine manufacturer. "Pierce, Fenner and Beane" is part of the original name of the Merrill Lynch stock brokerage. The "Taft-Hartley Act" is a 1947 law that deals with labor relations and labor unions.

The correct answer is the only one with the name **Colpitts** in it. Three oscillator circuits used in Amateur Radio equipment are **Colpitts, Hartley and Pierce**

E7H04 (C) How is positive feedback supplied in a Colpitts oscillator?
A. Through a tapped coil
B. Through link coupling
C. **Through a capacitive divider**
D. Through a neutralizing capacitor

Take a quick look at the simplified schematic for a typical Colpitts oscillator, Figure 46.1. We're not going to go through a whole analysis of this thing, but you should be able to quickly spot that it is an LC resonant circuit – a filter -- with a transistor amplifier.

Each of the oscillators we'll look at is a feedback oscillator. When power is applied to it, noise in the circuit gets filtered and amplified over and over and, much like that howling public address system, that noise coalesces into a sine wave at the resonant frequency of the circuit. We feed the filtered frequency into the amplifier, which amplifies it, and then it gets fed back into the filter, and round and round she goes. That's called "positive feedback."

What distinguishes the different sorts of oscillators on the exam is where that positive feedback comes from.

In the case of the Colpitts, it is from a spot between two capacitors, which are acting as a voltage divider. Just remember **Colpitts** oscillators get feedback from **Capacitors**.

(In the schematic, that circle with the downward pointing arrow is the symbol for a current source.)

E7H03 (A) How is positive feedback supplied in a Hartley oscillator?

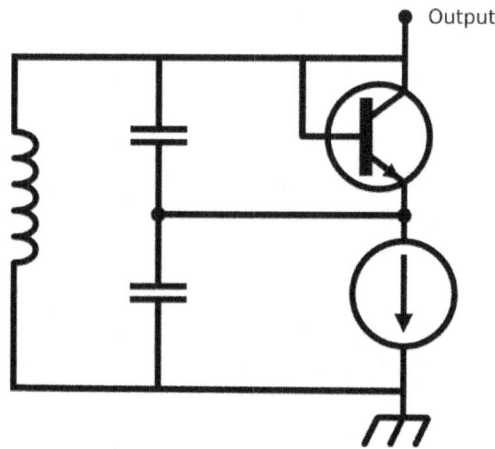

Figure 46.1: Colpitts Oscillator

A. **Through a tapped coil**
B. Through a capacitive divider
C. Through link coupling
D. Through a neutralizing capacitor

Figure 46.2 is a simplified schematic of a Hartley oscillator. It's something of a mirror image of the Colpitts. It gets its positive feedback from a tap on the inductor, rather than a spot between two capacitors.

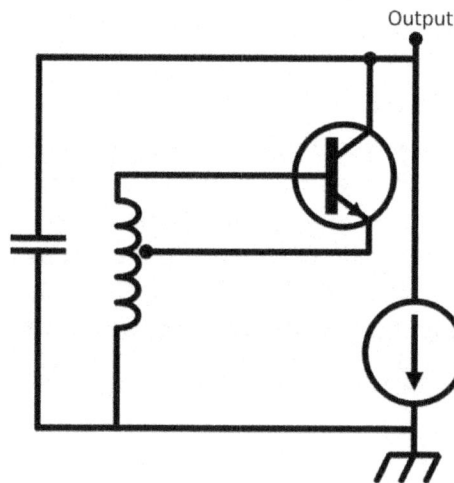

Figure 46.2: Hartley Oscillator

Maybe you can think of the **Hart**ley oscillator as getting its positive feedback from the **Heart** of an inductor, or remember that inductance is measured in **Henrys**.

E7H05 (D) How is positive feedback supplied in a Pierce oscillator?
A. Through a tapped coil
B. Through link coupling
C. Through a neutralizing capacitor
D. **Through a quartz crystal**

The Pierce oscillator uses a **quartz crystal** to provide the positive feedback – and it doesn't necessarily contain a transistor. Instead, many use an inverter – that same inverter you studied in Chapter 36 – that is wired up to act, in effect, as an amplifier in the circuit.

These Pierce oscillators are often used in computers as clocks – they're cheap as can be to add into an integrated circuit, and very accurate.

To remember the Pierce oscillator uses a quartz crystal, picture a crystal, like the one in Figure 46.4.

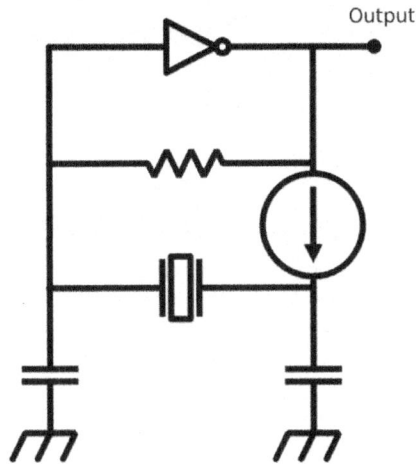

Figure 46.3: Pierce Oscillator

Figure 46.4: Pointy, Sharp, PIERCING! Crystal

It's very pointy and sharp, right? It could ...yes... **Pierce** you!

E7H06 (B) Which of the following oscillator circuits are commonly used in VFOs?

    A. Pierce and Zener

    B. **Colpitts and Hartley**

    C. Armstrong and deForest

    D. Negative feedback and balanced feedback

Colpitts and Hartley oscillator circuits are the oscillators commonly used in VFO's. Since a Pierce oscillator has a fixed, crystal-controlled frequency, it would be a terrible choice for a Variable Frequency Oscillator.

Ham history trivia: Back when the FCC administered ham exams, the examiners would often ask the candidate to draw a schematic of either a Colpitts or Hartley oscillator.

E7H02 (C) What is a microphonic?

    A. An IC used for amplifying microphone signals

    B. Distortion caused by RF pickup on the microphone cable

    C. **Changes in oscillator frequency due to mechanical vibration**

    D. Excess loading of the microphone by an oscillator

Crystals in particular are subject to microphonic interference. Some (cheap!) microphones use a crystal element to take advantage of this very effect. Crystals are, after all, electro-*mechanical* devices, so if they are jostled mechanically, their electrical properties change, and what we end up with is **changes in oscillator frequency due to mechanical vibration**.

Notice the correct answer to this question about microphonics is the only one that doesn't contain the word "microphone."

E7H07 (D) How can an oscillator's microphonic responses be reduced?

    A. Use NP0 capacitors

    B. Reduce noise on the oscillator's power supply

    C. Increase bias voltage

    D. **Mechanically isolate the oscillator circuitry from its enclosure**

When an electronic component responds to mechanical phenomena, such as impact or sound, that component is said to be "microphonic."

No electrical fix will repair microphonic responses. We need a mechanical fix, and that fix is **mechanically isolating the oscillator circuitry from its enclosure**. Typically, rubber or flexible plastic standoffs are used for this.

E7H08 (A) Which of the following components can be used to reduce thermal drift in crystal oscillators?

    A. **NP0 capacitors**

    B. Toroidal inductors

    C. Wirewound resistors

    D. Non-inductive resistors

"NP0" – that's a zero on the end, not the letter "O" – is a ceramic material used to make a particular flavor of ceramic capacitor.

Ceramic capacitors are, at least by some accounts, the single most manufactured electronic component. Inside, they consist of two metal plates separated by a ceramic disc. The characteristics of that ceramic disc govern many of the characteristics of the capacitor.

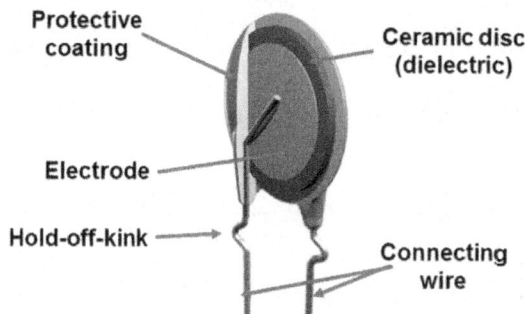

Figure 46.5: Ceramic Capacitor

You'll recall resistors have a temperature coefficient rating – it's how much they change value as they change temperature. So do capacitors, and for **NP0 capacitors**, that rating is zero – they don't change value with temperature. If you want to reduce thermal drift in a crystal oscillator, or any other circuit, go with NP0 capacitors.

E7H12 (B) Which of the following must be done to ensure that a crystal oscillator provides the frequency specified by the crystal manufacturer?

    A. Provide the crystal with a specified parallel inductance

    B. **Provide the crystal with a specified parallel capacitance**

    C. Bias the crystal at a specified voltage

    D. Bias the crystal at a specified current

In Chapter 32 there was a question about a crystal's "equivalent circuit" – it was shown Figure 33.4 on page 211.

For this question, it is important to be clear that C_2 in that schematic is an *equivalent* shunt capacitor – that imaginary circuit does not act like a parallel resonant circuit, it operates like a series resonant circuit.

Crystals can be used in series mode, but virtually all of our applications use crystals in parallel mode. Crystals in series mode have a different reactance characteristic than crystals in parallel mode. The difference looks like Figure 46.6.

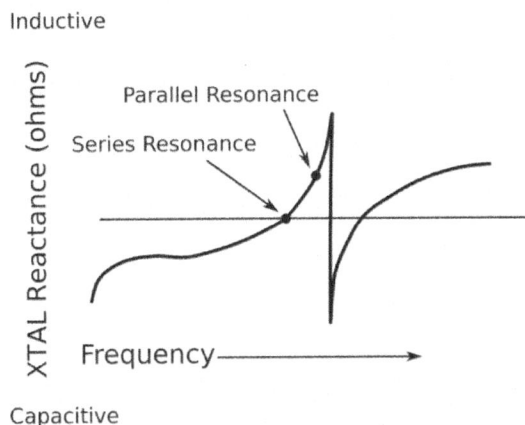

Figure 46.6: Crystal Resonance and Reactance in Series and Parallel

You can see there's a difference in both the frequency of the resonance depending on the mode and in the character of the reactance at resonance. At series resonance, a crystal has zero reactance – what we'd expect of a resonant series circuit. At the higher frequency of parallel resonance, the crystal becomes somewhat inductive, so we need to add a bit of capacitance in parallel to get that reactance back down to zero.

To assure the crystal resonates at the frequency specified by the manufacturer, we must **provide the crystal with a specified parallel capacitance.**

E7H13 (D) Which of the following is a technique for providing highly accurate and stable oscillators needed for microwave transmission and reception?
   A. Use a GPS signal reference
   B. Use a rubidium stabilized reference oscillator
   C. Use a temperature-controlled high Q dielectric resonator
   D. **All these choices are correct**

If we're designing an oscillator for, say, the 40-meter band, somewhere in the neighborhood of 7.5 MHz, and we have a little bit of thermal drift, and a bit of parasitic reactance here and there, and we manage to create a 0.01% frequency error, we're off frequency by 750 Hz. Not great, but not a disaster, and easy enough for which to compensate, so long as we're aware of it.

If we're designing for the 2 GHz range, and we create the same percentage of error, we're now off by 200,000 Hz! 200 kHz! That's ......not good.

Besides, ordinary quartz crystals don't go up in frequency beyond about 30 MHz – after that, we have to use what are called "overtones" – like harmonics, only not quite precisely the fundamental times 2, 3, 4, etc. Once we get into the microwaves, we need some more exotic ways of creating precise frequencies.

We can use a **rubidium stabilized reference oscillator**. That's your own personal atomic clock! It uses the subatomic activities of rubidium to create a very precise time standard. You can get one on e-bay for about $200.

We can also tap into the GPS system for a **GPS signal reference**, so long as we can pick up a GPS signal. Those GPS satellites up there are constantly sending down very precise time signals – that's how the GPS system works. So that's as official a standard as there is – quite literally.

Or we can use a **temperature-controlled high Q dielectric resonator**, which is nothing more than a high-tech version of a cavity filter that lives inside a thermostatically controlled oven.

Digital Frequency Synthesis

E7H09 (A) What type of frequency synthesizer circuit uses a phase accumulator, lookup table, digital to analog converter, and a low-pass anti-alias filter?

    A. **A direct digital synthesizer**

    B. A hybrid synthesizer

    C. A phase locked loop synthesizer

    D. A diode-switching matrix synthesizer

A **direct digital synthesizer** uses a phase accumulator, a lookup table, digital to analog converter, and a low-pass anti-alias filter.

"Lookup table," "digital to analog converter," and "low-pass anti-alias filter" tell you this must be some sort of digital thingamajig. The only answer with "digital" in it is the correct one.

The real giveaway is that "lookup table." A lookup table is sort of a digital version of a wizard's pantry. It contains the "ingredients" for the various waveforms a direct digital synthesizer can create.

E7H10 (B) What information is contained in the lookup table of a direct digital synthesizer (DDS)?

    A. The phase relationship between a reference oscillator and the output waveform

    B. **Amplitude values that represent the desired waveform**

    C. The phase relationship between a voltage-controlled oscillator and the output waveform

    D. Frequently used receiver and transmitter frequencies

The lookup table is analogous to a collection of MP3 files. You tell the radio "give me 144.280 MHz" and it plays a "song" from its library – the lookup table – that consists of one long 144.280 MHz note. Of course, it's not really an MP3 file, but it's similar in that it contains **the amplitude values that represent the desired waveform**.

E7H11 (C) What are the major spectral impurity components of direct digital synthesizers?

    A. Broadband noise

    B. Digital conversion noise

    C. **Spurious signals at discrete frequencies**

    D. Nyquist limit noise

Any time you see a square wave, think "harmonics." In fact, as we'll see shortly, a square wave consists of a sine wave plus all of the odd harmonics of that sine wave. If it looks like Figure 46.7, it is literally, by definition, loaded with harmonics, and odd order harmonics at that, the ones we worry about most.

A direct digital synthesizer can't really produce a sine wave directly. It's a digital device, it only knows about steps, not the smooth lines of analog devices, so it doesn't produce Figure

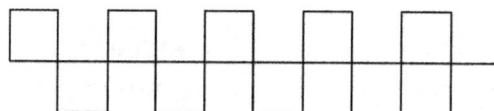

Figure 46.7: Square Wave

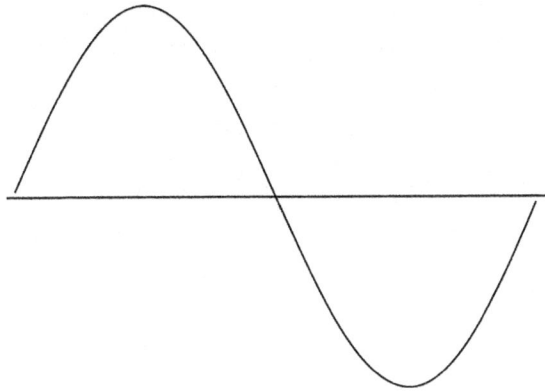

Figure 46.8: Sine Wave

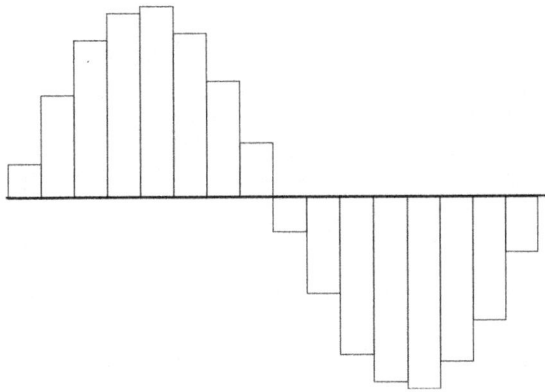

Figure 46.9: Sampled Sine Wave

46.8.

It produces some version of Figure 46.9.

That's nothing but a whole bunch of square waves, all at the same frequency, and that's why **spurious signals at discrete frequencies** are the major spectral impurity component of digital frequency synthesizers. It takes some serious filtering to manage all those spurious signals, but happily, they happen at **discrete**, predictable frequencies, so we know "where to find them."

By the way, if this seems a little familiar, this is the very same phenomenon we covered in the question about spurious mixer products back in Chapter 41.

Phase-Locked Loops

E7H14 (C) What is a phase-locked loop circuit?

A. An electronic servo loop consisting of a ratio detector, reactance modulator, and voltage-controlled oscillator

B. An electronic circuit also known as a monostable multivibrator

C. **An electronic servo loop consisting of a phase detector, a low-pass filter, a voltage-controlled oscillator, and a stable reference oscillator**

D. An electronic circuit consisting of a precision push-pull amplifier with a differential input

A phase-locked loop is a circuit that locks on to a signal's phase and frequency. It's a frequency synthesizer. Whenever we want to precisely match another frequency, a phase-locked loop often serves the purpose.

A phase-locked loop is **an electronic servo loop consisting of a phase detector, a low-pass filter, a voltage-controlled oscillator, and a stable reference oscillator**.

Figure 46.10 is a simplified look at the phase-locked loop.

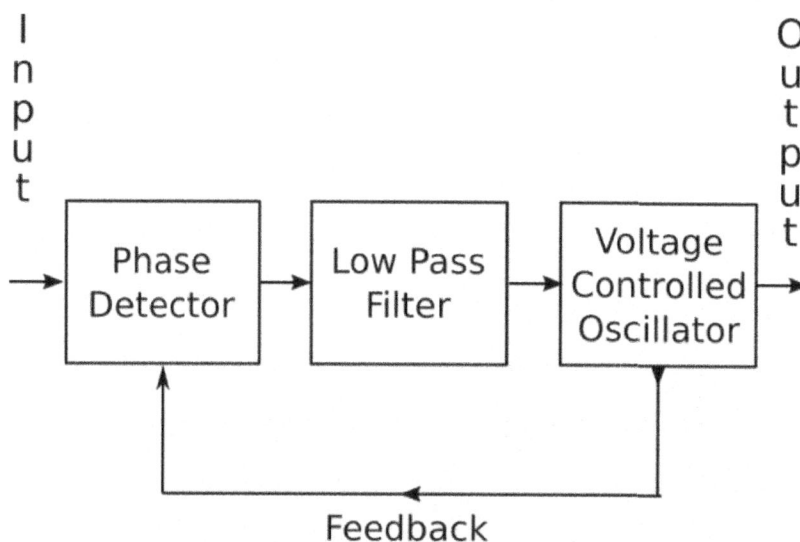

Input → Phase Detector → Low Pass Filter → Voltage Controlled Oscillator → Output

Feedback

Figure 46.10: Phase-locked Loop

The voltage controlled oscillator on the right is humming along, minding its own business, when an input comes in. If we're using the phase-locked loop as a frequency synthesizer, that input would come from the stable reference oscillator referred to in the question, but it could also be an unstable input, such as a frequency modulated radio signal. The phase detector compares the input to whatever the VCO was doing, and creates a control voltage for the VCO that represents the difference between the frequency of the input and the present frequency of the VCO. The low pass filter strips out the original input signal – since we don't really need or want that – and the DC voltage that's left changes the frequency of the VCO, and that output loops back to the phase detector. The phase detector compares the incoming signal from the VCO with the input signal, and that creates another voltage that readjusts the VCO, and around and around it goes until the VCO locks onto that input's phase. (Thus the name, phase-locked loop.)

E7H15 (D) Which of these functions can be performed by a phase-locked loop?

   A. Wide-band AF and RF power amplification

   B. Comparison of two digital input signals, digital pulse counter

   C. Photovoltaic conversion, optical coupling

   D. **Frequency synthesis, FM demodulation**

I've said the phase-locked loop is a frequency synthesizer, but we can also use one for FM demodulation. Here's the simplified version of how that's done.

In frequency modulation, the carrier's frequency is constantly varying, so the modulation envelope is equal to the amount the carrier deviates from the center frequency.

If we take our basic phase-locked loop block diagram ....

```
I                                                    O
n                                                    u
p                                                    t
u                                                    p
t                                                    u
                                                     t
        ┌──────────┐   ┌──────────┐   ┌──────────┐
   ─────│  Phase   │──▶│ Low Pass │──▶│  Voltage │────▶
        │ Detector │   │  Filter  │   │Controlled│
        │          │   │          │   │Oscillator│
        └──────────┘   └──────────┘   └──────────┘
             ▲                              │
             │                              │
             └──────────────◀───────────────┘
                        Feedback
```

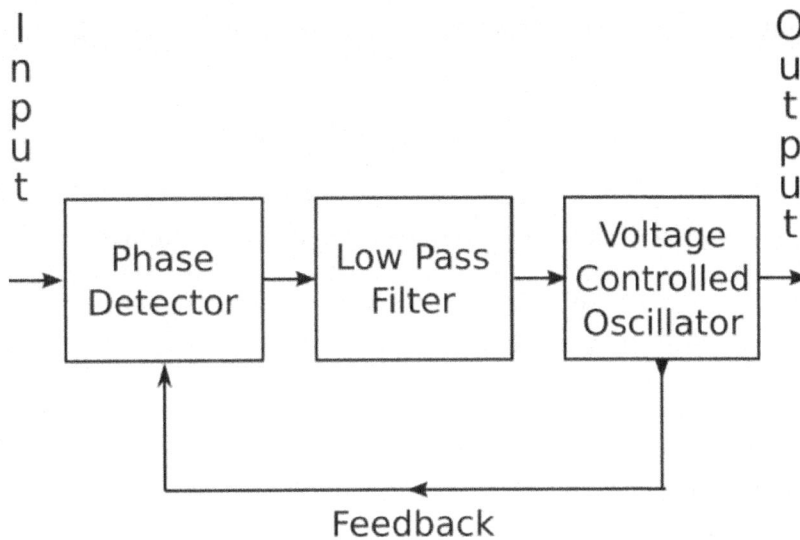

...use the incoming FM signal as the input, set the VCO to the center frequency of the desired signal and take an output from the low pass filter ...

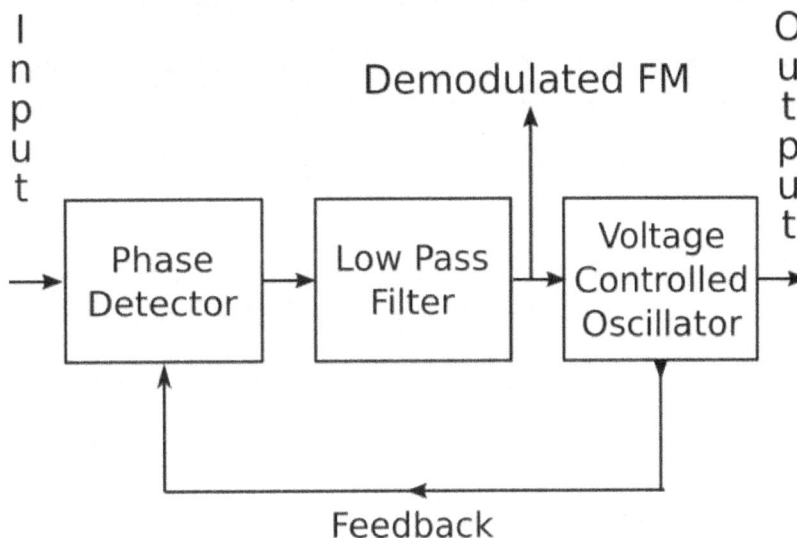

```
I                   Demodulated FM                   O
n                         ▲                          u
p                         │                          t
u                                                    p
t                                                    u
                                                     t
        ┌──────────┐   ┌──────────┐   ┌──────────┐
   ─────│  Phase   │──▶│ Low Pass │──▶│  Voltage │────▶
        │ Detector │   │  Filter  │   │Controlled│
        │          │   │          │   │Oscillator│
        └──────────┘   └──────────┘   └──────────┘
             ▲                              │
             │                              │
             └──────────────◀───────────────┘
                        Feedback
```

Figure 46.11: Phase-locked Loop Used as FM Demodulator

Boom! We have an electrical signal representing the deviation from the carrier frequency of the FM signal; we've recovered the original information. We have demodulated the FM. High fives!

Key Concepts in This Chapter

- Three oscillator circuits used in Amateur Radio equipment are the Colpitts, the Hartley, and the Pierce.

- Positive feedback in a Colpitts oscillator is supplied through a capacitive divider.

- Positive feedback in a Hartley oscillator is supplied through a tapped coil.

- Positive feedback in a Pierce oscillator is supplied through a quartz crystal.

- The oscillator circuits commonly used in VFO's are Colpitts and Hartley oscillators.

- A microphonic can be described as changes in oscillator frequency due to mechanical vibration.

- An oscillator's microphonic responses can be reduced by mechanically isolating the oscillator circuitry from its enclosure.

- NP0 capacitors can be used to reduce thermal drift in crystal oscillators.

- To insure that a crystal oscillator provides the frequency specified by the crystal manufacturer, it is important to provide the crystal with a specified parallel capacitance.

- Providing highly accurate and stable oscillators needed for microwave transmission and reception can be accomplished by using a GPS signal reference, by using a rubidium stabilized reference oscillator, or by using a temperature-controlled high Q dielectric resonator.

- The type of frequency synthesizer circuit that uses a phase accumulator, lookup table, digital to analog converter, and a low-pass anti-alias filter is a direct digital synthesizer.

- The information contained in the lookup table of a direct digital frequency synthesizer is the amplitude values that represent a sine-wave output.

- The major spectral impurity components of direct digital synthesizers are spurious signals at discrete frequencies.

- A phase-locked loop circuit is an electronic servo loop consisting of a phase detector, a low-pass filter, a voltage-controlled oscillator, and a stable reference oscillator.

- A phase-locked loop can perform frequency synthesis and FM demodulation functions.

Chapter 47

AC Waveforms & Measurements

Fourier Analysis of Waveforms

E8A01 (A) What is the name of the process that shows that a square wave is made up of a sine wave plus all its odd harmonics?

    A. **Fourier analysis**

    B. Vector analysis

    C. Numerical analysis

    D. Differential analysis

We've touched a few times on the idea that square waves are loaded with harmonics. Now we'll take a very light and brief look at the math behind that.

If we could create a true square wave – we can't, but if we could – it would go straight up from zero, turn a hard right 90 degree corner, go straight across, then make another hard right and head back to zero, then reverse the whole process on the negative side and we'd truly have a wave like Figure 47.1.

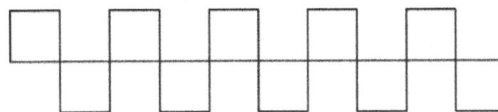

Figure 47.1: Square Waves

But that's not really what we get. Voltage cannot instantaneously change, because that would take an infinite amount of power. There's always a little bit of slope to those sides of the square and to the top (or bottom.) But, we get close enough to still create all those harmonics.

Just by its inherent nature, the creation of a square wave creates odd order harmonics. (In another sense, we could accurately say that creating a square wave *requires* all those odd order harmonics, because they are the "stuff" the square wave is made of.) So we don't just get, say a 1 kHz square wave and a smaller 3 kHz sine wave, which is riding on that flat portion.

We also get a 5 kHz sine wave, a 7 kHz, a 9, and 11, etc. In theory, that continues out to infinity, but of course we run out of bandwidth long before then. In the end, it's quite complex and just the first few harmonics leave us with something like Figure 47.2.

Plus, buried under there is a sine wave of the same frequency as the square wave, as in Figure 47.3.

So that's what a **Fourier analysis** of a square wave shows us and why square waves are often so very troublesome.

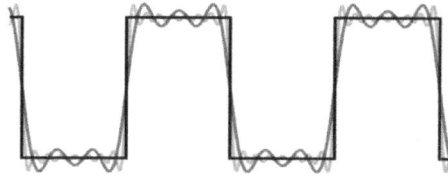

Figure 47.2: "Close-up" of a Square Wave

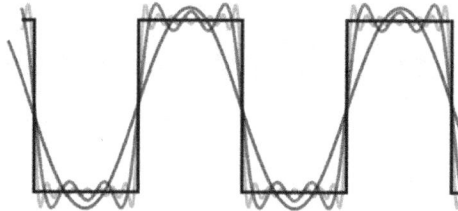

Figure 47.3: Square Wave Plus Fundamental Sine Wave

The Fourier analysis can be applied to *any* waveform and it will show that the waveform is made up of an infinite series of sine waves. The only exception is a sine wave. A pure sine wave has no harmonics.

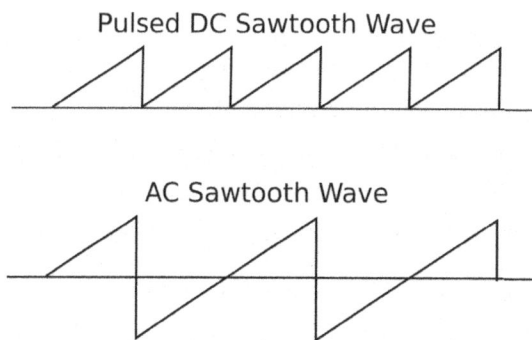

Figure 47.4: DC and AC Sawtooth Waves

E8A03 (A) What type of wave does a Fourier analysis show to be made up of sine waves of a given fundamental frequency plus all its harmonics?

    A. **A sawtooth wave**

    B. A square wave

    C. A sine wave

    D. A cosine wave

The square wave contains all the odd harmonics. A wave with *all* of the harmonics of its frequency – the odd ones and the even ones – is **a sawtooth wave**.

RMS Voltage

E8A05 (D) What of the following instruments would be the most accurate for measuring the RMS voltage of a complex waveform?

    A. A grid dip meter

    B. A D'Arsonval meter

C. An absorption wave meter

D. **A true-RMS calculating meter**

If you're trying to read the voltage of a simple waveform, like your household current under normal circumstances, your "I got it with a coupon!" inexpensive multimeter is probably more than adequate, and it will read the *average* voltage of that waveform, which will be fairly close to the RMS value. ($V_{RMS} = V_{Average} \times 1.11$.)

When you need more precision, or when you're trying to measure a complex waveform – for instance, the output of reactance modulator, or an audio signal – then you need **a true-RMS calculating meter.** You'll see them advertised as *TRMS* meters, and they're considerably more expensive than average reading meters.

PEP-to-Average Ratio

E8A06 (A) What is the approximate ratio of PEP-to-average power in a typical single-sideband phone signal?

    A. **2.5 to 1**

    B. 25 to 1

    C. 1 to 1

    D. 100 to 1

PEP is something we've been dealing with since the Technician level; it is Peak Envelope Power. PEP is the peak power of the signal at 100% modulation. CW has a PEP-to-average power ratio of 1 to 1 – any time it's "on" it's at 100%. SSB isn't like that, though; it has a lot of ups and downs in its waveform. The approximate ratio of PEP-to-average power for SSB is **2.5 to 1**.

E8A07 (B) What determines the PEP-to-average power ratio of a single-sideband phone signal?

    A. The frequency of the modulating signal

    B. **Speech characteristics**

    C. The degree of carrier suppression

    D. Amplifier gain

Of the possibilities listed, only **speech characteristics** affect the PEP-to-average power ratio of an SSB phone signal.

If the operator's voice is very consistently the same volume, then the PEP-to-average power ratio will be lower than if the operator's voice is constantly going from very loud to very soft.

Key Concepts in This Chapter

- The name of the process that shows that a square wave is made up of a sine wave plus all of its odd harmonics is the Fourier analysis.

- A Fourier analysis shows a sawtooth wave to be made up of sine waves of a given fundamental frequency plus all of its harmonics.

- The instrument that would be the most accurate for measuring the RMS voltage of a complex waveform is a true-RMS calculating meter.

- The approximate ratio of PEP-to-average power in a typical single-sideband phone signal is 2.5 to 1.

- What determines the PEP-to-average power ratio of a single-sideband phone signal is speech characteristics.

Chapter 48

Digital Signals

Symbol Rate

E8C02 (C) What is the definition of symbol rate in a digital transmission?

    A. The number of control characters in a message packet

    B. The duration of each bit in a message sent over the air

    C. **The rate at which the waveform changes to convey information**

    D. The number of characters carried per second by the station-to-station link

Symbol rate is **the rate at which the waveform of a transmitted signal changes to convey information**.

You can get this question right without knowing anything at all about digital transmissions, but by being a clever taker of tests. The question asks about the definition of some sort of rate. The answer must contain either the word "rate" or some synonym for rate. Sure enough, the correct answer is the only one with the word "rate" in it.

Let's get clear on this, though. Symbol rate has nothing to do with the number of control characters in a message packet, nor is it the duration of each bit in a message sent over the air. And the symbol rate isn't even "the number of characters carried per second," even though it sounds like it would be.

The symbol rate is about the waveform of **the transmitted signal and how rapidly it changes to convey information** – but remember, it might need to change state several times to transmit a single character.

We measure the symbol rate in baud. One baud is equal to one symbol per second.

E8C11 (A) What is the relationship between symbol rate and baud?

    A. **They are the same**

    B. Baud is twice the symbol rate

    C. Symbol rate is only used for packet-based modes

    D. Baud is only used for RTTY

Baud and symbol rate **are the same**. Just different words for the same thing.

Bandwidth of Digital Signals

E8C03 (A) Why should phase-shifting of a PSK signal be done at the zero crossing of the RF signal?

    A. **To minimize bandwidth**

    B. To simplify modulation

    C. To improve carrier suppression

    D. All these choices are correct

A quick review. PSK is phase shift keying. Simply put, in phase shift keying we're modulating the phase of the carrier and using that phase information to carry our information. Something like having our code key hooked up to switch the carrier from negative to positive instead of off and on. We end up, in a simple model, with a carrier that, ideally, looks like Figure 48.1.

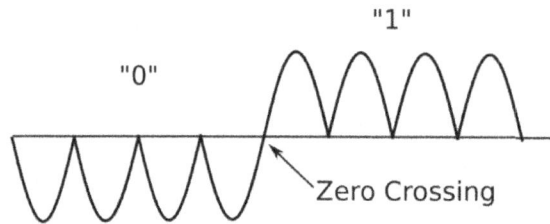

Figure 48.1: Ideal Phase Modulation

Notice that in this ideal – there's that word again – model, the keying happens right at the zero crossing. The effect, then, is to make that carrier one nice, smooth, continuous pseudo-sine wave – that wonderful wave with no harmonics.

Figure 48.2 shows an extreme example of how not to do it.

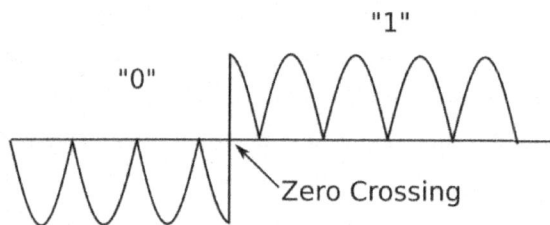

Figure 48.2: Non-ideal Phase Modulation

In this example, we've switched the phase right at the maximum negative swing of the carrier, as far from the zero-crossing as we could possibly be. A couple of things happen as a result, both unwanted. First, remember on our graph, the frequency of a wave shows up as the width of the wave relative to the X axis. In other words, that wave that suddenly switches in the middle is double the frequency of the carrier. So we've already produced a second harmonic, vastly widening our bandwidth. But wait, there's more! That vertical line sure looks like part of a square wave, doesn't it? We know from our extensive Fourier analysis of square waves that they contain *all* the odd harmonics of the frequency of the square wave – and that particular square wave is already double the carrier frequency, so now we're putting harmonics out there at 2x, 6x, 10x ---- oh, man, we're all over the spectrum now.

That's why shifting phase precisely at the zero crossing of the RF carrier **minimizes the bandwidth** of the transmitted signal.

E8C04 (C) What technique minimizes the bandwidth of a PSK31 signal?
    A. Zero-sum character encoding
    B. Reed-Solomon character encoding
    C. **Use of sinusoidal data pulses**
    D. Use of trapezoidal data pulses

PSK31 **uses sinusoidal data pulses** to minimize its bandwidth requirements. "Sinusoidal data pulses" means it uses sine waves – again, those waves without harmonics, and no harmonics means we're not taking up extra bandwidth.

E8C10 (C) How may data rate be increased without increasing bandwidth?
    A. It is impossible
    B. Increasing analog-to-digital conversion resolution
    C. **Using a more efficient digital code**
    D. Using forward error correction

Data rate may be increased without increasing bandwidth by **using a more efficient digital code.**

Throughout the 1980's and 1990's, modems – the devices that allowed computers to talk to each other over phone lines – steadily increased in speed from 300 bps to 33.6 kbps; that's an increase of over 100×. Very little of that increase was due to improvements in the telephone lines, which still have a bandwidth of about 3 kHz. All the improvements were due to **more efficient digital code**, which included more efficient encoding/decoding systems.

E8C05 (C) What is the approximate bandwidth of a 13-WPM International Morse Code transmission?

    A. 13 Hz

    B. 26 Hz

    C. **52 Hz**

    D. 104 Hz

What's a CW question doing in here with all this modern digital stuff? It can be argued that CW was the original digital mode, and apparently the question pool committee agrees.

There's a handy rule of thumb that will allow you to almost instantly calculate the bandwidth of a CW signal based on the words-per-minute being sent. Just multiply the WPM by 4, and that's the approximate bandwidth. In this case

$$Bandwidth = WPM \times 4 = 13 \times 4 = 52$$

However, let's understand the principles behind how we got to that number, because understanding that is going to be important for a couple of upcoming questions.

To calculate the bandwidth of a data signal, we need to know the frequency shift, the baud rate, and a constant known as K, about which more in a moment. If you were sending RTTY with frequency shift keying using mark and space frequencies 170 Hz apart, the frequency shift would be 170 Hz.

The formula for calculating the bandwidth occupied by a signal with a given frequency shift and a given baud rate is:

$$Bandwidth = (K \times shift) + B$$

In this formula, K is a constant, reflecting the amount of distortion we're willing to tolerate and the quality of the signal path. If you were an electrical engineer, you'd need to do a bunch of calculations to get a value of K for your problem, but you're a ham, and for our purposes, 1.2 is a perfectly useful and standard value of K, except for CW, when it's 4.8.

"B" is the baud rate of the transmission.

"Ah, but what about CW, my good sir, there's no frequency shift there!" Fair enough; we need a slightly different formula for CW:

$$Bandwidth = B \times K$$

Now, let's see how we got to **52 Hz** for the bandwidth of that 13 WPM signal.

First we need to convert WPM to baud. One baud equals one symbol per second, 60 per minute, so we need to know how many symbols there are in a Morse Code word. It turns out there's a standard Morse Code word used for this calculation, and it is PARIS. (Talk about obscure ham trivia! Wow!) A dot is a single symbol, as is the space between the dots and dashes of a single character. A dash is three dots long, as are the spaces between characters, so they count as three symbols. The space at the end of a word is seven symbols long. See Figure 48.3 for a graphic representation.

By the time you count up all the dots, dashes, and spaces in PARIS, there are 50 Morse Code symbols in that word, so if we were transmitting at 1 WPM we'd be managing 50 symbols per minute, 50/60ths of a baud or 0.83 baud. At 13 WPM we're sending at 13×0.83 baud.

Now we can do our formula.

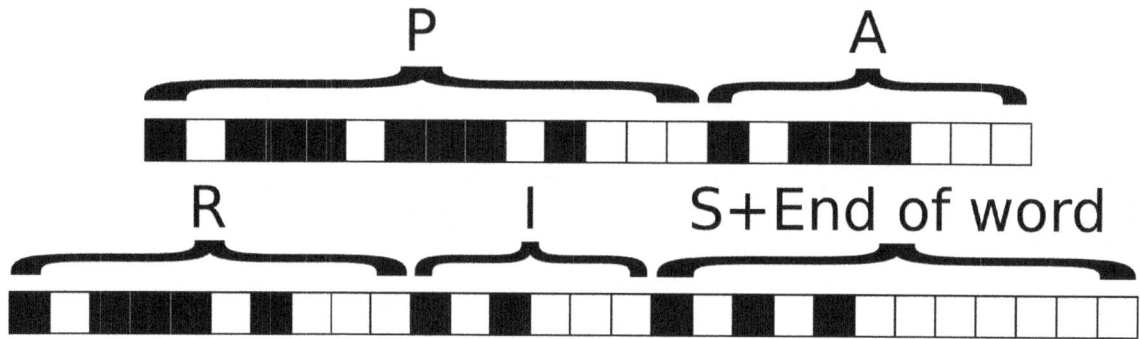

Figure 48.3: PARIS = 50 Symbols

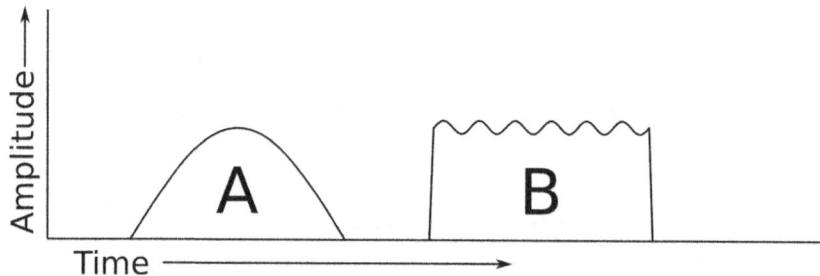

Figure 48.4: Two Extreme Keying Waveforms

$$Bandwidth = B \times K = (13\,WPM \times 0.83\,baud) \times 4.8 = 51.7 = 52\,Hz$$

Of course, this all assumes an ideal CW transmission system, with no ugly surprises in the transmitted waveform, as well as an extraordinarily skilled operator with perfect Morse Code rhythm, so it's really an approximation.

E8C12 (C) What factors affect the bandwidth of a transmitted CW signal?

A. IF bandwidth and Q

B. Modulation index and output power

C. **Keying speed and shape factor (rise and fall time)**

D. All these choices are correct

We've just seen that **keying speed** (words per minute) directly affects the bandwidth of a CW signal. Another influence is the shape of the waveform being sent.

Here's why I said that the bandwidth calculation for CW "assumes an ideal CW transmission system." A number of things can affect the waveforms or **shape factors** that make up the pulses of a CW signal. They range from the transmitter's power supply to, the type of key being used and even, in some cases, how hard the operator is hitting the key!

Compare the two extreme waveforms in Figure 48.4. You're looking at a single "dit" from two different stations, station A and station B.

Waveform A is half of a sine wave. You'll recall, pure sine waves contain no harmonics, so the bandwidth of that CW signal will be very narrow. The resulting tone, though, will be less distinct and more difficult to copy than signal B.

Waveform B is a real mess! It's almost a square wave, so those sharp leading and trailing edges are going to contain lots of harmonics, spreading that signal out across the dial. To top it off, there's some sort of modulation across the top of it, perhaps from a ringing filter or AC hum, and that's going to contribute to wider bandwidth as well. As you'll learn in Chapter 49, this shape factor also leads to key clicks.

E8C06 (C) What is the approximate bandwidth of a 170-hertz shift, 300-baud ASCII transmission?

    A. 0.1 Hz

    B. 0.3 kHz

    C. **0.5 kHz**

    D. 1.0 kHz

The formula for calculating bandwidth of a signal that is shifting frequency is:

$$Bandwidth = (K \times shift) + B$$

Here we're given the frequency shift, the baud rate, and we know that K is 1.2, since this isn't CW. It doesn't matter a bit that it is an ASCII transmission. We have everything we need to calculate this bandwidth.

$$Bandwidth = (K \times shift) + B = (1.2 \times 170\,Hz) + 300\,baud = 504\,Hz = 0.5\,kHz$$

E8C07 (A) What is the bandwidth of a 4800-Hz frequency shift, 9600-baud ASCII FM transmission?

    A. **15.36 kHz**

    B. 9.6 kHz

    C. 4.8 kHz

    D. 5.76 kHz

Once again, we have the frequency shift, and the baud, and we know that K is still 1.2. And, again, it doesn't matter that it is an ASCII transmission, nor does it matter that it is FM. It could be you trying to send a message by playing different notes on an electric guitar, the formula would still apply.

$$Bandwidth = (K \times shift) + B = (1.2 \times 4800\,Hz) + 9600\,baud = 15,360\,Hz = 15.36\,kHz$$

I think it's always valuable to be able to evaluate whether a possible answer to a question makes sense. So let's look at this problem a little differently.

The basics of Information Theory tell us that bandwidth must be *at least* equal to the baud rate of information being transmitted. If we're transmitting 9600 baud, we need a bare minimum of 9.6 kHz of bandwidth. That's one of those ideal numbers, so we'll never really fit 9600 baud into 9.6 kHz of bandwidth, but we might get in the ballpark, at least.

Look at the possible answers to this question. 4.8 kHz *can't* be right. There's just not enough room! It's like trying to pack twenty bologna sandwiches into little Jimmy's lunch sack. Neither can 5.76 kHz. In the real world, 9.6 kHz isn't possible either. That leaves **15.36 kHz** as the only answer that makes any sense at all, and, yeah, seems about right.

Digital Error Correction

E2E02 (A) What do the letters FEC mean as they relate to digital operation?

    A. **Forward Error Correction**

    B. First Error Correction

    C. Fatal Error Correction

    D. Final Error Correction

Some digital modes incorporate some sort of system for error correction for assuring that the data sent gets received correctly. Error correction systems are broadly categorized as either *forward* or *backward*. **Forward Error Correction** (FEC) means that, at least in theory, all the information needed to correct errors in the received data arrives with that data. This requires that "extra" data be transmitted with the original message, and that extra data can be used at the receiving end to correct any errors detected. A crude version of FEC would be to send each packet three times in succession. The receiver would then average or otherwise compare the three packets to extract the "correct" information. If all of the packets are completely messed up, the receiver is just out of luck. In pure forward error correction, all the communication is moving "forward", from the sender to the recipient. As you can probably guess, modern forward error correction schemes are far more sophisticated than what I described, but they still require a lot of extra data to be sent, and that's one of the disadvantages of forward error correction.

Backward Error Correction sends some very basic error checking information along with the original message, such as a checksum. The basic idea of a checksum is to assign a numeric value to each possible letter and symbol, add up all the numbers represented by the symbols in the message, and send along that number. On the receiving end, a similar process "checks" the "sum." If things don't add up, the receiver sends an NAK message– a Not Acknowledged message – and the transmitter repeats the transmission. Now we have communication moving "backward" from the recipient to the sender. Assuming a reasonably reliable connection, we save bandwidth, because we don't have to send all that redundant data, but communication might slow down if there are lots of NAK's and retransmissions. Also, if you think about it, backward error correction over the airwaves is not practical with multiple receivers – if, say, 10 receivers are all sending back retransmission requests, and all requesting different packets, it would create so much interference the system would break down.

E8C01 (C) How is Forward Error Correction implemented?

    A. By the receiving station repeating each block of three data characters

    B. By transmitting a special algorithm to the receiving station along with the data characters

    C. **By transmitting extra data that may be used to detect and correct transmission errors**

    D. By varying the frequency shift of the transmitted signal according to a predefined algorithm

Forward error correction operates only in the "forward" direction, relative to the transmitter. In other words, all the error correction is in the outgoing message, in the form of repeated data. The burden of error correction is on the receiver.

Forward error correction has been around since the first person sending a message repeated the message to be sure it got through. "Do not cut the blue wire! I repeat! DO NOT CUT THE BLUE WIRE!" is an example of forward error correction.

    At the receiving end, we combine all the received symbols back into, we hope, the original message.

E8C08 (D) How does ARQ accomplish error correction?

    A. Special binary codes provide automatic correction

    B. Special polynomial codes provide automatic correction

    C. If errors are detected, redundant data is substituted

    D. **If errors are detected, a retransmission is requested**

ARQ, Automatic Repeat reQuest, or Automatic Repeat Query, is backwards error correction. It works "backwards" relative to the transmitter, with the burden of error correction now resting with the transmitter. In "pure" ARQ, the transmitter sends part of the message once, then the receiver applies some sort of analysis to the message to see if it "makes sense." **If errors are detected, a retransmission is requested**. "Hey, send that last part again, please."

Gray Code

E8C09 (D) Which digital code allows only one bit to change between sequential code values?
   A. Binary Coded Decimal Code
   B. Extended Binary Coded Decimal Interchange Code
   C. Excess 3 code
   D.**Gray code**

Gray code is named for Frank Gray, who was a Bell Labs physicist and researcher. He also made early contributions to the invention of television.

The original application of Gray code was in mechanical switch systems. It addressed some very real problems, but I'll paint you a silly but simpler example of the sort of problem this solved.

Imagine you have a complex set of four switches that can be either on or off. We'll say they control some complicated machine, and there are 10 valid possible patterns of on and off. The switches are controlled by a round mechanical cam that opens and closes them as it rotates underneath them. When the switches are in one pattern, the machine does something, in a different pattern it does something else. If it does the wrong thing at the wrong time, it instantly explodes. The switches need to switch from one state to another at just the right time, and always forward or back only one state. How should we design the on/off patterns of the switches?

We could use a mechanical form of standard binary, using off to represent 0 and on to represent 1, as in Table 48.1.

Standard Binary System				
Position 0=0000	Position 2=0010	Position 4=0100	Position 6=0110	Position 8=1000
Position 1=0001	Position 3=0011	Position 5=0101	Position 7=0111	Position 9=1001

Table 48.1: Standard Binary System

Should work, right? Ah, but the switches have to all change position at the same time and send just the right signal at just the right moment. Look at the transition from Position 7 to Position 8. All four switches must change at exactly the same time or something horrible will happen. What if the timing's just a little bit off? Disaster!

Gray code addresses this by tossing binary out the window. There's no rule that says we must use 0111 to represent "7", after all. Gray code arranges things like Table 48.2.

Gray Code				
Position 0=0000	Position 2=0011	Position 4=0110	Position 6=0101	Position 8=1100
Position 1=0001	Position 3=0010	Position 5=0111	Position 7=0100	Position 9=1101

Table 48.2: Gray Code

Now no more than one switch changes state at a time. We could even design another machine to check for that behavior. "Did more than one switch change? BAD COMMAND! IGNORE IT! TURN ON THE WARNING LIGHT AND SIREN!"

The Gray code is even circular – after position 15 it just takes one switch change to get back to 0. (15 is 1000 in Gray code, 1111 in binary.)

This coding turns out to be very useful in some digital error checking algorithms, as well, having nothing at all to do with mechanical switches, but everything to do with on/off patterns.

Gray code patterns are often used in the more complex optical shaft encoders, as well.

Key Concepts in This Chapter

- The symbol rate of a digital transmission is the rate at which the waveform of a transmitted signal changes to convey information.

- Symbol rate and baud are the same.

- The phase shifting of a PSK signal should be done at the zero crossing of the RF signal to minimize bandwidth.

- The technique used to minimize the bandwidth requirements of a PSK31 signal is the use of sinusoidal data pulses.

- Symbol rate may be increased without increasing bandwidth by using a more efficient digital code.

- The necessary bandwidth of a 13-WPM International Morse Code transmission is approximately 52 Hz. The way to calculate that is to multiply the WPM by 4.

- The necessary bandwidth of a 170 Hz shift, 300 baud ASCII transmission is 0.5 kHz. The way to calculate that is to multiply the frequency shift by a standard K factor of 1.2, then add the baud rate.

$$Bandwidth = (K \times shift) + B$$

- The necessary bandwidth of a 4800 Hz frequency shift, 9600 baud ASCII FM transmission is 15.36 kHz.

- The letters FEC, as related to digital operation, mean Forward Error Correction.

- Forward error correction is implemented by transmitting extra data that may be used to detect and correct transmission errors.

- If a system is using ARQ error correction, if errors are detected, a retransmission is requested.

- The name of a digital code where each preceding or following character changes by only one bit is Gray code.

Chapter 49

Miscellaneous Signals & Emissions

Spread Spectrum

E8D01 (A) Why are received spread spectrum signals resistant to interference?

    A. **Signals not using the spread spectrum algorithm are suppressed in the receiver**

    B. The high power used by a spread spectrum transmitter keeps its signal from being easily overpowered

    C. The receiver is always equipped with a digital blanker

    D. If interference is detected by the receiver it will signal the transmitter to change frequencies

Know who invented spread spectrum? It was a film star, Hedy Lamarr.

Figure 49.1: Hedy Lamarr

    Hedy was quite intelligent and loved to invent and tinker with machinery. (Spiritually, I like to think she was one of us.) She and a composer, George Antheil, came up with a way for US radio-controlled torpedoes to defeat German jamming interference to those torpedoes. They got a patent for their system in 1942. She wanted to join the National Inventor's Council that was designed to serve as a clearinghouse for inventions that would aid the war effort. They told her to go put on a show and sell war bonds. It only took the US Navy 20 years to decide spread spectrum was a good idea and start to develop it in earnest.

    Where were we? Oh, yes. Why are received spread spectrum signals resistant to interference? Because each time the transmitter shifts to a new frequency, it sends a key along with the message, and the spread spectrum receiver will only respond to messages with that key attached. **Signals not using the spread spectrum algorithm are suppressed in the receiver.**

E8D02 (B) What spread spectrum communications technique uses a high-speed binary bit stream to shift the phase of an RF carrier?

    A. Frequency hopping

    B. **Direct sequence**

    C. Binary phase-shift keying

    D. Phase compandored spread spectrum

There are two major types of spread spectrum modulation techniques. One is **direct sequence**, the other is frequency hopping. Other types are hybrids of these two. Both end up spreading our signal across a wide spectrum, but they do it in very different ways.

Spread spectrum **direct sequence** modulation uses a very – *very* – wideband carrier that, if you saw it on an oscilloscope or spectrum analyzer, would look almost exactly like white noise. The key word there is *almost*. The "noise" is a bunch of phase shifts caused by first phase modulating an RF signal with that binary bit stream, called the "chip", then phase modulating that signal with the message. This double modulation creates a signal that is far wider than we "need" for the amount of real data it is carrying, but that wideband characteristic is essential to the working of this form of spread spectrum. At the receiving end, the noise – that binary number "chip" -- is removed, because the receiver "knows" that number. Put simply, direct sequence adds a whole lot of wideband "magic noise" to the original signal at the transmit end, then removes the magic noise at the receive end.

For any normal radio, that very wideband carrier just shows up as a bit of background noise – normal radios are way too narrowband to make any sense of that signal.

Direct sequence spread spectrum is widely used in WiFi routers, GPS satellites, and 900 MHz cordless phones. Your cell phone might use it as well.

The other two wrong answers besides "frequency hopping" are techniques we might use in the modulation/demodulation process, but neither uses a **high-speed binary bit stream**. (That bit stream is the "magic noise.")

E8D03 (D) How does the spread spectrum technique of frequency hopping work?

    A. If interference is detected by the receiver it will signal the transmitter to change frequencies

    B. If interference is detected by the receiver it will signal the transmitter to wait until the frequency is clear

    C. A binary bit stream is used to shift the phase of an RF carrier very rapidly in a pseudo-random sequence

    D. **The frequency of the transmitted signal is changed very rapidly according to a pseudorandom sequence also used by the receiving station**

Frequency hopping is more along the lines of how we usually think of spread spectrum, and it is the technology that runs your Bluetooth devices. It's the system that keeps changing the transmitter and receiver frequencies at the same (it has to be *precisely* the same) moment so that the frequency they are using keeps "hopping" around the spectrum.

CW Keying Defects

E8D04 (C) What is the primary effect of extremely short rise or fall time on a CW signal?

    A. More difficult to copy

B. The generation of RF harmonics

C. **The generation of key clicks**

D. Limits data speed

If your CW signal looks like this on an oscilloscope....

You are key clicking! The primary effect of those vertical rise and fall times – called "hard keying" – is **the generation of key clicks**.

E8D05 (A) What is the most common method of reducing key clicks?

A. **Increase keying waveform rise and fall times**

B. Low-pass filters at the transmitter output

C. Reduce keying waveform rise and fall times

D. High-pass filters at the transmitter output

Well, if the cause of key clicks is extremely short rise and fall times on a CW signal, the cure is to **increase keying waveform rise and fall times**. (And maybe stop banging on the code key with a hammer....)

Overmodulation of Digital Signals

E8D07 (D) What is a common cause of overmodulation of AFSK signals?

A. Excessive numbers of retries

B. Ground loops

C. Bit errors in the modem

D. **Excessive transmit audio levels**

AFSK is Audio Frequency Shift Keying. That annoying tone that triggers the Emergency Alert System is AFSK. If you can remember back to the days when computers were made of wood, your 1200 bit-per-second telephone line modem worked on AFSK. It is a modulation system that transmits two audio tones, one to signify "mark" and the other to signify "space."

These days common practice is to use the audio output of a computer to provide the audio that makes up an AFSK signal, and it's quite possible to crank up the volume to **excessive transmit audio levels** that cause overmodulation of the AFSK signal.

E8D08 (D) What parameter evaluates distortion of an AFSK signal caused by excessive input audio levels?

A. Signal to noise ratio

B. Baud rate

C. Repeat Request Rate (RRR)

D. **Intermodulation Distortion (IMD)**

When we crank the volume up to 11 and overdrive the transmitter, somewhere down the line the signal is going to get clipped. All the round tops of the sine waves will be shaved off and turned into our nemesis, square waves. One of the recurring themes of the Extra exam is *where we find square waves we find unwanted harmonics*, and it's the theme of this question as well.

Any time we have high levels of unwanted frequencies, we're barreling down the road to **intermodulation distortion (IMD)**. It's a very dependable indicator of overmodulation.

Notice that the correct answer is the only one that even mentions **distortion**.

E8D09 (D) What is considered an acceptable maximum IMD level for an idling PSK signal?

A. +10 dB

B. +15 dB

C. -20 dB

D. **-30 dB**

We've changed modes now, to PSK, Phase Shift Keying, but this maximum IMD – intermodulation distortion -- level of **-30 dB** is a valid number for any digital mode.

Most digital mode software will give you a readout of the IMD level of the received signal.

Digital Codes

E8D10 (B) What are some of the differences between the Baudot digital code and ASCII?

A. Baudot uses 4 data bits per character, ASCII uses 7 or 8; Baudot uses 1 character as a letters/figures shift code, ASCII has no letters/figures code

B. **Baudot uses 5 data bits per character, ASCII uses 7 or 8; Baudot uses 2 characters as letters/figures shift codes, ASCII has no letters/figures shift code**

C. Baudot uses 6 data bits per character, ASCII uses 7 or 8; Baudot has no letters/figures shift code, ASCII uses 2 letters/figures shift codes

D. Baudot uses 7 data bits per character, ASCII uses 8; Baudot has no letters/figures shift code, ASCII uses 2 letters/figures shift codes

Baudot uses **5 data bits per character**. Remember that – it was on your General exam -- and you have this question nailed! **Baudot uses 5 data bits per character, ASCII uses 7 or 8; Baudot uses 2 characters as letters/figures shift codes, ASCII has no letters/figures shift code**.

E8D11 (C) What is one advantage of using ASCII code for data communications?

A. It includes built in error correction features

B. It contains fewer information bits per character than any other code

C. **It is possible to transmit both upper and lower case text**

D. It uses one character as a shift code to send numeric and special characters

BAUDOT CODE CAN ONLY SEND UPPER CASE LETTERS. Using ASCII, **it is possible to transmit both upper and lower case text**, but it costs us some transmission speed because it is a less efficient code.

E8D06 (D) What is the advantage of including parity bits in ASCII characters?

A. Faster transmission rate

B. The signal can overpower interfering signals

C. Foreign language characters can be sent

D. **Some types of errors can be detected**

A parity bit is an "extra" bit tagged onto the end of a byte, i.e., a character. The parity bit makes the number of 1's in the byte an even number or an odd number, depending on whether we're using an even or odd parity system.

For instance, say we want to send 10010110 and add an even parity bit. We count the total of 1's in the byte – 4 – so the parity bit is 0, since the number of 1's is already even. We send 100101100. On the receiving end, the software counts the 1's, comes up with an even number, and gives it a digital thumbs-up. If we want to send 10010111, now there's an odd number of 1's, so the parity bit will be a 1 to bring the total to 6. Thumbs up again! But, say somehow one of those 0's in the middle gets sent as a 1: That's seven 1's, an odd number, so that's an error.

Key Concepts in This Chapter

- Received spread spectrum signals are resistant to interference because signals not using the spread spectrum algorithm are suppressed in the receiver.

- The spread spectrum technique that uses a high-speed binary bit stream to shift the phase of an RF carrier is direct sequence.

- In the spread spectrum technique of frequency hopping the frequency of the transmitted signal is changed very rapidly according to a particular sequence also used by the receiving station.

- The primary effect of extremely short rise or fall time on a CW signal is the generation of key clicks. The most common method of reducing key clicks is to increase keying waveform rise and fall times.

- A common cause of overmodulation of AFSK signals is excessive transmit audio levels.

- The parameter that might indicate that excessively high input levels are causing distortion in an AFSK signal is intermodulation distortion.

- A good minimum IMD level for an idling PSK signal is -30 dB.

- Some of the differences between the Baudot digital code and ASCII are; Baudot uses 5 data bits per character, ASCII uses 7 or 8; Baudot uses 2 characters as letters/figures shift codes, and; ASCII has no letters/figures shift code.

- One advantage of using ASCII code for data communications is that with ASCII it is possible to transmit both upper and lower case text.

- The advantage of using a parity bit with an ASCII character stream is that some types of errors can be detected.

At this point in the program, if you can score 100% on Progress Check #10 consistently, you should score at least 82% on the real exam – well above the minimum passing score of 74%. If you can score 91% consistently, you'd still almost certainly have a passing score on the final exam. We'd urge you to have more of a margin of safety than that, but you're closing in on your goal!

Progress Check #10

Chapter 50

Basic Antenna Parameters

Antenna Gain

E9A01 (C) What is an isotropic antenna?

    A. A grounded antenna used to measure Earth conductivity

    B. A horizontally polarized antenna used to compare Yagi antennas

    C. **A theoretical omnidirectional antenna used as a reference for antenna gain**

    D. A spacecraft antenna used to direct signals toward Earth

An isotropic antenna is another of those handy, non-existent fantasy objects physicists and engineers love so much. It's **a theoretical antenna used as a reference for antenna gain**. It's a mathematical convenience. Isotropic is a physics word. It means "having the same properties in all directions." An isotropic antenna radiates signal absolutely equally in all directions, so it has a gain of 0 dB.

    There are many reasons an isotropic antenna is impossible to construct, one being that the isotropic antenna is lossless – precisely none of the signal fed into it gets turned into heat.

E9H03 (D) What is Receiving Directivity Factor (RDF)?

    A. Forward gain compared to the gain in the reverse direction

    B. Relative directivity compared to isotropic

    C. Relative directivity compared to a dipole

    D. **Forward gain compared to average gain over the entire hemisphere**

Receiving Directivity Factor (RDF) is a parameter of receiving antennas. Technically, it is **forward gain compared to average gain over the entire hemisphere.** Simplified, let's say a receiving antenna had forward gain of $10dB$ and an average gain in every direction of $3dB$. That would make this antenna's Receiving Directivity Factor $7dB$.

    Charles Rauch, W8JI, has experimented extensively with high-gain receiving antennas and his highest RDF figures are in the $13dB$ range. (W8JI.com.) According to his tests, RDF is a better indicator of receiving performance than simple gain numbers for antennas designed for frequencies below UHF.

E9A12 (A) How much gain does an antenna have compared to a 1/2-wavelength dipole when it has 6 dB gain over an isotropic antenna?

    A. **3.85 dB**

    B. 6.0 dB

    C. 8.15 dB

    D. 2.79 dB

Remember; a dipole has 2.15 dB of gain relative to an isotropic antenna.

The antenna in this question has 6 dB of gain over an isotropic antenna. We subtract 2.15 from 6 and the answer is, the antenna has **3.85 dB** of gain over a dipole antenna.

Antenna Efficiency

E9A05 (D) What is included in the total resistance of an antenna system?
    A. Radiation resistance plus space impedance
    B. Radiation resistance plus transmission resistance
    C. Transmission-line resistance plus radiation resistance
    D. **Radiation resistance plus loss resistance**

The total resistance of an antenna system consists of **radiation resistance plus loss resistance**. You're not going to be able to whip out your rusty, trusty ohm meter and measure either of these values.

It takes some energy to jostle the electrons in the antenna enough to get them to emit electromagnetic waves, and radiation resistance is a measure of that energy. It's actually radiation resistance that creates the electromagnetic waves. Radiation resistance is a calculated value, and to calculate it even somewhat precisely you need to know the "far field" field strength of the whole radiation pattern of the antenna. However, a good rule-of-thumb estimate for a dipole's radiation resistance is around 73 ohms. The value of the radiation resistance is the value of a resistor that would dissipate the same amount of power as is radiated by the antenna. (That's why we need to know just how much *is* being radiated before we can calculate it.) Ponder that concept of radiation resistance for a moment, because that's going to be an important concept in an upcoming question.

Loss resistance, sometimes known as *ohmic resistance*, is the resistance that is loss in the structure of the antenna – it's a measure of how much of your transmitter's signal is being turned into heat. Loss resistance of an efficient antenna is considerably lower than the radiation resistance. The value of the loss resistance is equal to a resistor that would dissipate the amount of the signal's power *not* being radiated by the antenna.

E9A03 (C) What is the radiation resistance of an antenna?
    A. The combined losses of the antenna elements and feed line
    B. The specific impedance of the antenna
    C. **The value of a resistance that would dissipate the same amount of power as that radiated from an antenna**
    D. The resistance in the atmosphere that an antenna must overcome to be able to radiate a signal

It's like deja vu all over again! Radiation resistance is **the value of a resistance that would dissipate the same amount of power as that radiated from an antenna**.

Hmmm ... I wonder why they're asking us about these radiation and loss resistance thingamajigs ....

E9A09 (B) What is antenna efficiency?
    A. Radiation resistance divided by transmission resistance
    B. **Radiation resistance divided by total resistance**
    C. Total resistance divided by radiation resistance
    D. Fffective radiated power divided by transmitter output

Aha. Here's the point of loss and radiation resistance. They tell us the antenna's efficiency – how much of what we put in with our transmitter is coming out as useful radio waves?

To calculate that, we divide the radiation resistance by the total resistance; radiation resistance plus loss resistance. If we want, we can multiply the result by 100 to get a percentage.

$$Antenna\ Efficiency = \frac{Radiation\ Resistance}{Total\ Resistance} \times 100\%$$

Let's make up some numbers and see how this works. Let's say we have an antenna with a radiation resistance of 73 ohms and a loss resistance of 7 ohms.

$$Antenna\ Efficiency = \frac{73\ \Omega}{80\ \Omega} \times 100\% = 91.25\%$$

Now, let's make the radiation resistance 73 ohms and the loss resistance 50 ohms.

$$Antenna\ Efficiency = \frac{73\ \Omega}{123\ \Omega} \times 100\% = 59.3\%$$

Ouch. If we were pumping 100 watts into the first antenna, we'd be getting 91.25 watts out as radio waves. But with the second antenna, we're only getting 59.3 watts out. That's why we want high radiation resistance relative to the loss resistance.

E9B07 (C) How does the total amount of radiation emitted by a directional gain antenna compare with the total amount of radiation emitted from a theoretical isotropic antenna, assuming each is driven by the same amount of power?

    A. The total amount of radiation from the directional antenna is increased by the gain of the antenna

    B. The total amount of radiation from the directional antenna is stronger by its front-to-back ratio

    C. **They are the same**

    D. The radiation from the isotropic antenna is 2.15 dB stronger than that from the directional antenna

Grrrrrrrr ...... trick question.

    Big hint: Heat is a form of radiation.

    This question is asking about the *total* amount of radiation emitted by an antenna; both radio waves and heat. The isotropic antenna will radiate all the input as radio waves while any real-world antenna will radiate some of the input as radio waves and the rest as heat. They give us some wrong answers that are quite creative! The total amount of radiation can't be increased by the gain of the antenna – that would mean the antenna was creating energy from nothing. It's not "stronger by its front-to-back ratio", which makes no sense. And not only is the isotropic antenna irrelevant to this question, they have the dipole/isotropic 2.15 dB ratio backwards!

    The Laws of Thermodynamics decree that the *total* amount of radiation emitted by *any* antenna *exactly* equals the power put into it. The radiation either gets radiated as radio waves or as heat, regardless of any other factor.

E9A10 (A) Which of the following improves the efficiency of a ground-mounted quarter-wave vertical antenna?

    A. **Installing a radial system**

    B. Isolating the coax shield from ground

    C. Shorten the radiating element

    D. All these choices are correct

Ground-mounted quarter-wave vertical antennas *love* a good radial system. That radial system acts like a mirror under the antenna, and makes it look, electrically speaking, like it is twice as tall. A *good* radial system is usually buried in the ground, with wires the length of the antenna's height strung out radially from the base of the antenna. (If your antenna is in the middle of your lawn, you can just hold down the radial wires with turf staples and let the lawn grow up around it.) More wires is better, though I've seen a portable antenna system that uses a single wire laying on the ground under the antenna. It worked.

Note – these radial wires are not for electrically connecting to the earth. In fact, they are insulated wires. They *are* the ground system, not a connection to ground. We specifically do not want the earth being our antenna's ground system because, especially at the frequencies we use, that leads to lots of ground losses.

Virtually every commercial AM radio station in the US uses one or more ground-mounted quarter-wave vertical antennas, and they usually have ground systems buried in the ground around the antenna with a radial every 3° or so. That's a lot of copper, and overkill for our purposes, but it illustrates the importance of radials.

E9A11 (C) Which of the following factors determines ground losses for a ground-mounted vertical antenna operating in the 3 MHz to 30 MHz range?
  A. The standing wave ratio
  B. Distance from the transmitter
  C. **Soil conductivity**
  D. Take-off angle

The more conductive the soil, the more it sucks up our radio waves – especially in the 3 MHz to 30 MHz range, or, say, 80 meters to 10 meters. (Things are a little different for the very long waves of 160 meters. At frequencies below 3 MHz, we often *want* to create a strong ground wave and soil conductivity helps with that.) **Soil conductivity** determines ground losses for a ground-mounted vertical antenna.

Feed Point Impedance

E9A04 (B) Which of the following factors affect the feed point impedance of an antenna?
  A. Transmission-line length
  B. **Antenna height**
  C. The settings of an antenna tuner at the transmitter
  D. The input power level

Be careful with this one. Everything they list could certainly affect your *transmissions*, but they're asking specifically about which will affect the feed point impedance of the antenna, and that's **antenna height.**

The feed point impedance of a wire dipole antenna, for instance, climbs dramatically as you bring it close to the ground.

E9D10 (B) What happens to feed point impedance at the base of a fixed length HF mobile antenna when operated below its resonant frequency?
  A. The radiation resistance decreases and the capacitive reactance decreases
  B. **The radiation resistance decreases and the capacitive reactance increases**
  C. The radiation resistance increases and the capacitive reactance decreases
  D. The radiation resistance increases and the capacitive reactance increases

As the frequency of operation is lowered on a fixed length antenna, **the radiation resistance decreases and the capacitive reactance increases**. In other words, two bad things happen. The antenna looks too short, so capacitive reactance goes up. And radiation resistance goes *down.* Remember, we *like* radiation resistance – that's the resistance that is the result of pushing radio waves out the antenna.

Antenna Bandwidth

E9A08 (B) What is antenna bandwidth?
  A. Antenna length divided by the number of elements
  B. **The frequency range over which an antenna satisfies a performance requirement**
  C. The angle between the half-power radiation points
  D. The angle formed between two imaginary lines drawn through the element ends

We have two antenna-related terms that are easily confused. Beamwidth and bandwidth. Beamwidth is how wide or narrow the forward lobe of the antenna's radiation pattern is. High gain antennas have a narrow beamwidth.

This question isn't asking about that, though, it's asking about the bandwidth of an antenna, and that's simply **the frequency range over which an antenna satisfies a performance requirement**.

Let's be honest, our "performance requirement" is usually not very formal. It's something like, "Can I work 40 meters with this thing or not?" or, even better, "Can I work 80 through 10 meters on this thing without having an SWR so high it sets my transceiver on fire?" The answer to either of those questions lies in the answer to another question, which is, "Well, Betty Lou, what's the bandwidth of that whatchamacallit?"

E9D08 (B) What happens as the Q of an antenna increases?
  A. SWR bandwidth increases
  B. **SWR bandwidth decreases**
  C. Gain is reduced
  D. More common-mode current is present on the feed line

Think of an antenna as a resonant circuit. They aren't all resonant – there are non-resonant designs – but mostly, they are. That tells you that as the Q of an antenna increases, **bandwidth decreases**, just as in a resonant circuit bandwidth decreases as Q increases.

"SWR bandwidth" is shorthand for "how wide is the bandwidth with acceptable SWR?" Once you get outside the antenna's bandwidth, you'll see increased SWR.

Effective Radiated Power

E9A13 (C) What term describes station output, taking into account all gains and losses?
  A. Power factor
  B. Half-power bandwidth
  C. **Effective radiated power**
  D. Apparent power

Most of the power regulations for ham radio specify the power limit as PEP – Peak Envelope Power, which is the number of watts coming out of your transmitter. However, a few situations require us to calculate our ERP – **Effective Radiated Power**. That's the watts coming out of your transmitter, minus the losses in everything between your transmitter and the antenna, then plus the gain of your antenna.

If we're operating on the 60-meter band, we're restricted to not more than 100 watts **effective radiated power.** If we add up all the gains and losses of our feed line, antenna, and any other gear between our transmitter and the antenna, and we come up with 3 dB of gain, we can't have a PEP of any more than 50 watts, because that would give us 100 watts of ERP.

On the other hand, if we're operating in the 60-meter band and we add up those gains and losses for a total of -3 dB, we can crank up the amplifier to 200 watts PEP to get to 100 watts ERP.

E9A02 (D) What is the effective radiated power relative to a dipole of a repeater station with 150 watts transmitter power output, 2 dB feed line loss, 2.2 dB duplexer loss, and 7 dBd antenna gain?

    A. 1977 watts

    B. 78.7 watts

    C. 420 watts

    D. 286 watts

I'll show you how to calculate this precisely, but we can "eyeball estimate" our way to the correct answer. Here's the estimation method.

$$-2dB + -2.2dB + 7dB = 2.8dB\,(of\,gain)$$

We know 3 dB equals a doubling of power. 2.8 is pretty close to 3, so the answer is going to be just a little less than 150 watts times 2, 300 watts; only one of the possible answers is even close. **286 watts**.

Now, for those who like their numbers to be exactly right ...

First we calculate the multiplier for 2.8 dB. To do that, it's 10 raised to the power of the dB divided by 10. ' $10^{\frac{2.8}{10}}$. That's 1.905. Then we multiply 150 watts by 1.905.

$$10^{\frac{2.8\,dB}{10}} = 1.905\times$$

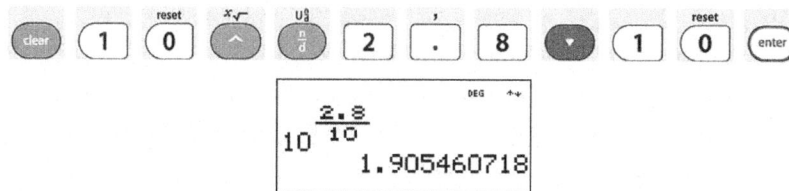

$$1.905 \times 150\,watts = 285.81\,watts \approx 286\,watts$$

Whoa, but wait! That question asked about "power relative to a dipole!" Do we have to do something about that?

Nope. They're being a bit clever with this – notice that in the question, the gain of the antenna is given in **dBd**. That's already "dB of gain relative to a dipole" so it's all handled in the setup of the question. If we subtract the 2.15 dB of gain of a dipole relative to an isotropic antenna, we end up with gain of 0.65 dB, and if we run that through our calculations we'll end up, thank goodness, at an answer not included in the possible answers.

E9A06 (A) What is the effective radiated power relative to a dipole of a repeater station with 200 watts transmitter power output, 4 dB feed line loss, 3.2 dB duplexer loss, 0.8 dB circulator loss, and 10 dBd antenna gain?

    A. 317 watts

    B. 2000 watts

    C. 126 watts

    D. 300 watts

We can guesstimate this one, too. $-4dB + -3.2dB + -0.8dB + 10dBd = 2dBd\,gain$. 2 dB is about 60% gain. 60% of 200 is 120, and $200 + 120 = 320$ Let's see, do we have something close to 320? Yep, **317 watts**.

Let's double-check that.

$$10^{\frac{2}{10}} = 1.584$$

$$200\,watts \times 1.584 = 316.8\,watts \approx 317\,watts$$

E9A07 (B) What is the effective isotropic radiated power of a repeater station with 200 watts transmitter power output, 2 dB feed line loss, 2.8 dB duplexer loss, 1.2 dB circulator loss, and 7 dBi antenna gain?

    A. 159 watts

    B. **252 watts**

    C. 632 watts

    D. 63.2 watts

We have a gigantic total of 1 dB of gain in this system. Thrilling. To estimate this one, remember that 1 dB is about a 25% gain in power. 25% of 200 is 50 watts – we're looking for an answer around 250 watts and there it is! **252 watts**.

To calculate it, it's:

$$10^{\frac{1}{10}} = 1.258$$

$$1.258 \times 200 = 251.78 \approx 252\,watts$$

If you're sharp-eyed, you spotted that they gave this antenna's gain as 7 dBi, "dB relative to an isotropic antenna." It's all still an apples to apples to apples comparison, though, because they didn't ask anything about "relative to a dipole", so no need to worry about that.

Key Concepts in This Chapter

- An isotropic antenna is a theoretical omnidirectional antenna used as a reference for antenna gain.

- Receiving Directivity Factor, or RDF, is a receiving antenna parameter equal to forward gain compared to average gain over the entire hemisphere.

- If an antenna has 6 db gain over an isotropic antenna, it has 3.85 dB of gain over a 1/2 wavelength dipole. The difference is 2.15 dB.

- The total resistance of an antenna system includes the radiation resistance plus loss resistance.

- Radiation resistance is the value of a resistance that would dissipate the same amount of power as that radiated from an antenna.

- The total amount of radiation emitted by a directional gain antenna and the total amount of radiation emitted from an isotropic antenna is exactly the same.

- One good way to improve the efficiency of a ground-mounted quarter-wave vertical antenna is to install a good radial system.

- In the 3 MHz to 30 MHz range, the most significant factor that determines ground losses for a ground-mounted vertical antenna is soil conductivity.

- The term antenna bandwidth means the frequency range over which an antenna satisfies a performance requirement.

- As the Q of an antenna increases, SWR bandwidth decreases.

- The term that describes station output, taking into account all gains and losses, is *effective radiated power.*

- The effective radiated power relative to a dipole of a repeater station with 150 watts transmitter power output, 2 dB feed line loss, 2.2 dB duplexer loss, and 7 dBd antenna gain is 286 watts.

- The effective radiated power relative to a dipole of a repeater station with 200 watts transmitter power output, 4 dB feed line loss, 3.2 dB duplexer loss, 0.8 dB circulator loss, and 10 dBd antenna gain is 317 watts.

- The effective radiated power of a repeater station with 200 watts transmitter power output, 2 dB feed line loss, 2.8 dB duplexer loss, 1.2 dB circulator loss, and 7 dBi antenna gain is 252 watts.

Chapter 51

Antenna Patterns

Reading Azimuth & Elevation Plots

E9B01 (B) In the antenna radiation pattern shown in Figure E9-1, what is the beamwidth?
  A. 75 degrees
  B. **50 degrees**
  C. 25 degrees
  D. 30 degrees

Figure E9-1

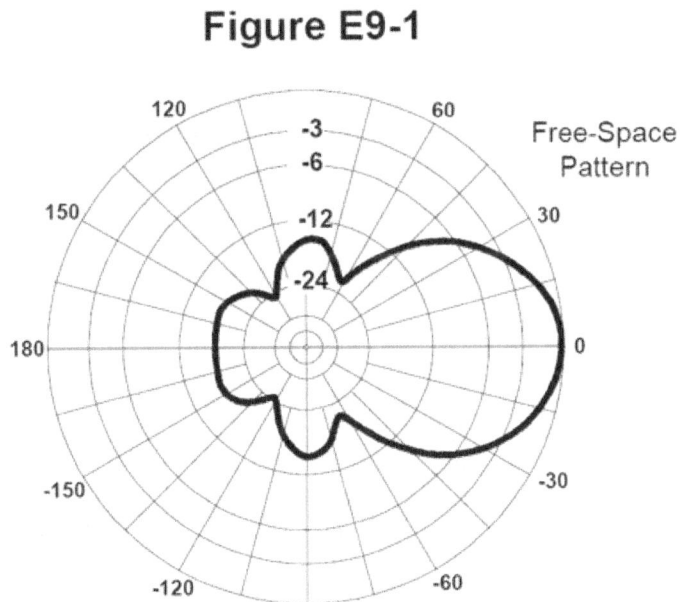

Figure 51.1: Figure E9-1 in the question pool

What you see in Figure E9-1 is a map of an antenna's radiation pattern. It's known as an "azimuth plane pattern." It's a bird's-eye view, looking down from the top with the antenna represented by the center of the circle. From our adventures in impedance studies you'll no doubt recognize that it is a polar coordinates plot with circular axes. The chart plots signal strength in dB against direction. That zero point on the far right, on the outer ring, represents the maximum signal strength and some arbitrary direction designated as zero. The numbers around the outside represent degrees from zero – something like degrees on a compass. We call $0°$ the "front" of the antenna, while the "back" is at $180°$ and the "sides" are at $+90°$ and $-90°$. Concentric rings represent signal strengths of $-3\,dB$, $-6\,dB$, $-12\,dB$ and $-24\,dB$, all relative to the maximum strength of $0\,dB$, which is the outer circle.

The question asks us to read the beamwidth; in other words, how wide is the beam, in degrees? What they don't specify is "at what signal strength" nor "in which direction."

Common usage when referring to antenna patterns is that it is always "beamwidth in the forward direction," since that's our direction of interest, and "beamwidth at $-3dB$" since that's the half-power point. The question committee has gone with common usage on this one.

You can see how to read this in Figure 51.2.

Looking at Figure E9-1, we can see the plot crosses the $-3dB$ mark somewhere between +15 and +30 degrees, and somewhere between -15 and -30. Call it +25 and -25. We're 25 from zero on both sides. $25 + 25 = 50$. That means the $3\,dB$ beamwidth of this antenna is **50 degrees**.

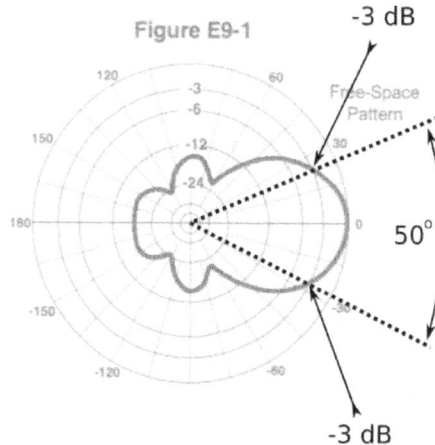

Figure 51.2: Reading 3 dB Beamwidth

E9B02 (B) In the antenna radiation pattern shown in Figure E9-1, what is the front-to-back ratio?

    A. 36 dB

    B. **18 dB**

    C. 24 dB

    D. 14 dB

The front-to-back ratio of an antenna is the difference, in dB, between the signal strength on the 0-degree axis and the signal strength on the 180-degree axis. There's no complicated math, we just need to read the numbers, as in Figure 51.3.

The 0-degree axis is, of course 0 dB, and the signal strength on the 180-degree axis is about halfway between -12 and -24 dB – call it -18 dB, so the front-to-back ratio is **18 dB.**

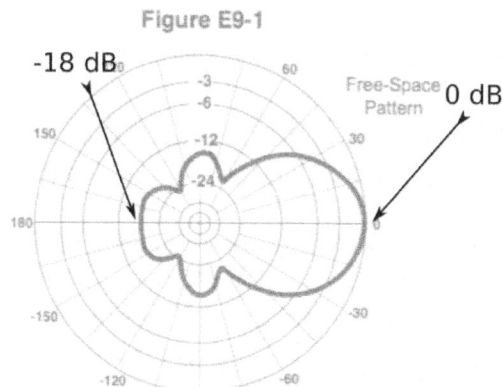

Figure 51.3: Reading Front-to-back Ratio

E9B03 (B) In the antenna radiation pattern shown in Figure E9-1, what is the front-to-side ratio?

    A. 12 dB

    B. **14 dB**

    C. 18 dB

    D. 24 dB

Now we're looking for the front-to-side ratio of the antenna. We know the front is at $0°$ and the sides are at $+90°$ and $-90°$. Happily for us this antenna's pattern is perfectly symmetrical – it must be from the Ideal™Antenna Company! – so it's simple enough to read right off the chart, as shown in Figure 51.4. On both sides the signal strength is down just a bit more than -12 dB, at -14 dB, giving us a front-to-side ratio of **14 dB**.

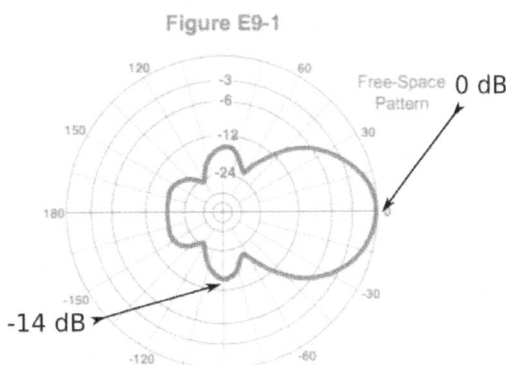

Figure 51.4: Reading Front-to-side Ratio

I'd suggest you memorize both the front-to-back and front-to-side ratio answers. Even though the question committee spread out the numbers a bit on this edition of the exam, both answers require you to estimate a value off a chart that isn't exactly a masterpiece of precision chart making, and both questions give you possible wrong answers that are fairly close to correct.

E9B08 (D) What is the far field of an antenna?

    A. The region of the ionosphere where radiated power is not refracted

    B. The region where radiated power dissipates over a specified time period

    C. The region where radiated field strengths are constant

    D. **The region where the shape of the antenna pattern is independent of distance**

If you could see radio waves, and you were looking at the area around an operating transmitting antenna, you'd most likely see chaos. Waves are reflecting off of stuff, being absorbed and re-radiated by other stuff, they're interacting with anything and everything that conducts electricity in the nearby area, and the nearby radiated pattern of the antenna is really nothing at all like that neat azimuth pattern they show us in figure E9-1. As we get some distance from the antenna, all that extra stuff fades away into the distance and we're left with the "real" pattern of the antenna.

How far away is the "far field?" There's some fancy math you can use to determine that, but for our purposes and for most of our antennas, it works out to a minimum of two wavelengths. So if you have an antenna for 40 meters, you need to be at least 80 meters away to be in the far field.

In the practical world, when we actually go out to measure the field strengths created by an antenna to create a real-world plot of the antenna's pattern, we back off at least a couple of *miles* to a **region where the shape of the antenna pattern is independent of distance**.

E9B05 (A) What type of antenna pattern is shown in Figure E9-2?

A. **Elevation**
B. Azimuth
C. Radiation resistance
D. Polarization

Figure E9-2

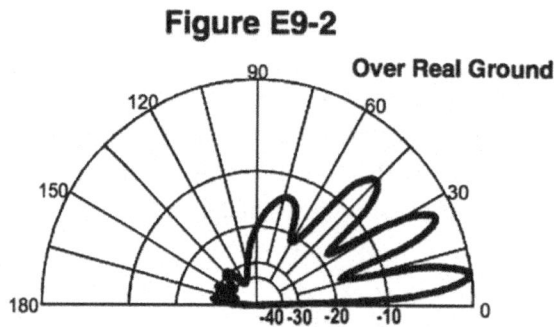

Figure 51.5: Figure E9-2 in the Question Pool

Figure E9-2 is an **elevation** pattern. Now we're looking at the antenna's radiation pattern from the side, with the antenna located right at the center of the axis that runs from 180 to 0. 0 still represents the front of the antenna and 180 the back, but now the degrees are rising up into the air. So, if you stood on the antenna and pointed straight ahead with your arm raised 60° above the plane of the ground, you'd be pointing at the 60-degree mark. (Standing on the antenna is not a recommended procedure.)

E9B06 (C) What is the elevation angle of peak response in the antenna radiation pattern shown in Figure E9-2?

A. 45 degrees
B. 75 degrees
C. **7.5 degrees**
D. 25 degrees

To find the elevation angle of peak response, we need to find where the 0 dB plot point touches the degree axis. That spot is halfway between 0° and 15°, as shown in Figure 51.6. The elevation angle of peak response is **7.5 degrees**.

Figure E9-2

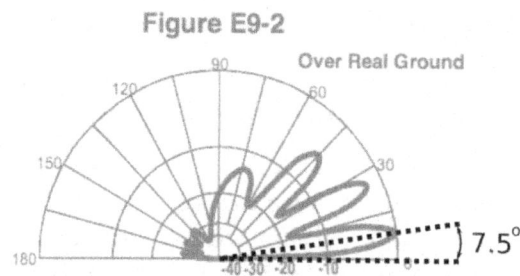

Figure 51.6: Reading Elevation Angle of Peak Response

E9B04 (B) What is the front-to-back ratio of the radiation pattern shown in Figure E9-2?

A. 15 dB
B. **28 dB**
C. 3 dB
D. 38 dB

To get the front-to-back ratio from an elevation chart we read the difference, in dB, between the highest reading at the front and the reading at the back, just the way we did when reading the front-to-back ratio from the azimuth chart. The front, remember, is toward the 0-degree mark, the back toward the 180-degree mark.

We're looking at the point shown in Figure 51.7.

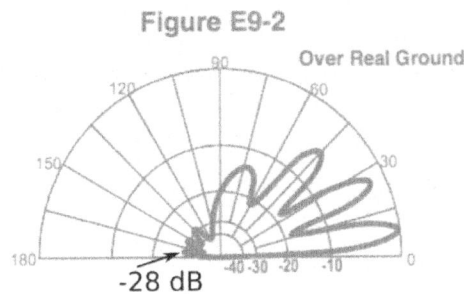

Figure 51.7: Reading Front-to-back Ratio

That lobe just tops the -30 dB axis, so it's at -28 dB. That makes our front-to-back ratio **28 dB**.

Antenna Modeling

E9B09 (B) What type of computer program technique is commonly used for modeling antennas?
   A. Graphical analysis
   B. **Method of Moments**
   C. Mutual impedance analysis
   D. Calculus differentiation with respect to physical properties

As you might imagine, the mathematics of directional antennas, with all those interactions between elements, can get fiendishly complex. Antenna modeling programs simplify things by using a method that comes from the science of statistics, called **Method of Moments**.

The driving force of any antenna is the current flow through the elements. That creates all the other effects. Every antenna carries a continually changing amount of current relative to the distance from the feed point. At the feed point, the current is at maximum. At the far end of the antenna, or of that half of the dipole, the current is zero. The challenge is, the current flow is different at every single infinitely small point of each element, and varies continuously. We're faced with completing an infinite number of calculations, which, if my estimate is correct, will take an infinite amount of time.

To break this all down to something manageable, the computer models the antenna as a bunch of discrete segments of wire, with equal current through the individual segment. Segments as large as about 1/20th of a wavelength – 10 per half wavelength -- will work.

Then the computer basically runs a limited number of calculations, plots the results, then connects the dots to create, say, the azimuth pattern, or the feed point impedance. It's much the same idea as measuring the behavior of a large number of people by sampling the behavior of a small number of people.

E9B10 (A) What is the principle of a Method of Moments analysis?
   A. **A wire is modeled as a series of segments, each having a uniform value of current**
   B. A wire is modeled as a single sine-wave current generator
   C. A wire is modeled as a single sine-wave voltage source
   D. A wire is modeled as a series of segments, each having a distinct value of voltage across it

The principle of a Method of Moments analysis is that **a wire is modeled as a series of segments, each having a uniform value of current**.

Remember that phrase, **uniform value of current**. Voltage is all well and good, but in an antenna, it's the current that gets the work done.

E9B11 (C) What is a disadvantage of decreasing the number of wire segments in an antenna model below 10 segments per half-wavelength?
A. Ground conductivity will not be accurately modeled
B. The resulting design will favor radiation of harmonic energy
C. **The computed feed point impedance may be incorrect**
D. The antenna will become mechanically unstable

Decreasing the number of "wire segments" in an antenna model is just like reducing the sample size in a statistical analysis. The smaller the sample, the more likely the results will be inaccurate. If we reduce the number of wire segments below 10 segments per half-wavelength, one result is that **the computed feed point impedance may be incorrect.**

Antenna modeling programs predict the electrical characteristics of an antenna, so we're looking for an answer that's about an antenna's electrical characteristics. Ground conductivity is not modeled by the program, nor is the mechanical stability of the antenna. The resulting design *might* favor radiation of "harmonic energy", but it also might not – that's not a result of reducing the number of wire segments used by the model.

Key Concepts in This Chapter

- In the antenna radiation pattern shown in Figure E9-1, the beamwidth is 50 degrees. The front-to-back ratio is 18 dB. The front-to-side ratio is 14 dB.

- The far field of an antenna is the region where the shape of the antenna pattern is independent of distance.

- The type of antenna pattern shown in Figure E9-2 is "elevation."

- The elevation angle of peak response in the antenna radiation pattern shown in Figure E9-2 is 7.5 degrees. The front-to-back ratio is 28 dB.

- The type of computer program technique commonly used for modeling antennas is called Method of Moments.

- The principle behind a Method of Moments analysis is that a wire is modeled as a series of segments, each having a uniform value of current.

- A disadvantage of decreasing the number of wire segments in an antenna model below 10 segments per half-wavelength is that the computed feed point impedance may be incorrect.

Chapter 52

Phased Arrays & Wire Antennas

Phased Arrays

This section starts with three questions about pairs of ¼ wavelength vertical antennas.

We have three different combinations to remember. Each question gives us a spacing distance between the pair, and a phase relationship between the pair. The answer is the pattern created by that combination.

An antenna complex like this is called a "phased array". Phased arrays can be quite elaborate, but the exam keeps things reasonably simple by focusing only on two-antenna arrays.

Let's see if we can get this organized in some way that will help make it memorable enough to pass this one should it show up on your exam, then we'll tear into how these combinations work.

Spacing	Phase Relationship	Pattern
½ Wavelength	180° Out of Phase	Figure 8 along the axis of the array.
½ Wavelength	In Phase	Figure 8 broadside to the array
¼ Wavelength	90° Out of Phase	Cardioid

When we say two antennas are out of phase, it means one is getting the signal later than the other. When we combine this delay, or lack of delay, with the space between the antennas, we get different radiation patterns, much the way two stones would make different wave patterns if you dropped them into a pond at different times and distances from each other.

Notice that if the antennas are ½ wavelength apart, we get a figure 8 pattern. If they are 180° out of phase, the figure 8 lines up with the axis of the array – remember the "8" is parallel to the "1" in 180, since it's the only array that's 180° out of phase.

If the antennas are ½ wavelength apart and in phase, the 8 rotates 90°, so now it's across the axis of the array.

If we space the antennas ¼ wavelength apart and feed them 90° out of phase, we get a cardioid, or heart shaped pattern. The "notch" in the cardioid pattern is centered on the antenna that gets the signal first.

E9C01 (D) What is the radiation pattern of two 1/4-wavelength vertical antennas spaced 1/2-wavelength apart and fed 180 degrees out of phase?

    A. Cardioid

    B. Omni-directional

    C. A figure-8 broadside to the axis of the array

    D. **A figure-8 oriented along the axis of the array**

337

Our vertical antennas spaced ½ wavelength apart and fed 180° out of phase create a pattern that is **a figure-8 oriented along the axis of the array.**

1/2 Wavelength Apart

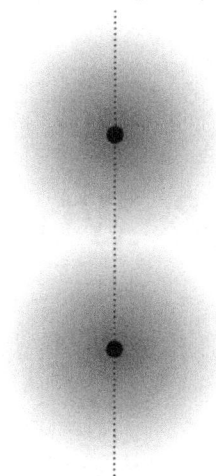

180 Degrees Out of Phase

In these diagrams, the black dots represent a top view of the antenna, the dotted line is the "axis of the array," and the fuzzy clouds are the radiation pattern. We'll call the top antenna Antenna A and the bottom antenna Antenna B.

Let's sort out what's happening here. antennas A and B get signals that are 180° out of phase. This all gets a lot easier to understand if we also think of Antenna A and B as being, not ½ wavelength apart, but 180° of phase apart – and that's exactly what ½ wavelength means.

So the signals leave Antenna A and Antenna B in this relationship:

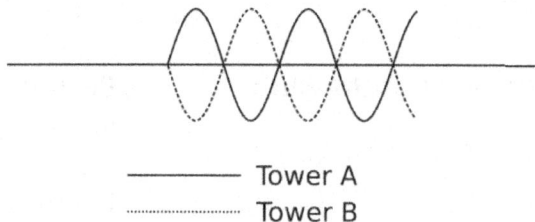

——————— Tower A
·················· Tower B

They travel through space. When they meet at the point in middle of the antennas, they've both traveled the same distance, so they're still out of phase and cancel each other out. That pulls in the sides of that figure 8.

Once Antenna A's signal reaches Antenna B, though, it has traveled through 180° and now the signals are in phase.

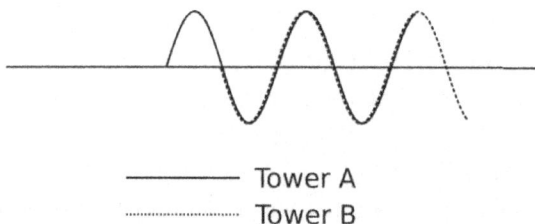

——————— Tower A
·················· Tower B

The same thing happens when Antenna B's signal reaches Antenna A, so the lobes expand around the antennas.

E9C03 (C) What is the radiation pattern of two 1/4-wavelength vertical antennas spaced 1/2-wavelength apart and fed in phase?

    A. Omni-directional

    B. Cardioid

    C. **A Figure-8 broadside to the axis of the array**

    D. A Figure-8 end-fire along the axis of the array

The signals leave the antennas in phase and travel a distance equal to 180° of phase.

338

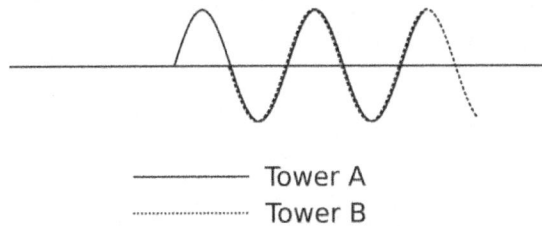

──────── Tower A
·············· Tower B

Then just the opposite of what happened in the previous question happens. The waves are still *in* phase when they meet in the middle, and they're out of phase when they reach that other antenna.

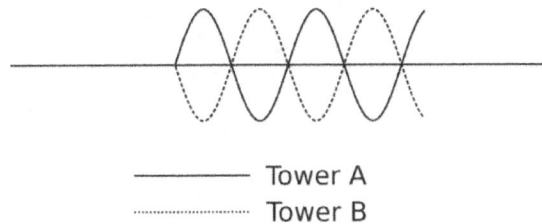

──────── Tower A
·············· Tower B

The result is a figure 8 rotated so that now it is going across the axis of those two antennas.

1/2 Wavelength Apart

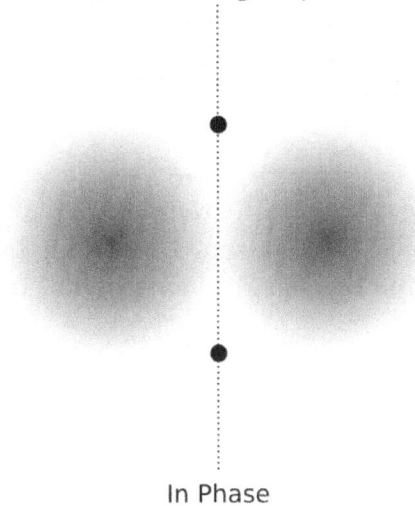

In Phase

Because the waves headed in the direction of Antenna B from Antenna A get cancelled out, and vice versa between B and A, the energy gets pushed out the sides and we get **a figure-8 broadside to the axis of the array**.

If you think about it, we're not stuck with just 0° or 180° out of phase for those feeds. If we could control the phase on one of the antennas, we could pick any combination, and as those relationships change the figure 8 will rotate around the center of the axis of the antennas. Since we have 90° of rotation of our figure 8 available to us, we could point in any direction we want by varying the phase relationship of those two antennas. Well, any two directions at a time that we want, to be perfectly accurate.

E9C02 (A) What is the radiation pattern of two 1/4-wavelength vertical antennas spaced 1/4 wavelength apart and fed 90 degrees out of phase?

   A. **Cardioid**

   B. A figure-8 end-fire along the axis of the array

   C. A figure-8 broadside to the axis of the array

   D. Omni-directional

Well, we've called in the crane and the other heavy equipment, picked up one of our antennas, and moved it ¼ wavelength toward the other, so now they're ¼ wavelength, or 90° of phase, apart. We'll feed them 90° out of phase and see what happens.

The signals leave Antenna A and B in this relationship.

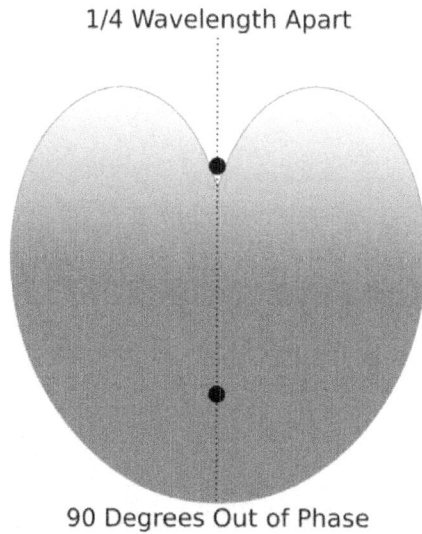

1/4 Wavelength Apart

90 Degrees Out of Phase

Figure 52.1: Cardioid Pattern

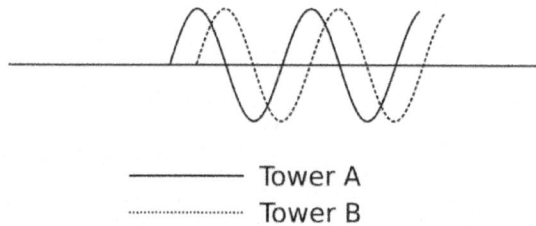

——— Tower A
·············· Tower B

The Antenna A signal takes off through space, rocketing toward Antenna B. Just as that signal gets to Antenna B, Antenna B sends the same signal. In this direction, the signals are now in phase.

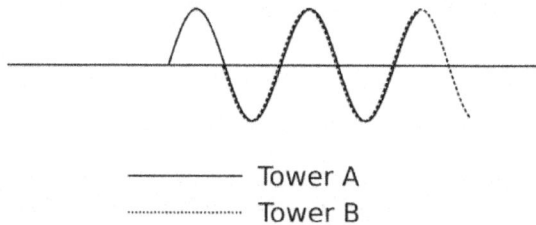

——— Tower A
·············· Tower B

The signal gets a big push in the direction marked by a line from Antenna A to Antenna B.

The signal from Antenna B heads off in the direction of Antenna A. It's 90° out of phase and has to cover 90 phase degrees worth of ground ... 90° + 90° = 180° ...so at Antenna A the signals are canceling each other.

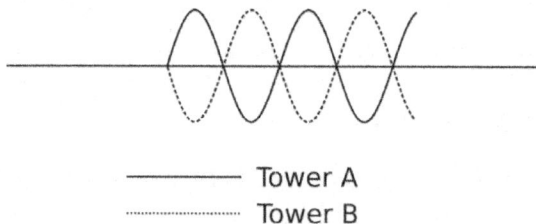

——— Tower A
·············· Tower B

And we get this what you see in Figure 52.1.

That's called a cardioid pattern – you can see it somewhat resembles a heart. It puts a lot of power in the direction from A to B and not much power at all in the direction from B to A.

Now, of course, in all these scenarios, and any other directional antenna, the waves are mixing and matching throughout the whole field, which is why the patterns are rounded, with no squared off corners.

Over the years, I think these three questions are, by far, the questions I've most often heard as the subject of, "I only missed one, and it was about those #&*%!! antenna patterns." I offer the following as a way of keeping them straight.

- First, "quarter" and "cardioid" sound somewhat alike, and the only quarter-wave spacing in the set of three is the one that produces a cardioid pattern. Quarter = Cardioid.

- Now, imagine the number 180, for the array that is 180° out of phase.

    Throw out the zero; it's nothing! That leaves you with the 1 and the 8. Slide them together, and you have a picture of a **figure-8 oriented along the axis of the array.**

- If you get one of these questions on your exam and "quarter=cardioid" doesn't fit, and you have no 8 to slide over a 1, you know it's the pattern that is **a figure-8 broadside to the axis of the array.**

    Sure, it's a little corny, but if generations of anatomy students could remember the bones of the ankle with "Tiger Cubs Need MILC," you can afford to munch a little corn for the sake of this question.

Wire Antennas

E9C04 (B) What happens to the radiation pattern of an unterminated long wire antenna as the wire length is increased?
    A. The lobes become more perpendicular to the wire
    B. **The lobes align more in the direction of the wire**
    C. The vertical angle increases
    D. The front-to-back ratio decreases

An unterminated long wire antenna is just what it sounds like – you string out wire until you run out of wire or trees, hook up one end to the transmitter, stick a ground stake in the ground for your transmitter ground, and fire up. Okay, yeah, there's a little more to it than that. You'll need to match impedances somehow, and you'll need a feed line up to the near end of the long wire – a piece of ladder line with only one side hooked up will probably work.

An unterminated long wire antenna behaves a bit like a Beverage antenna, in that it gets more and more directional as it gets longer and longer. **The lobes align more in the direction of the wire**, so it radiates less and less like a dipole, where the lobes come off the sides of the wire.

E9C10 (B) Which of the following describes a Zepp antenna?
    A. A dipole constructed from zip cord
    B. **An end-fed dipole antenna**
    C. An omni-directional antenna commonly used for satellite communications
    D. A vertical array capable of quickly changing the direction of maximum radiation by changing phasing lines

The Zepp antenna is an **end-fed dipole antenna**. It's a ½ wavelength antenna, fed with ladder-line, not coax, and, depending on construction, can be a multiband antenna.

The Zepp is more formally known as a Zeppelin antenna and, yes, it was originally designed to hang off a Zeppelin lighter-than-air craft while the craft was in flight – which accounts for the end fed design.

E9C12 (C) Which of the following describes an Extended Double Zepp antenna?
    A. A wideband vertical antenna constructed from precisely tapered aluminum tubing
    B. A portable antenna erected using two push support poles
    C. **A center-fed 1.25-wavelength antenna (two 5/8-wave elements in phase)**
    D. An end fed folded dipole antenna

An Extended Double Zepp antenna is **a center fed 1.25 wavelength antenna (two 5/8-wave elements in phase.)**

Figure 52.2: USS Macon; Navy Zeppelin Over New York, 1933

If you looked at a quick sketch of an Extended Double Zepp antenna, you'd probably scratch your head and think, "Looks like a garden variety dipole to me." The big difference is in the length – each half is 5/8 wavelength long.

With an antenna tuner, an Extended Double Zepp can work multiple bands.

E9C05 (A) Which of the following is a type of OCFD antenna?

A. **A dipole fed approximately 1/3 the way from one end with a 4:1 balun to provide multiband operation**

B. A remotely tunable dipole antenna using orthogonally controlled frequency diversity

C. A folded dipole center-fed with 300-ohm transmission line

D. A multiband dipole antenna using one-way circular polarization for frequency diversity

I guess if they had just said, "Off-Center Fed Dipole" instead of "OCFD" it would have given away the answer to this one, right? As it is, they certainly came up with some artful gibberish for the wrong answers.

An OCFD is a dipole fed off-center! It's **a dipole fed approximately 1/3 of the way from one end with a 4:1 balun to provide multiband operation**. That balun serves as an impedance matching device, since the farther we move our feed point from the center of a dipole, the higher the antenna's impedance.

The benefit of an OCFD is that it is a multiband antenna. Not "all band", by any means, but multiband. A standard dipole is generally a one band antenna (with some we can work the third harmonic, too, such as working 40 and 15 meters.) The downside of an OCFD is slightly higher losses than the standard dipole.

E9C08 (C) What is a folded dipole antenna?

A. A dipole one-quarter wavelength long

B. A type of ground-plane antenna

C. **A half-wave dipole with an additional parallel wire connecting its two ends**

D. A dipole configured to provide forward gain

You say you have room for a half-wave dipole? How about just folding a full-wave dipole in half and putting twice the antenna in the same amount of space? Would that work? Yes, it would. That's a folded dipole antenna – **a half-wave dipole with an additional parallel wire connecting its two ends.**

The impedance of your antenna feed point is going to quadruple to about 300 ohms, so you'll need to find some 300-ohm feedline, and build or buy a matching network of some sort, but you'll get wider bandwidth than a half-wave dipole, and even a tiny bit more gain.

While it's perfectly possible to build a folded dipole for HF, you're more likely to see them used on frequencies from the 2-meter band up. They usually look something like Figure 52.3.

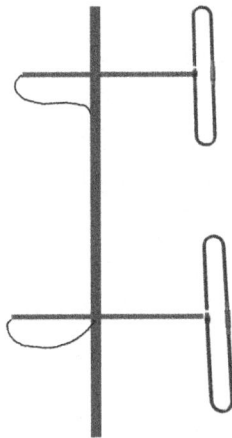

Figure 52.3: VHF Folded Dipole

E9C07 (A) What is the approximate feed point impedance at the center of a two-wire folded dipole antenna?

    A. **300 ohms**

    B. 72 ohms

    C. 50 ohms

    D. 450 ohms

The nominal feed point impedance of a standard dipole is ≈ 75 ohms. Turning it into a two-wire folded dipole multiplies that feed point impedance by four, to roughly **300 ohms**.

E9C09 (A) Which of the following describes a G5RV antenna?

    A. **A multi-band dipole antenna fed with coax and a balun through a selected length of open wire transmission line**

    B. A multi-band trap antenna

    C. A phased array antenna consisting of multiple loops

    D. A wide band dipole using shorted coaxial cable for the radiating elements and fed with a 4:1 balun

The very popular G5RV antenna is named for the British ham who invented it, Louis Varney, call sign G5RV. It is a **multi-band dipole antenna fed with coax and a balun through a selected length of open wire transmission line**. Put more simply, it's a dipole fed by a piece of ladder-line.

The legs of the dipole are 15.55 meters – about 51 feet – long. The ladder-line is precisely 8.84 meters long. It, in turn, is fed by a balun hooked to 50-ohm coax. An antenna tuner is required.

That magical combination creates an antenna that is claimed to work any band from 80 meters through 10 meters.

Effects of Ground

The next few questions deal with the effects of the ground under an antenna. In reality, the interaction of radio waves with the ground is a complex subject, to say the least. For purposes of the exam, though, we'll treat the ground as though the rocks on and in it are reflectors for radio waves. Lots of rocks and irregularities means more reflectors ricocheting our watts off to the sky. Fewer rocks means our signal stays closer to the earth.

E9C13 (B) How does the radiation pattern of a horizontally polarized 3-element beam antenna vary with increasing height above ground?

    A. The takeoff angle of the lowest elevation lobe increases

    B. **The takeoff angle of the lowest elevation lobe decreases**

    C. The horizontal beam width increases with height

    D. The horizontal beam width decreases with height

As we get that antenna higher, the ground has less and less effect, so **the main lobe takeoff angle decreases with increasing height.**

    Usually, low takeoff angles are very desirable, especially for DX, since they send our signals to the ionosphere at oblique angles, leading to better propagation.

E9C14 (B) How does the performance of a horizontally polarized antenna mounted on the side of a hill compare with the same antenna mounted on flat ground?

    A. The main lobe takeoff angle increases in the downhill direction

    B. **The main lobe takeoff angle decreases in the downhill direction**

    C. The horizontal beamwidth decreases in the downhill direction

    D. The horizontal beamwidth increases in the uphill direction

If ground reflection tends to push our signal toward the sky, then if the ground slopes down around some part of our antenna, that signal is going to go less straight up and more to the side, right? **The main lobe takeoff angle decreases in the downhill direction**. In other words, it tilts down.

    Of course, on the uphill side, the main lobe is going to tilt up, increasing the takeoff angle.

E9C11 (D) How is the far-field elevation pattern of a vertically polarized antenna affected by being mounted over seawater versus soil?

    A. The low-angle radiation decreases

    B. Additional higher vertical lobes will appear

    C. Fewer vertical lobes will be present

    D. **The low-angle radiation increases**

Rocky ground is going to provide lots of reflectors to bounce our signal skyward. Seawater is about as flat and rock-free a surface as one can imagine, unless one is in the midst of a storm, so not much scattering skyward.

    That's a good way to remember these for the exam; in reality, this has to do with the conductivity of sea water, which is much greater than that of soil.

Key Concepts in This Chapter

- The radiation pattern of two 1/4 wavelength antennas spaced 1/2 wavelength apart and fed 180 degrees out of phase is a figure-8 oriented along the axis of the array.

- The radiation pattern of two 1/4 wavelength vertical antennas spaced 1/2 wavelength apart and fed in phase is a figure-8 broadside to the axis of the array.

- The radiation pattern of two 1/4 wavelength vertical antennas spaced 1/4 wavelength apart and fed 90 degrees out of phase is cardioid.

- If the length of an unterminated long wire antenna is increased the lobes align more in the direction of the wire.

- A Zepp antenna is an end fed dipole antenna.

- An extended double Zepp antenna is a center fed 1.25 wavelength antenna (two 5/8-wave elements in phase.)

- An OCFD antenna is a dipole fed approximately 1/3 the way from one end with a 4:1 balun to provide multiband operation.

- A folded dipole antenna is a half-wave dipole with an additional parallel wire connecting its two ends.

- The approximate feed point impedance at the center of a two-wire folded dipole antenna is 300 ohms.

- A G5RV antenna is a multi-band dipole fed with coax and a balun through a selected length of open wire transmission line.

- For a horizontally polarized 3-element beam antenna, the main lobe takeoff angle decreases with increasing height.

- Compared with the same antenna mounted on flat ground, if a horizontally polarized antenna is mounted on the side of a hill, the main lobe takeoff angle decreases in the downhill direction.

- If a vertically polarized antenna is mounted over seawater versus rocky ground, the low-angle radiation increases in the far-field elevation pattern.

Chapter 53

Directional Antennas

Rhombic Antennas

E9C06 (B) What is the effect of adding a terminating resistor to a rhombic antenna?

    A. It reflects the standing waves on the antenna elements back to the transmitter

    B. It changes the radiation pattern from bidirectional to unidirectional

    C. It changes the radiation pattern from horizontal to vertical polarization

    D. It decreases the ground loss

Feed Line

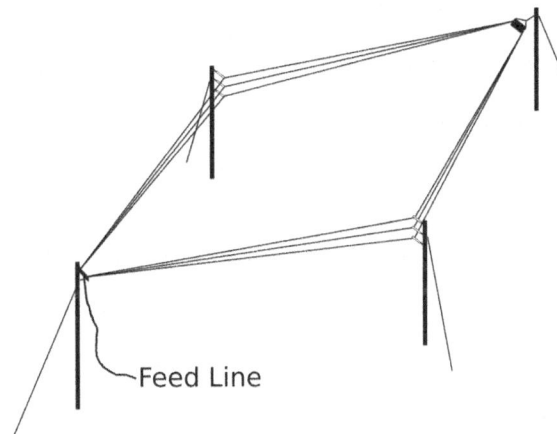

Figure 53.1: Rhombic Antenna

A rhombic antenna is a diamond shaped antenna, mounted on poles and parallel to the ground.

If we just make a diamond of wire that runs straight around the poles we get a radiation pattern something like what you see in Figure 53.2.

It's a high-gain bidirectional pattern. If we install a terminating resistor at the end of the diamond opposite the feed point, **it changes the radiation pattern from bidirectional to unidirectional**, pointing in the direction of the line between the feed point and the terminating resistor, as you see in Figure 53.3.

For a lot of years, big unidirectional rhombic antennas were the antenna of choice for

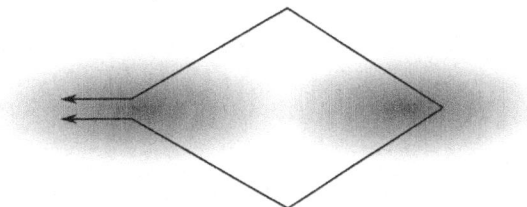

Figure 53.2: Pattern of Unterminated Rhombic Antenna

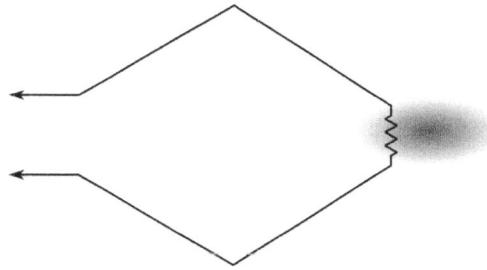

Figure 53.3: Pattern of Terminated Rhombic Antenna

long-distance point-to-point communication. When the phone company needed to put a radio signal into one particular spot overseas, they'd put up a big unidirectional rhombic with that resistor at the far end, and get the job done.

Rhombics are even wide bandwidth antennas, usually useful over multiple bands.

The downside? Each of those sides of the diamond is 1 to 2 wavelengths long. To put up a rhombic for 40 meters, you need at least a half-acre of land, probably in a secluded rural setting unless you have neighbors with good senses of humor.

Parabolic Dish Antennas

E9D01 (C) How much does the gain of an ideal parabolic dish antenna change when the operating frequency is doubled?

    A. 2 dB
    B. 3 dB
    C. 4 dB
    D. **6 dB**

The nature of a parabolic reflector is that if something radiates from the focal point, no matter where it impacts the parabola, it will be reflected in a line parallel to all the other reflections. (See Figure 53.4.)

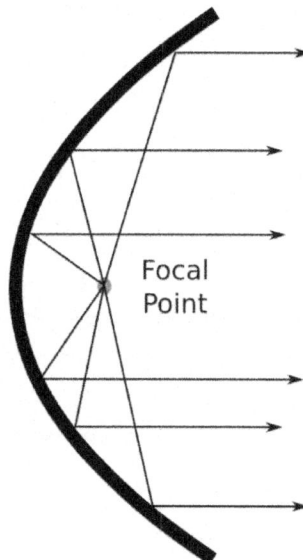

Focal Point

Figure 53.4: Parabolic Dish Antenna

This makes them very useful for highly directional antennas. The catch is, the parabola must be wider than one wavelength of what is being reflected. Bigger is better, but smaller is no good.

348

Given that, you can see that as the operating frequency – the frequency it is transmitting, in other words -- increases to shorter wavelengths, the parabolic dish will be more and more reflective. The precise amount that the effectiveness increases is 6 dB per octave – every time the frequency doubles, **gain increases by 6 dB.** Why 6 dB? Basic geometry. By doubling the frequency, we've effectively doubled the diameter of the parabolic dish relative to the wavelength. When a circle's diameter is doubled, the area increases 4 times. 6 dB is a 4 times increase.

Yagi Antennas

E9D05 (B) What usually occurs if a Yagi antenna is designed solely for maximum forward gain?

   A. The front-to-back ratio increases
   B. **The front-to-back ratio decreases**
   C. The frequency response is widened over the whole frequency band
   D. The SWR is reduced

As hams, we learn early that antennas are all about compromise. With Yagi antennas, it is easy to get caught up in the quest for Maximum Forward Gain. Unfortunately, if that's all we design for, we can go from a fairly typical Yagi pattern with a quite good front-to-back ratio and decent gain, as shown in Figure 53.5, to something where we've narrowed down the front lobe, and squeezed those side lobes in very tight, but they've sort of shot out the back end of the antenna, ruining our front-to-back ratio, as shown in Figure 53.6.

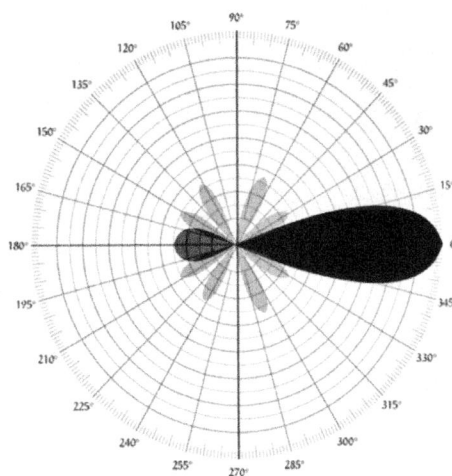

Figure 53.5: Typical Yagi Pattern, Azimuth View

At least these days we can make those mistakes in an antenna modeling program instead of real life!

E9D02 (C) How can linearly polarized Yagi antennas be used to produce circular polarization?

   A. Stack two Yagis fed 90 degrees out of phase to form an array with the respective elements in parallel planes
   B. Stack two Yagis fed in phase to form an array with the respective elements in parallel planes
   C. **Arrange two Yagis perpendicular to each other with the driven elements at the same point on the boom fed 90 degrees out of phase**
   D. Arrange two Yagis collinear to each other with the driven elements fed 180 degrees out of phase

To produce circular polarization from a pair of Yagi antennas we arrange them perpendicular to each other with the driven elements at the same point on the boom fed 90° out of phase.

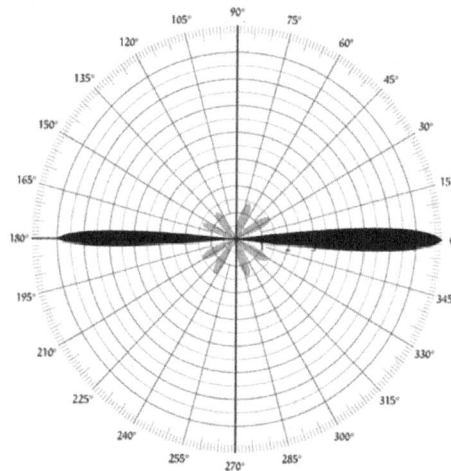

Figure 53.6: Yagi Over-Designed for Maximum Forward Gain

Figure 53.7 shows a USMC Radioman setting up a portable Yagi with perpendicular elements, probably preparing to transmit to a satellite.

Figure 53.7: US Marine Radioman Setting up Portable Circularly Polarized Yagi

The key word to look for in the answers is "perpendicular." Most circularly polarized antennas have elements at right angles to each other, and only one answer contains that word.

Loading Coils

E9D03 (A) Where should a high Q loading coil be placed to minimize losses in a shortened vertical antenna?
    A. **Near the center of the vertical radiator**
    B. As low as possible on the vertical radiator
    C. As close to the transmitter as possible
    D. At a voltage node

For this question, we're almost certainly talking about a mobile "whip" style antenna that needs a loading coil to make the antenna look electrically longer to the transmitter.

When an antenna is too short for a frequency, it presents a capacitive reactance to the transmitter, and to balance that we need to add some inductance.

We can place a coil at the base of the antenna, but there are some advantages to be gained by placing the coil toward the center of the antenna, having to do with the distribution of current in the antenna and our old friend radiation resistance. However, if we place a coil **near the center of the vertical radiator**, we need a higher value inductance than one placed at the base, and we end up gaining efficiency at the expense of bandwidth. So, despite the wording of the question, it's really less a matter of "where should we place a high Q coil" than of "if we choose to place a coil **near the center of the vertical radiator**, what Q coil do we end up with?"

E9D04 (C) Why should an HF mobile antenna loading coil have a high ratio of reactance to resistance?

    A. To swamp out harmonics

    B. To lower the radiation angle

    C. **To minimize losses**

    D. To minimize the Q

Why? Because we don't want to burn up our expensive watts heating a dumb resistor! We want our loading coils to be high reactance, low resistance **to minimize losses.**

E9D06 (B) What happens to the SWR bandwidth when one or more loading coils are used to resonate an electrically short antenna?

    A. It is increased

    B. **It is decreased**

    C. It is unchanged if the loading coil is located at the feed point

    D. It is unchanged if the loading coil is located at a voltage maximum point

The reason we can fit that HF antenna on our car's bumper is that big loading coil in the design, which electrically lengthens the antenna without physically lengthening it.

What we give up in the process is bandwidth. When loading coils are added, **bandwidth is decreased**.

Figure 53.8: Miniature Antenna Loading Coil (1912, New Jersey)

E9D09 (D) What is the function of a loading coil used as part of an HF mobile antenna?

    A. To increase the SWR bandwidth

    B. To lower the losses

    C. To lower the Q

    D. **To cancel capacitive reactance**

The reactance of an antenna that is too short for the chosen frequency is capacitive. **To cancel capacitive reactance** we add some inductance.

E9D07 (D) What is an advantage of using top loading in a shortened HF vertical antenna?

    A. Lower Q

    B. Greater structural strength

    C. Higher losses

    D. **Improved radiation efficiency**

To top load a shortened antenna, we cap it with a "capacitance hat." That has the same effect as putting a loading coil at the base or the center, but with the advantage of minimizing losses. By minimizing losses, we get **improved radiation efficiency**.

We also get a physically challenging design for some applications, particularly mobile HF antennas – that's a mighty big wind-catcher to put at the top of an antenna that's going to go 60 mph.

RF Grounding

E9D11 (B) Which of the following conductors would be best for minimizing losses in a station's RF ground system?
   A. Resistive wire, such as spark plug wire
   B. **Wide flat copper strap**
   C. Stranded wire
   D. Solid wire

We covered this pretty extensively in the General course, but to quickly review: to minimize losses in a station's RF ground system, **wide flat copper strap** is the stuff we want, at least out of the choices given. (Ground braid works, too, but they didn't offer us that.)

Remember, compared to household current, even our 160-meter waves are incredibly high frequencies, so the only place they'll be conducted will be on the surface of the conductors – that's the skin effect. We also want nice, hefty conductors that will present minimal resistance. We certainly don't want a resistive wire. That would be insane. We don't want a puny little piece of stranded wire, nor do we want a single solid wire.

I know **wide flat copper strap** is expensive, as is ground braid, but suck it up, buttercup; there's no substitute. It will last you a lifetime and get the job done.

E9D12 (C) Which of the following would provide the best RF ground for your station?
   A. A 50-ohm resistor connected to ground
   B. An electrically short connection to a metal water pipe
   C. **An electrically short connection to 3 or 4 interconnected ground rods driven into the earth**
   D. An electrically short connection to 3 or 4 interconnected ground rods via a series RF choke

No, we don't want a 50-ohm resistor in our ground connection, nor do we want to try to ground RF through an RF choke that won't let it go to ground. And, yes, I know we've all done it at some point in our radio adventures, but grounding to a metal water pipe really is not a good idea, if for no other reason than those water pipes and the water in them go to the rest of your house – and your neighbor's house.

The best RF ground for your station is **an electrically short connection to 3 or 4 interconnected ground rods driven into the earth**. Those rods need to be bonded – electrically connected – and that bond needs to go all the way to the house ground rod. For more details on station grounding please see the *Fast Track to Your General Class Ham Radio License, The Fast Track to (Finally!) Getting on the Air With Ham Radio*, or Ward Silver's excellent book *Grounding and Bonding for the Radio Amateur* from the ARRL.

Key Concepts in This Chapter

- The effect of a terminating resistor on a rhombic antenna is to change the radiation pattern from bidirectional to unidirectional.

- When the operating frequency of an ideal parabolic dish antenna is doubled, the gain increases by 6 dB.

- If a Yagi antenna is designed solely for maximum forward gain, the front to back ratio decreases.

352

- Linearly polarized Yagi antennas can be used to produce circular polarization. Arrange two Yagis perpendicular to each other with the driven elements at the same point on the boom fed 90° out of phase.

- High Q loading coils for shortened vertical antennas should be placed near the center of the vertical radiator and should have a high ratio of reactance to resistance to minimize losses. When one or more loading coils are used to resonate an electrically short antenna, SWR bandwidth is decreased.

- The function of a loading coil used as part of an HF mobile antenna is to cancel capacitive reactance.

- An advantage of using top loading in a shortened HF vertical antenna is improved radiation efficiency.

- The best type of conductor for minimizing losses in a station's RF ground system is a wide flat copper strap forming an electrically short connection to 3 or 4 interconnected ground rods driven into the earth.

Once you've mastered the Practice Exam for this chapter, you'll be ready for the next-to-last Progress Check, which is #11. Have you set the date for your exam?

Chapter 54

Antenna Systems

Matching Networks

This chapter asks you to have some familiarity with several types of antenna matching networks – often known as "matches" – and throws in a couple of other doodads you might attach to an antenna.

Matches connect feed lines and antennas. They might be designed to match impedances, to connect a balanced feed line to an unbalanced antenna or vice versa, or even to do both.

We're not going to do a deep exploration of the math and physics involved in these – that's at least a whole 'nother volume of information, and much more than the exam requires.

The matching networks to know a little about include:

- The delta match
- The gamma match
- The stub match
- The hairpin match

Delta Match

E9E01 (B) What system matches a higher-impedance transmission line to a lower-impedance antenna by connecting the line to the driven element in two places spaced a fraction of a wavelength each side of element center?

    A. The gamma matching system

    B. The delta matching system

    C. The omega matching system

    D. The stub matching system

Typical Yagi antenna feed point impedances are in the 20 to 25 ohm range. Typical ham transmitter outputs and feed lines are 50 ohms. Unless we're willing to tolerate a minimum of 2:1 SWR, we'll need to match the feed point impedance.

Many Yagi installations use a gamma match, which was covered briefly in your General exam, and which we will touch on again, shortly. However, there is another way to match a higher impedance transmission line to a lower impedance antenna, and it is one that does offer some advantages over the gamma match. It's called the **delta matching system**.

Figure 54.1 shows a classic Yagi antenna.

Reflector

Radiator   Directors

Yagi Antenna
(Top View)

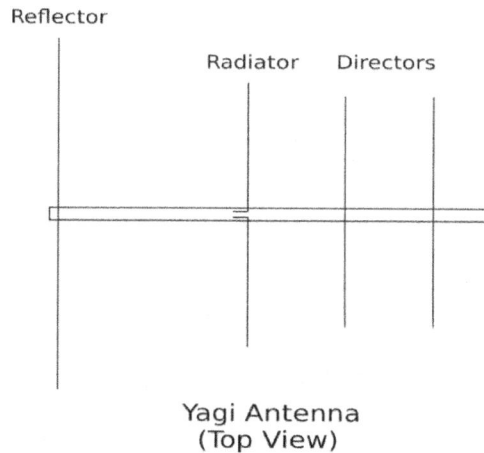

Figure 54.1: Yagi Antenna

Notice the driven radiator element is split in the middle. That's where the feed line usually connects, if it is a balanced Yagi. If we build the antenna to use a delta match, the radiator is just one solid piece. Advantage #1 for the delta, because that's easier to fabricate.

To connect to the radiator element, we fabricate two arms that form a triangle. The Greek letter delta (Δ) is shaped like a triangle, and that's where this match got its name. We attach one end of each arm to the supporting boom, the other to the radiator. Those arms are going to conduct the signal to the radiator, and they also serve as mechanical braces. Advantage #2.

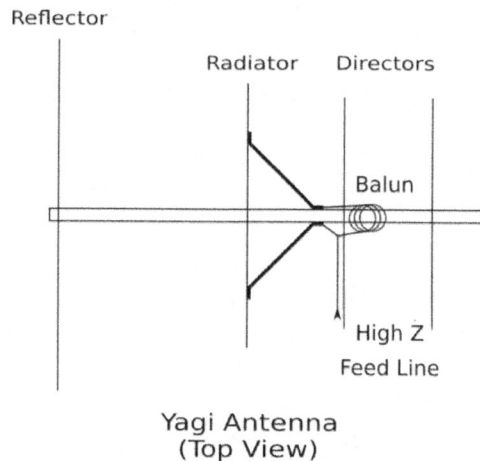

Reflector

Radiator   Directors

Balun

High Z
Feed Line

Yagi Antenna
(Top View)

Figure 54.2: Yagi Antenna with Delta Match

The actual impedance matching is accomplished both by the spacing of the contact points of the arms on the radiator and by the balun matching transformer, constructed from ½ wavelength of coax. (All the coax shields are electrically connected to the boom.)

The delta match is a low-Q matching network, so it offers more bandwidth than a gamma match. Advantage #3.

However, it can get cumbersomely large when designed for antennas with lower frequencies. That ½ wavelength of coax is going to be a coil of about 40 feet of coax at 40 meters. Disadvantage #1.

For this question on the exam, remember in the **delta matching system** *both* connection points are *a fraction of a wavelength each side of element center* – the delta is the only matching system mentioned in the exam that is described that way.

356

Gamma Match

E9E02 (A) What is the name of an antenna matching system that matches an unbalanced feed line to an antenna by feeding the driven element both at the center of the element and at a fraction of a wavelength to one side of center?

    A. **The gamma match**
    B. The delta match
    C. The epsilon match
    D. The stub match

Like the delta match, the **gamma match** gets its name from its physical resemblance to a Greek letter; in this case gamma (Γ).

    The gamma match is a conductor inside, and electrically isolated from, a conductive tube, usually copper pipe. The inner conductor can be another piece of smaller copper tubing or just the inner conductor and insulator of a piece of coax. The inner conductor is fed by the center conductor of the feed line. The feed line's shield is grounded to the supporting beam.

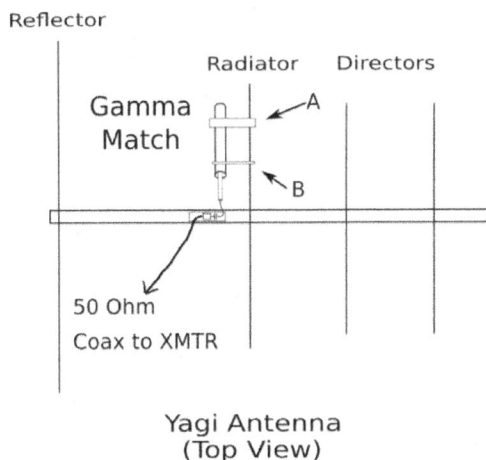

Figure 54.3: Yagi Antenna with Gamma Match

    The standoff that attaches at point A is a conductor and the standoff that attaches at point B is an insulator. (If you ignore the insulator you can see the resemblance to the Greek letter gamma, Γ. Kinda sorta ...)

    That arrangement of parts serves both as an impedance matching network and as a match between the unbalanced feed line and the antenna. Think of it as a matching transformer. Do note there is a small series capacitor connected ahead of the gamma match – that's to cancel the inherently inductive reactance of the rest of the gamma match.

    The key phrase to remember for the gamma match is *unbalanced feed line* – it's the only thing in this section with that phrase.

E9E04 (B) What is the purpose of the series capacitor in a gamma-type antenna matching network?

    A. To provide DC isolation between the feed line and the antenna
    B. **To cancel the inductive reactance of the matching network**
    C. To provide a rejection notch that prevents the radiation of harmonics
    D. To transform the antenna impedance to a higher value

Capacitive reactance cancels inductive reactance and to RF that "tube inside a tube" of the gamma network looks like an inductance. To balance that out, a small value capacitor is inserted in series **to cancel the inductive reactance of the matching network**.

E9E09 (C) Which of the following is used to shunt-feed a grounded tower at its base?

    A. Double-bazooka match

    B. Hairpin match

    C. **Gamma match**

    D. All these choices are correct

The **gamma match** is an effective method of connecting a 50-ohm coaxial cable feed line to a grounded tower so the tower itself can be used as a vertical antenna. For an example, see Figure 54.4.

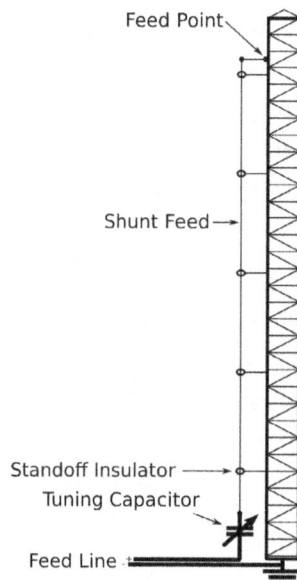

Figure 54.4: Shunt-Fed Tower

One side of the feed line is attached to ground, as is the base of the tower. The other side travels up the tower to the feed point, which is calculated to the be point with impedance that matches the transmitter. The tuning capacitor is used to tune out any remaining inductive reactance.

There's no such thing as a "double-bazooka match", though there is an antenna design called a double-bazooka.

The hairpin match is not a workable choice. As you'll learn soon, the hairpin's coil is grounded on one side, and in this case, the tower is also grounded. That won't work – all the signal will just go to ground.

Stub Match

E9E03 (D) What is the name of the matching system that uses a section of transmission line connected in parallel with the feed line at or near the feed point?

    A. The gamma match

    B. The delta match

    C. The omega match

    D. **The stub match**

At radio frequencies, we can use a stub of coax as an inductor or a capacitor to achieve impedance matching. When we do that, it's called, simply enough, a **"stub match."**

It can be as simple as inserting a "T" splitter in your coax and screwing on the stub. The trick is that splitter has to be at a point just the right distance from the connection to the antenna and the stub has to be just the right length. It will only work at precisely one frequency.

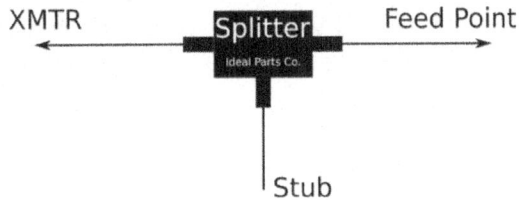

Figure 54.5: Stub Match

The stub of coax can be shorted at the end or open, depending on the design. We generally prefer shorted stubs because they don't become an extra radiator of radio waves.

How in the world can a shorted chunk of coax act as a matching network? At the right frequency, it's a resonant circuit, and it has reactance. Depending on the length and whether the unconnected end is shorted or open, the reactance can be capacitive or inductive.

In the next chapter we'll cover something called a Smith chart; this is the sort of problem for which Smith charts can be used.

E9E10 (C) Which of these choices is an effective way to match an antenna with a 100-ohm feed point impedance to a 50-ohm coaxial cable feed line?

A. Connect a 1/4-wavelength open stub of 300-ohm twin-lead in parallel with the coaxial feed line where it connects to the antenna

B. Insert a 1/2 wavelength piece of 300-ohm twin-lead in series between the antenna terminals and the 50-ohm feed cable

C. Insert a 1/4-wavelength piece of 75-ohm coaxial cable transmission line in series between the antenna terminals and the 50-ohm feed cable

D. Connect 1/2 wavelength shorted stub of 75-ohm cable in parallel with the 50-ohm cable where it attaches to the antenna

This is another form of stub match; it uses coaxial cable as an impedance matching device. There are two ways to approach this question. One is, memorize "50-ohm feed line, 100-ohm feed point impedance, ¼ wavelength, series."

The other way is with the actual formula for an impedance matching transformer, which is what that ¼ wavelength piece of 75-ohm coax will become. Specifically, it becomes what is known as a synchronous transformer.

The formula is;

$$Z_{Transformer} = \sqrt{Z_1 \times Z_2}$$

We just multiply the two mismatched impedances together, take the square root of the product, and that's the impedance we want for the matching transformer.

$$Z_{Transformer} = \sqrt{Z_1 \times Z_2} = \sqrt{50\Omega \times 100\Omega} = \sqrt{5000\Omega} = 70.71\Omega \approx 75\Omega$$

You can see we came up with 70.71 ohms, but the ham radio store was fresh out of 70.71 ohm coax, so 75 ohms is, as we say, close enough for radio!

Why ¼ wavelength? I promised we wouldn't go too deeply into the physics, so suffice it to say that at ¼ wavelength, that "mismatched" piece of coax creates reflecting waves, the same way any impedance mismatch creates reflecting waves. But in this special case, the waves reflecting back toward the 50-ohm feed line and the waves reflecting toward the load precisely cancel each other, effectively matching the impedances.

E9E06 (C) Which of these feed line impedances would be suitable for constructing a quarter-wave Q-section for matching a 100-ohm loop to 50-ohm feed line?

A. 50 ohms

B. 62 ohms

C. 75 ohms

D. 450 ohms

The quarter-wave length of coax referred to in question E9E10 above is sometimes called a *Q section*. Aside from that terminology, this is essentially the same question as above with the same method of reaching the same answer.

Hairpin Match

E9E05 (A) How must an antenna's driven element be tuned to use a hairpin matching system?

    A. **The driven element reactance must be capacitive**

    B. The driven element reactance must be inductive

    C. The driven element resonance must be lower than the operating frequency

    D. The driven element radiation resistance must be higher than the characteristic impedance of the transmission line

Electrically, a hairpin matching system is simply an inductance connected in parallel across the feed line at the antenna feed point. For bottom-fed vertical antennas, it really is a big copper coil sitting at the base of the antenna and connected to each leg of the feed line, with one side then going to the base of the antenna – which must be insulated from ground -- and the other going to the radial system.

    Up in the VHF and UHF world, where Yagis are smaller and thus more common, that coil doesn't look a lot like a coil. It looks something like Figure 54.6.

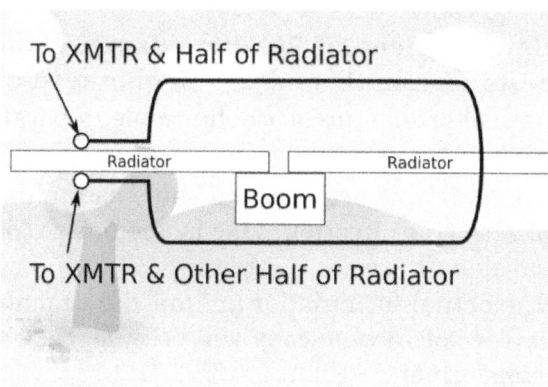

Figure 54.6: Hairpin Match

    In that illustration you're looking right down the boom, toward the front of the antenna. That loop is the "hairpin" that serves as an inductor. (Remember, impedance is a function of frequency, and at high frequencies, a little bit of inductance goes a long way.) We're back to a split radiator configuration on the Yagi, with one terminal of the hairpin connected to one half, and the other to the other.

    Since we're using an inductor to match impedance, the Yagi's radiator must be cut a little short for the frequency, so it becomes a capacitive reactance. In this setup, **the driven element reactance must be capacitive** so the inductance and capacitance balance out.

Wilkinson Divider

E9E08 (C) What is a use for a Wilkinson divider?

    A. It divides the operating frequency of a transmitter signal so it can be used on a lower frequency band

    B. It is used to feed high-impedance antennas from a low-impedance source

    C. **It is used to divide power equally between two 50-ohm loads while maintaining 50-ohm input impedance**

    D. It is used to feed low-impedance loads from a high-impedance source

If we want to **divide power equally between two 50-ohm loads while maintaining 50-ohm input impedance**, we have a challenge. If we hook the loads up in series, now our load impedance is 100-ohms. If we put them in parallel, it's 25 ohms.

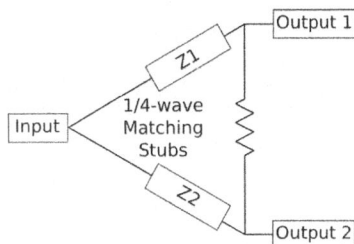

Figure 54.7: Wilkinson Divider

The Wilkinson divider meets this challenge by splitting the input through two series impedances with a resistor across the outputs of the impedances. The impedances are typically 1/4-wave lengths of coax.

Figure 54.7 shows the equivalent circuit.

Where might this come into play? How about if we have two driven elements on an antenna, each with the same impedance? Yep, one of these Wilkinson dividers would be just the ticket.

Reflected Power

E9E07 (B) What parameter describes the interactions at the load end of a mismatched transmission line?

    A. Characteristic impedance

    B. **Reflection coefficient**

    C. Velocity factor

    D. Dielectric constant

If we mismatch the load and the transmission line, we get reflected power, and the amount we get is the **reflection coefficient**. That's just another way to say "reflectance", which is what creates Standing Wave Ratios greater than 1:1.

The reflection coefficient, Γ, is equal to the reflected voltage, V^-, divided by the forward voltage, V^+.

$$\Gamma = \frac{V^-}{V^+}$$

The reflection coefficient converts directly to SWR:

$$SWR = \frac{1 + |\Gamma|}{1 - |\Gamma|}$$

Phasing Lines

E9E11 (A) What is the primary purpose of phasing lines when used with an antenna having multiple driven elements?

    A. **It ensures that each driven element operates in concert with the others to create the desired antenna pattern**

    B. It prevents reflected power from traveling back down the feed line and causing harmonic radiation from the transmitter

    C. It allows single-band antennas to operate on other bands

    D. It creates a low-angle radiation pattern

A phasing line can be as simple as a piece of coax cut to a length that corresponds to the amount of phase difference you want to create between two elements. Its purpose is to **ensure that each driven element operates in concert with the others to create the desired antenna pattern**.

Remember our twin vertical antennas spaced ¼ wavelength apart back on page 339? In principle, if we strung feed line to the first tower, split our feed, then ran "ideal" feed line straight across the field to the second tower, that would create the 90° of phase delay we want for that second tower, automatically. And that would be an example of a phasing line. (In the real world we'd control the phasing with reactances, making the directionality of the system adjustable.)

More commonly, we think of phasing lines being used in a multiple driven element beam antenna for VHF and higher frequencies, just because of size and weight considerations.

Key Concepts in This Chapter

- The system that matches a higher impedance transmission line to a lower impedance antenna by connecting the line to the driven element in two places spaced a fraction of a wavelength each side of element center is the delta matching system.

- The name of an antenna matching system that matches an unbalanced feed line to an antenna by feeding the driven element both at the center of the element and at a fraction of a wavelength to one side of center is the gamma match.

- The purpose of the series capacitor in a gamma-type antenna matching network is to cancel the inductive reactance of the matching network.

- A gamma match can be used to shunt-feed a grounded tower at its base.

- The name of the matching system that uses a section of transmission line connected in parallel with the feed line at or near the feed point is the stub match.

- An effective way to match an antenna with a 100-ohm feed point impedance to a 50-ohm coaxial cable feed line is to insert a 1/4 wavelength piece of 75-ohm coaxial cable transmission line in series between the antenna terminals and the 50-ohm feed cable.

- The correct feed line impedance for constructing a quarter-wave Q-section for matching a 100-ohm loop to a 50-ohm feed line is 75 ohms.

- To use a hairpin matching system, the driven element of a 3-element Yagi must be tuned to be capacitive.

- One use for a Wilkinson divider would be to divide power equally between two 50-ohm loads while maintaining 50-ohm input impedance.

- The term that best describes the interactions at the load end of a mismatched transmission line is "reflection coefficient."

- The primary purpose of phasing lines when used with an antenna having multiple driven elements is to ensure that each driven element operates in concert with the others to create the desired antenna pattern.

Chapter 55

The Smith Chart

When we start calculating things like, "If my antenna has a complex impedance of (250, +j 58) at frequency X, and my feed line is 75 ohms, what kind of SWR will I have?" and "what impedance does a ¼ wavelength long piece of transmission line present if the far end is shorted?" or even, "What components should I use to build a matching network?", we quickly find the math gets, to use a very technical math term, "freakin' hairy." That's why we have Smith charts.

Phillip Smith was a researcher at Bell Labs, where he went to work in 1928. He sounds like quite an interesting guy. When he was going through Electrical Engineering school, he commuted in a reconstructed Model T, then later on his Harley-Davidson motorcycle. He was a ham radio operator – call sign 1ANB, and a private pilot. He did some of the early development work on directional antenna arrays for commercial AM radio stations. He had a special interest in transmission lines. Transmission line measurements were tedious at best in those days, involving numerous physically challenging measurements and enormous quantities of calculations. He just knew there had to be a better way.

The Smith chart was the better way. In 1936, or thereabouts, he had developed the beginnings of what we now know as the Smith chart as an easy way to see, on one chart, what would be the results of lots of calculations.

Around 1940 came the real breakthrough for the chart. Smith tossed his rectangular coordinate graph paper out the window, then sent his polar coordinate graph paper out right behind it. Instead he plotted all those values on graph paper with axes that were a series of circles offset from each other – and that let him represent resistance, capacitive reactance, inductive reactance, impedance, SWR, and their interrelationships all on one chart. Solving complex transmission line problems went from hours of slide rule calculations – no computers in those days! -- to tracing a few lines on a chart, and maybe using a compass to draw a circle.

Smith went on to contribute to the development of radar, and you can thank him for high-power coaxial lines and the adjustable stub tuner, among other things.

Okay, here comes the only scary part of the chapter – your first look at an actual Smith chart in Figure 55.1 on page 364. Don't panic – remember, it makes things *easier*!

The actual chart used in the exam is a vastly simplified version of the illustration – but I want you to see what the real graph paper looks like. At the size we can reproduce in print or on e-readers, the numbers on this one probably won't be legible – don't sweat that. The important things to learn are the parts of the chart and what the chart represents – there isn't a single question that asks you to solve a numerical problem with a Smith chart on the exam.

Let's take this thing apart piece by piece. The Smith Chart is a coordinate system that consists of *resistance circles* and *reactance arcs*.

We'll start with the big circle around the outside as shown in Figure 55.2 on page 365.

That's the *reactance axis*, and any point on it represents a particular reactance. Think of it as the vertical axis of a phasor diagram bent into a circle.

Then, in Figure 55.3 we'll add an axis right across the center that represents resistance. No capacitive reactance, no inductive reactance, just pure resistance. It runs from zero ohms on

363

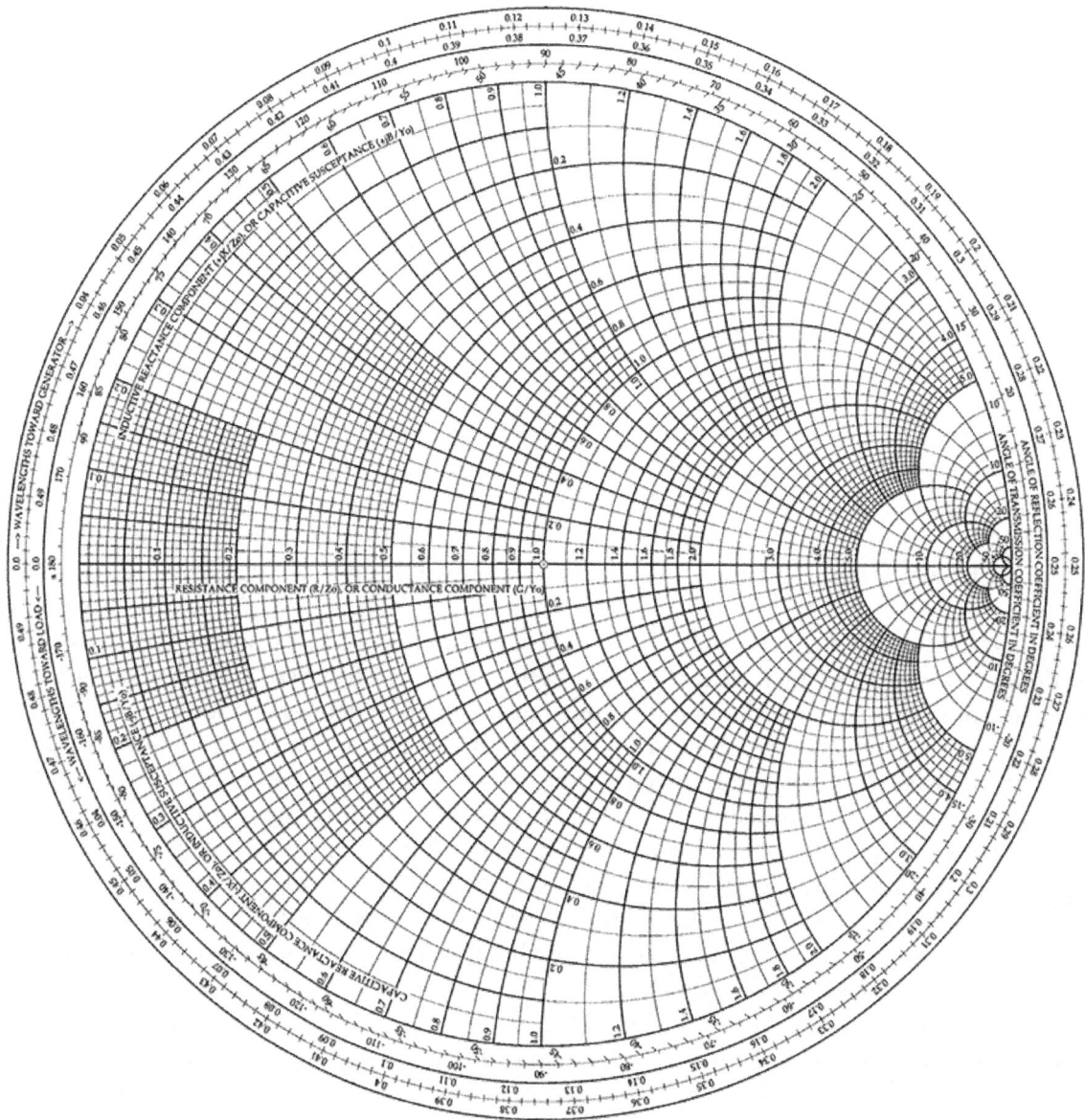

Figure 55.1: Smith Chart

the left – a short circuit -- to infinite ohms on the far right – an open circuit. It's the horizontal axis of a phasor diagram.

Now, that's all well and good except we seldom have circuits with zero or infinite resistance, so there are more circles to represent constant real-world values of resistance. They all intersect that infinite resistance point on the right. (They don't look like they do because other circles get in the way, but they do.) You can see a few of them highlighted in Figure 55.4 (page 366.)

Each spot on a resistance circle represents the value of resistance indicated by the spot where it intersects that horizontal resistance axis. (The "not infinity" spot where it intersects.)

There's another set of circles on the chart, shown in Figure 55.5 but we only see part of the arcs of those circles, because the rest of them are impossible.

Let's get rid of the parts that are, literally, off the chart ....and make a little more room ....as in Figure 55.6, page 366.

Those are the *reactance arcs*. The reactance arcs in the top half of the chart – above the resistance axis – represent inductive reactance, the arcs in the bottom half are capacitive

364

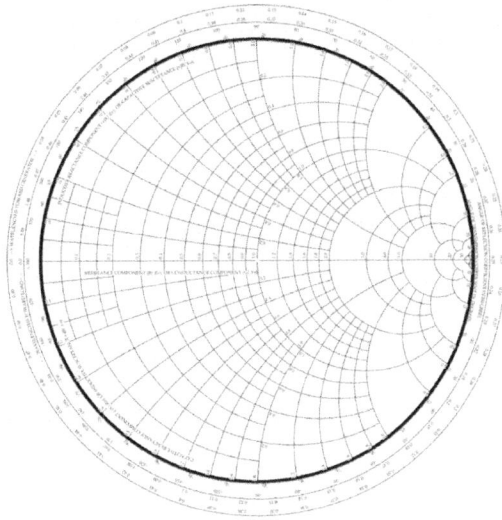

Figure 55.2: Reactance Axis

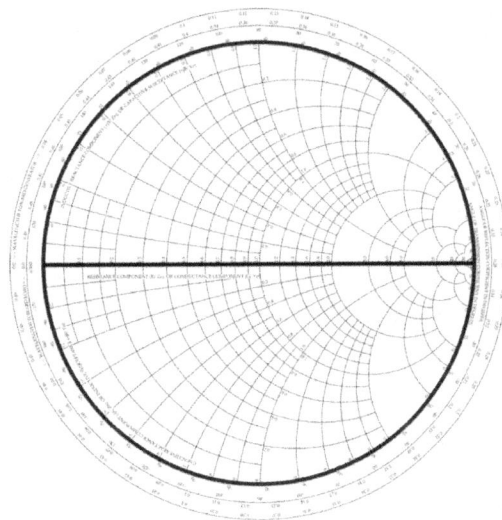

Figure 55.3: Resistance Axis

reactance. The value they represent is the value where they intersect the circular reactance axis.

The "shape" and the spatial relationships of the lines on the Smith chart are universal – they'll apply no matter what the values. If we could computerize the paper the Smith chart is printed on (which is precisely what some network analyzers do) then the basic values could be set to whatever we want – but to make it work on paper, we have to have a common starting point. In effect, we set the chart up with the assumption that one part of the system has a fixed value. That value is the nominal impedance of the transmission line in the system, and on this chart – and most Smith charts -- it's 50 ohms. That value goes at the "prime center" of the chart, dead center on the horizontal resistance axis. That process of setting that fixed value is called "normalizing the chart."

Because the chart is normalized to a 50-ohm transmission line, we also have to normalize all the values we plug into it.

Very simple – divide everything by 50. I'll demonstrate in a bit.

With values plugged in, the Smith chart becomes something very much like an old-fashioned slide rule.

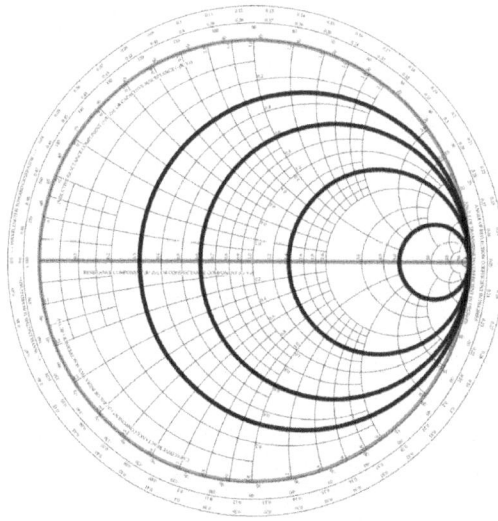

Figure 55.4: Resistance Circles

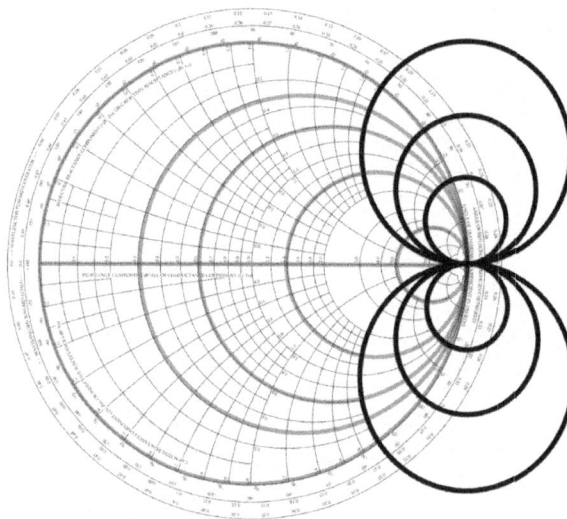

Figure 55.5: Reactance Circles

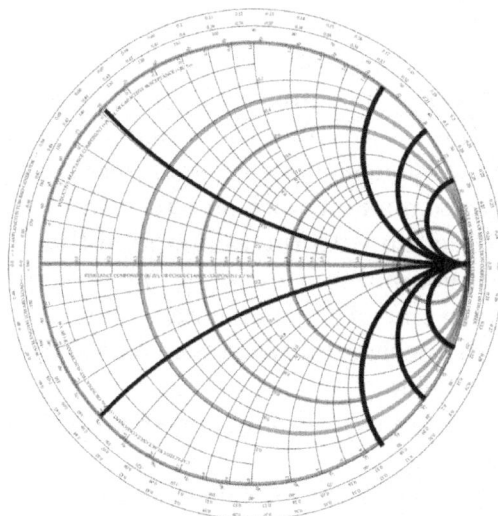

Figure 55.6: Reactance Arcs

Beginning to Use the Smith Chart

You are now armed with at least 90% of the knowledge about Smith charts you need to pass your Extra exam, but we've come this far – we might as well take a quick look at how the contraption actually works, then you can use that knowledge when we come to the section in the next chapter on the characteristics of various coax stubs.

We'll start with a simple example. We want to attach a load to our 50-ohm coax. We know the resistance and the reactance of the load – those same values we used to create our phasor diagram. (One way to get those values would be with an antenna analyzer.)

What we want to know is what the SWR will be with that setup.

To keep numbers simple, let's say our load has a 50-ohm resistance and 50 ohms of inductive reactance. Formally, those are rectangular coordinates of (50, +j50.)

Step 1 is to normalize our values. We divide the resistance and the reactance by the normalizing number of the chart, which is 50. That gives us (1, +j1.) The exam refers to this as "reassigning impedance values with regard to the prime center."

Next, we plot the resistance point on the resistance axis. It's easy enough to find, 1.0 is smack dab in the center of the resistance axis. (Figure 55.7.)

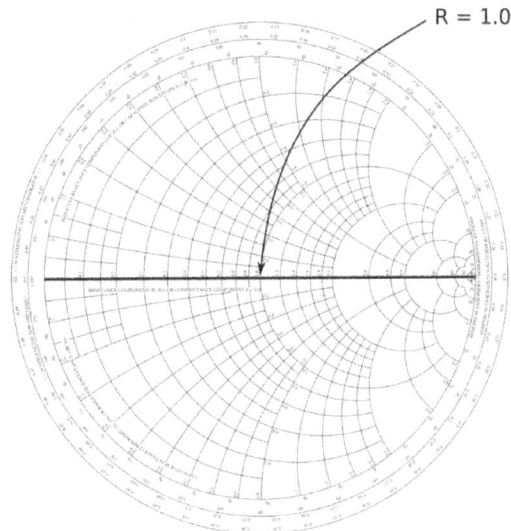

Figure 55.7: Plot the Resistance

Next, we grab our gigantic #2 pencil and trace the resistance circle that crosses the resistance axis at that 1.0 point.

That circle shows a resistance of 1.0 (normalized) ohms at every point.

Next, we add the reactance point on the reactance axis. The reactance axis is that circle that marks the outside border of the chart. You probably can't see the numbers, but there are (tiny) numbers showing the values just inside that outer border. Negative numbers are on the bottom half of the chart – those are the capacitive reactances, the –j numbers. Our reactance is a +j, so we'll look on the top half of the chart....and there it is. 1.0.

Next, we trace that reactance arc with our gigantic #2 pencil .....

Every point on that arc has a reactance of +j1. That spot where the resistance circle and the reactance arc meet is the (normalized) impedance of our load, which we can read on the chart as .... as .....well, the chart doesn't tell us the impedance in ohms. But we know that's a simple calculation, anyway, and that wasn't our question. We wanted to know something practical; what will the SWR be?

We grab our handy compass. Not the find your way in the woods kind; the make a circle kind.

We put the point of the compass right in the center of the chart, at 1.0 on the resistance

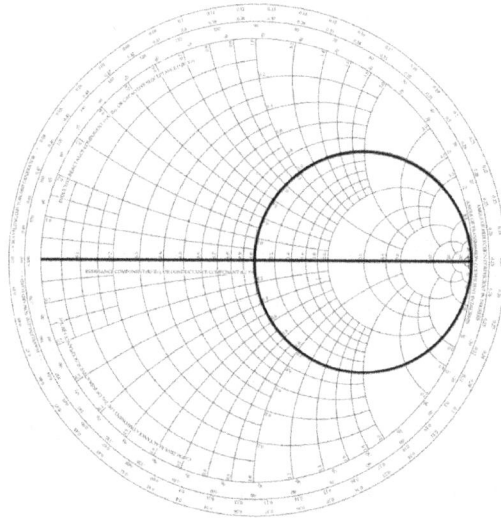

Figure 55.8: Trace Resistance Circle

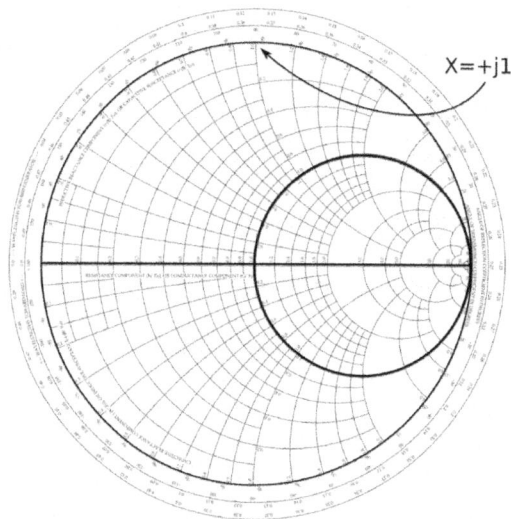

Figure 55.9: Find the Reactance on the Reactance Axis

scale. Then we set the pencil on the impedance point, and draw a circle. What we have added is called a standing wave ratio circle. We read the SWR on the resistance scale. (We read it on the right hand side, to the right of 1.0, because there's no such thing as a SWR of less than 1:1.) I read it as 2.6:1.

Just like a slide rule, there are *many* more things you can calculate using a Smith chart. You can buy whole books on using the Smith chart to solve far more complex problems than the one we just saw. If you'd like to explore a bit more, or just get another grounding in the basics, a fellow named Carl Oliver has a great set of videos on the topic on youtube. Get started with this one:

https://www.youtube.com/watch?v=hmqM8PnUkmo

There's also a whole site devoted to the Smith chart created by the Smith Chart Amateur Radio Society:

http://smithchart.org/phsmith.shtml

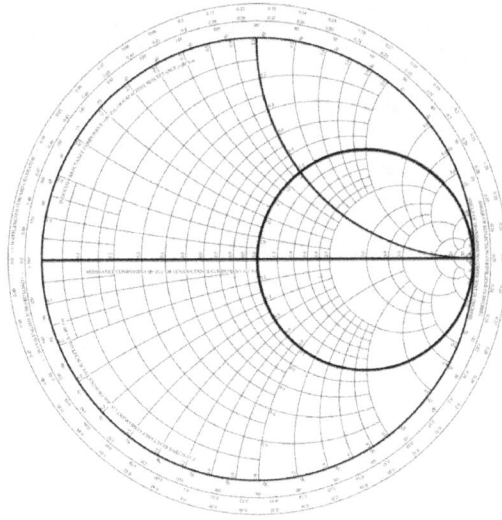

Figure 55.10: Trace the Reactance Arc

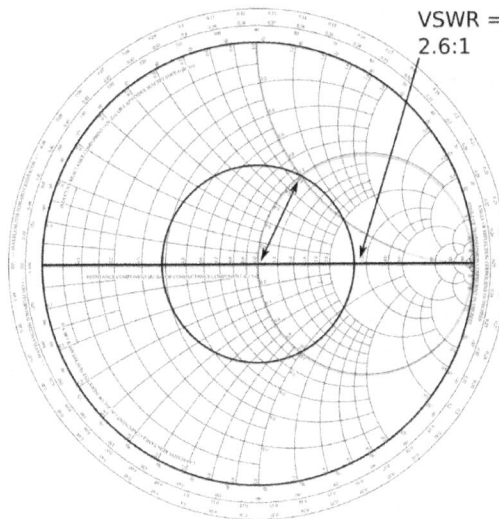

Figure 55.11: Draw the SWR Circle and Read Answer

Smith Chart Questions

E9G01 (A) Which of the following can be calculated using a Smith chart?
  A. **Impedance along transmission lines**
  B. Radiation resistance
  C. Antenna radiation pattern
  D. Radio propagation

You may be shocked – shocked, I say! – to learn that transmission line science is a bit more complex than we teach you at the Technician and General Class levels. In fact, the impedance of a transmission line changes along the line, especially at certain fractions and multiples of the wavelength they are carrying.

The Smith chart can help you calculate **impedance along transmission lines**.

E9G05 (A) Which of the following is a common use for a Smith chart?
  A. **Determine the length and position of an impedance matching stub**
  B. Determine the impedance of a transmission line, given the physical dimensions
  C. Determine the gain of an antenna given the physical and electrical parameters

D. Determine the loss/100 feet of a transmission line, given the velocity factor and conductor materials

Smith charts won't help you calculate the impedance of a transmission line from its physical dimensions, they don't relate to the gain of antennas, and there's nowhere on a Smith chart to calculate the loss of a transmission line based on the conductor material. They are, however, very useful for solving any problem that relates to questions like, "What impedance is presented by a one-quarter wavelength stub of transmission line that is shorted at the far end?"

In the next chapter I'll show you how to use a primitive version of the Smith chart to do some simple problem solving.

E9G02 (B) What type of coordinate system is used in a Smith chart?
    A. Voltage circles and current arcs
    B. **Resistance circles and reactance arcs**
    C. Voltage lines and current chords
    D. Resistance lines and reactance chords

The Smith chart uses coordinates on **resistance circles and reactance arcs** to define values.

E9G04 (C) What are the two families of circles and arcs that make up a Smith chart?
    A. Resistance and voltage
    B. Reactance and voltage
    C. **Resistance and reactance**
    D. Voltage and impedance

The two families of circles and arcs that make up a Smith chart are the same values we used to plot a phasor diagram – **resistance and reactance**.

E9G03 (C) Which of the following is often determined using a Smith chart?
    A. Beam headings and radiation patterns
    B. Satellite azimuth and elevation bearings
    C. **Impedance and SWR values in transmission lines**
    D. Trigonometric functions

The Smith chart is very useful for solving for impedance and SWR values in transmission lines.

E9G06 (B) On the Smith chart shown in Figure E9-3, what is the name for the large outer circle on which the reactance arcs terminate?
    A. Prime axis
    B. **Reactance axis**
    C. Impedance axis
    D. Polar axis

The big circle around the outside of the Smith chart is the **reactance axis**.

E9G07 (D) On the Smith chart shown in Figure E9-3, what is the only straight line shown?
    A. The reactance axis
    B. The current axis
    C. The voltage axis
    D. **The resistance axis**

Exam figure E9-3 is on page 371. I have no idea why the illustrator for the committee decided to rotate the exam's Smith chart 90° clockwise from every other Smith chart ever printed – it may be a leftover from when the Smith chart trademark was still in force, since it only expired in 2015. In any case, you can tell by the circles and arcs that that is a Smith chart.

The only straight line on the whole Smith Chart is the horizontal line in the middle of the chart; **the resistance axis**.

E9G10 (D) What do the arcs on a Smith chart represent?
    A. Frequency
    B. SWR
    C. Points with constant resistance

D. Points with constant reactance

Every point along any reactance arc on a Smith chart has the same reactance. The lines represent **points with constant reactance**.

Figure E9-3

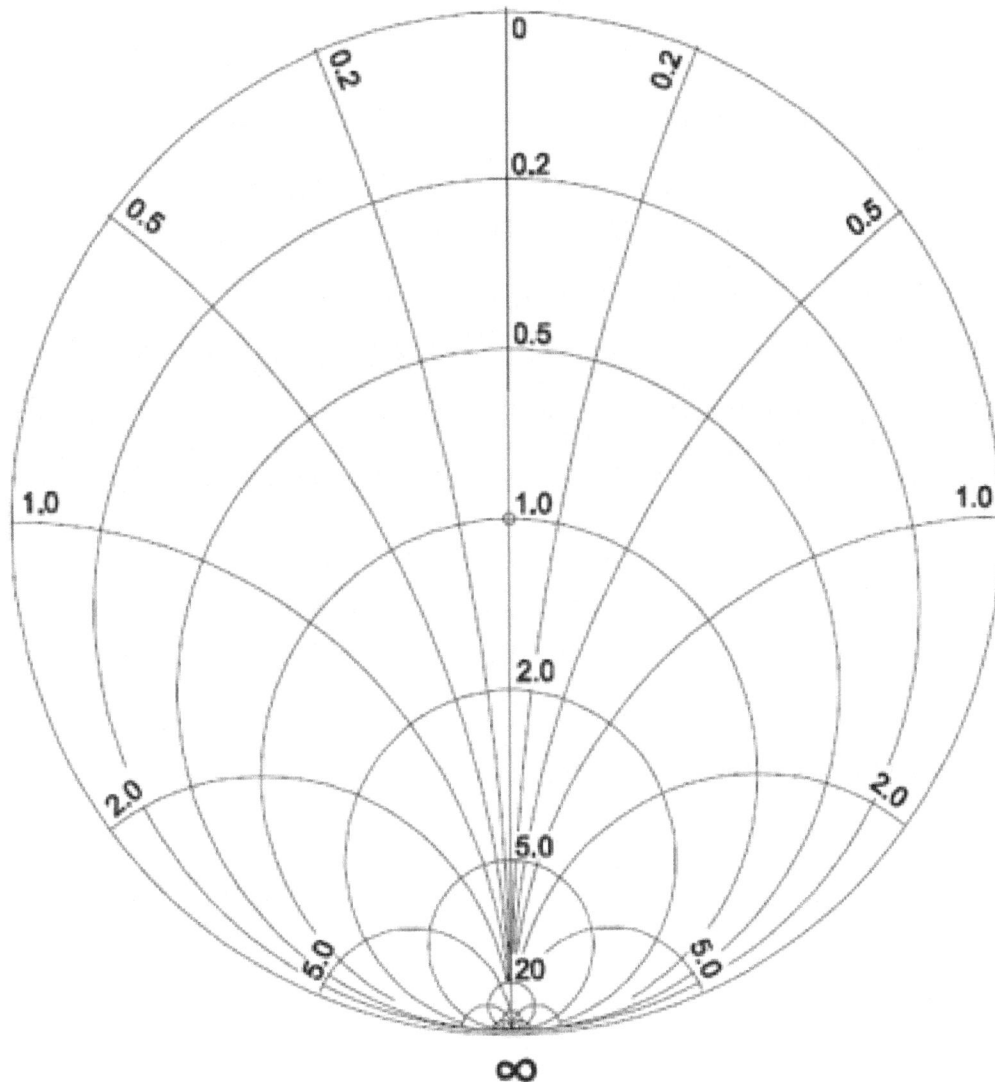

Figure 55.12: Figure E9-3 in the Question Pool

E9G08 (C) What is the process of normalization with regard to a Smith chart?
   A. Reassigning resistance values with regard to the reactance axis
   B. Reassigning reactance values with regard to the resistance axis
   C. **Reassigning impedance values with regard to the prime center**
   D. Reassigning prime center with regard to the reactance axis

Normalization is the process of **reassigning impedance values with regard to the prime center**.

   Since 50-ohm transmission line is quite common, most charts are normalized to 50-ohm transmission line – so that 1.0 right in the "prime center" represents 50 ohms of resistance. (That 50-ohm choice also makes the chart readable and useful at its standard size of 8½ x 11 inches.)

For us users, normalization is the process of correcting our resistance and reactance values to the chart's normalization of 1.0 = 50 ohms. We simply divide all our values by 50.

E9G09 (A) What third family of circles is often added to a Smith chart during the process of solving problems?

    A. **Standing wave ratio circles**
    B. Antenna-length circles
    C. Coaxial-length circles
    D. Radiation-pattern circles

The **standing wave ratio circle** was the one we drew with our compass to get the SWR in the example above.

E9G11 (B) How are the wavelength scales on a Smith chart calibrated?

    A. In fractions of transmission line electrical frequency
    B. **In fractions of transmission line electrical wavelength**
    C. In fractions of antenna electrical wavelength
    D. In fractions of antenna electrical frequency

Just know there is a wavelength scale on the Smith chart – it goes around the outside of the reactance axis – and it is calibrated in **fractions of transmission line electrical wavelength**.

I just have to share one last thing about Smith charts. Smith charts are such a "thing" with RF engineers, for Valentine's Day Microwaves101.com offers a free download of an Excel spreadsheet filled in with a set of S parameters that when plotted on a Smith chart creates a series of concentric hearts.

If the love of your life is electronically inclined, you can navigate to

https://www.microwaves101.com/59-downloads

and use Ctrl-F to search for the word "heart."

Key Concepts in This Chapter

- The coordinate system used in a Smith chart is resistance circles and reactance arcs. The two families of circles and arcs that make up a Smith chart are resistance and reactance.

- Impedance and SWR values in transmission lines are often determined using a Smith chart.

- The name for the large outer circle on which the reactance arcs terminate on a Smith chart is the reactance axis.

- The only straight line on a Smith chart is the resistance axis.

- The arcs on a Smith chart represent points with constant reactance.

- The process of normalization with regard to a Smith chart is reassigning impedance values with regard to the prime center.

- A third family of circles is often added to a Smith chart during the process of solving problems -- standing wave ratio circles.

- The wavelength scales on a Smith chart are calibrated in fractions of transmission line electrical wavelength.

Chapter 56

Transmission Lines

Velocity Factor

E9F01 (D) What is the velocity factor of a transmission line?
   A. The ratio of the characteristic impedance of the line to the terminating impedance
   B. The index of shielding for coaxial cable
   C. The velocity of the wave in the transmission line multiplied by the velocity of light in a vacuum
   D. **The velocity of the wave in the transmission line divided by the velocity of light in a vacuum**

You know how for two grades of license we've been saying "the speed of electricity equals the speed of light?" Heh. Not always. Sorry about that.

In uninsulated copper wire, yes, electricity propagates at very, very close to the speed of light in a vacuum. Even in insulated copper wire, it's about 99% of the speed of light in a vacuum.

In coaxial cable – not so much. In polyethylene dielectric coaxial transmission line, such as RG-8, for example, the waves are only traveling at about 66% of the speed of light in a vacuum. That's a number you'll want to remember for the exam. In the vast majority of our uses for coax, this makes absolutely no difference whatsoever. Who cares if the signal from your transmitter takes an extra nanosecond to get to the antenna? However, if we're cutting a ¼ wavelength of coax for a tuning stub, that's quite a significant difference. If we use the "divide into 300" rule, and we're cutting a stub of RG-8 for the 7.150 MHz range, we'll cut it at about 10.5 meters and – it won't even get close to working. Because of the velocity factor of RG-8, that stub only needs to be about 7 meters long.

We express this difference between the speed of light in a vacuum and the speed of a wavefront in a conductor with a factor which is **the velocity of the wave in the transmission line divided by the velocity of light in a vacuum.**

E9F02 (C) Which of the following has the biggest effect on the velocity factor of a transmission line?
   A. The termination impedance
   B. The line length
   C. **Dielectric materials used in the line**
   D. The center conductor resistivity

What sets the velocity factor for a piece of coax is not the resistivity, it's not the length of the line, and it has nothing to do with the termination impedance.

It's the **dielectric materials used in the line**.

Up till now in your ham radio education, we've often used "dielectric" and "insulator" interchangeably. For the most part, that works fine, but they're not precisely the same – at least in theory. An insulator is a substance that will not conduct electricity *and will not take a static charge*. A dielectric is a substance that will not conduct electricity and will take a static charge. In practice, there's no such thing as a "pure insulator" with absolutely zero dielectric properties, but – close enough. A pure insulator is electrically inert. A dielectric isn't – it interacts with the electricity near it by absorbing and releasing charges.

Physics tells us the speed of electricity in a given medium is dependent on several factors. In the conductors we typically use, the most important is the "permittivity" of the conductor. "Permittivity" may be one of the worst choices of words in all of physics – it sounds like it would "permit" something, but, in fact, it is a measure of a substance's *resistance* to being charged. As permittivity increases, the wave slows down.

The permittivity of our coax is determined almost 100% by the choice of **dielectric materials used in the line** – that "insulator" around the center conductor.

Consider – the speed of electricity is the speed of a wave, not of the electrons involved in the wave. When you see a wave on the ocean moving, say, 20 mph, that doesn't mean the whole ocean is moving toward you at 20 mph, nor even the water that is in the wave. The water is staying more or less in place, but the wave is moving. Now imagine the ocean suddenly turned into maple syrup – the waves would be much slower. The ocean's "permittivity" would have increased. A dielectric has the same effect on electrical waves.

E9F03 (D) Why is the physical length of a coaxial cable transmission line shorter than its electrical length?
   A. Skin effect is less pronounced in the coaxial cable
   B. The characteristic impedance is higher in a parallel feed line
   C. The surge impedance is higher in a parallel feed line
   D. **Electrical signals move more slowly in a coaxial cable than in air**

The physical length of a coaxial cable is shorter than its electrical length simply because of the velocity factor – **electrical signals move more slowly in a coaxial cable than in air.**

The effect isn't limited to coaxial cables, by the way. Velocity factor gets to be quite a consideration in complex computer networks, since Cat5 cable – that network cable that plugs into the back of your computer – has about the same velocity factor as RG-8 coax.

E9F05 (D) What is the approximate physical length of a solid polyethylene dielectric coaxial transmission line that is electrically 1/4 wavelength long at 14.1 MHz?
   A. 10.6 meters
   B. 5.3 meters
   C. 4.3 meters
   D. **3.5 meters**

Here's why you memorized that velocity factor of **0.66**. Let's work this out. We know the wavelength of 14.1 MHz in a vacuum is:

$$\frac{300}{14.1} = 21.28 \; meters$$

¼ wavelength would be

$$\frac{21.28 \; meters}{4} = 5.32 meters.$$

We also know that in a polyethylene dielectric coaxial transmission line, the velocity factor is 0.66, so we'll want an answer that's shorter than 5.32 meters. We'll multiply 5.32 meters by the velocity factor of 0.66.

$$5.32 \; meters \times 0.66 \; velocity \; factor = 3.51 \; meters \approx 3.5 \; meters$$

E9F09 (B) What is the approximate physical length of a foam polyethylene dielectric coaxial transmission line that is electrically 1/4 wavelength long at 7.2 MHz?

    A. 10.4 meters

    B. **8.3 meters**

    C. 6.9 meters

    D. 5.2 meters

Notice we've changed dielectric material with this question; it's asking about *foam polyethylene* dielectric. There's a range of values for the velocity factor of foam polyethylene coaxial, but they hover right around 0.80.

Again, we'll start by calculating the wavelength of this 7.2 MHz signal:

$$\frac{300}{7.2\ MHz} = 41.67\ meters$$

1/4 wavelength would be

$$\frac{41.67\ meters}{4} = 10.42\ meters$$

Now we take the final step of correcting that length for the velocity factor of this foam polyethylene dielectric coaxial cable, 0.80.

$$10.42\ meters \times 0.80\ velocity\ factor = 8.33\ meters \approx \textbf{8.3 meters}$$

E9F06 (C) What is the approximate physical length of an air-insulated, parallel conductor transmission line that is electrically 1/2 wavelength long at 14.10 MHz?

    A. 7.0 meters

    B. 8.5 meters

    C. **10.6 meters**

    D. 13.3 meters

An "air-insulated, parallel conductor transmission line" is not coax, but ladder line or window line. (The exam committee are probably the only people on planet earth who have ever called ladder line "air-insulated, parallel conductor transmission line.")

Air-insulated, parallel conductor transmission line has a velocity factor that is very close to 1.0; depending on the particular configuration, it's between 0.95 and 0.99.

We'll calculate the wavelength of 14.1 MHz.

$$\frac{300\ meters}{14.1\ MHz} = 21.28\ meters$$

Then we'll calculate the length of 1/2 wavelength;

$$\frac{1}{2}\ wavelength = \frac{21.28}{2} = 10.64\ meters$$

Hmmm ... I think I see the correct answer already ..... but let's be super-safe and work in that velocity factor of, say, 0.99 ....

$$10.64\ meters \times 0.99\ velocity\ factor = 10.53\ meters \approx 10.6\ meters$$

Transmission Line Characteristics

E9F08 (D) Which of the following is a significant difference between foam dielectric coaxial cable and solid dielectric cable, assuming all other parameters are the same?

    A. Foam dielectric has lower safe operating voltage limits

    B. Foam dielectric has lower loss per unit of length

    C. Foam dielectric has a higher velocity factor

    D. **All these choices are correct**

Not all coax uses solid polyethylene for a dielectric. There's also foam dielectric coax cable. It has significantly lower loss per unit of length than solid polyethylene, and a higher velocity factor, but it also has lower safe operating voltage limits compared to the same diameter of solid polyethylene cable. **All these choices are correct**.

E9F07 (A) How does ladder line compare to small-diameter coaxial cable such as RG-58 at 50 MHz?

    A. **Lower loss**

    B. Higher SWR

    C. Smaller reflection coefficient

    D. Lower velocity factor

You'd almost think that in the context of all the velocity factor questions, they'd be asking about ladder line's velocity factor, but the only choice they give us about that is incorrect. Ladder line's velocity factor is *higher* than coax. SWR is not really dependent on the particular type of transmission line, it's about the interaction of the transmitter with the transmission line and the transmission line with the antenna feed point, so that answer isn't right, and since reflection coefficient is related to SWR, that can't be the answer either.

    Ladder line does have **lower loss** than small-diameter coaxial cable at 50 MHz, though.

Coaxial Cable Stubs

E9F10 (C) What impedance does a 1/8-wavelength transmission line present to a generator when the line is shorted at the far end?

    A. A capacitive reactance

    B. The same as the characteristic impedance of the line

    C. **An inductive reactance**

    D. Zero

A 1/8 wavelength transmission line that is shorted at the far end presents **an inductive reactance**. Repeat after us: "Shorted 1/8 = inductive. Shorted 1/8 = inductive ..." Seems simple enough. However ...

    There are five of these questions, and that's a lot of wavelengths and opens and shorteds to remember. However, to figure out any of them really quickly, all you need is a very, very simple version of the Smith chart in your head, like the one in Figure 56.1. You don't need any of the arcs, nor any of the numbers; just the left and right ends of the resistance axis, and the big reactance circle around the outside, plus the knowledge that one-half wavelength equals one full trip around the circle, so 1/8 of a wavelength equals ¼ of the way around that circle going, for all these problems, in a clockwise direction. It's also important to know that all impedances in the top half of the circle have inductive reactances and all impedances in the bottom half of the circle have capacitive reactances. We don't even need to normalize any values; we'll start either at the zero end or the infinity end of the resistance axis, depending on the given resistance value, then travel around the reactance axis a distance equal to the length of the coax stub in wavelengths.

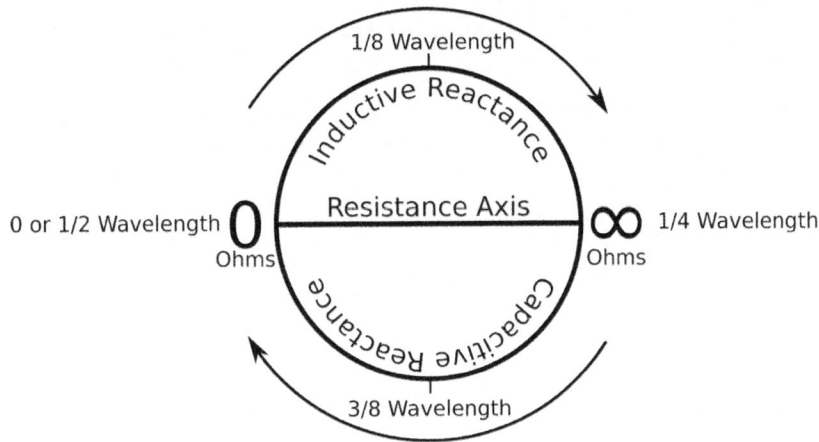

Figure 56.1: Super-Simple Smith Chart

We know the resistance of the piece of coax is zero ohms – it's shorted at the end. If we put our ohm meter on it, it will show zero ohms. (Or close enough for practical purposes.)

That puts our impedance somewhere on the reactance axis – the big circle around the outside, since it intersects the resistance axis at zero and, since it will come into play shortly, at infinity as well.

Now we can figure out the relative reactance. Follow along on Figure 56.2.

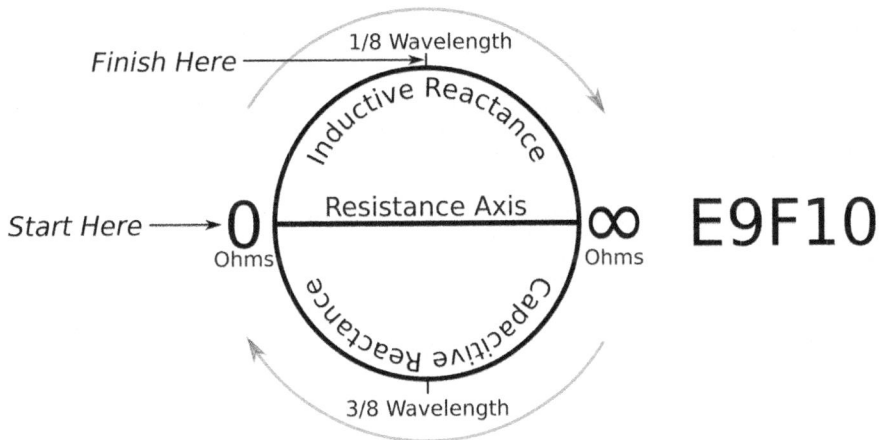

Figure 56.2: Solution for E9F10

The wavelength scale goes around the outside of the chart. On a real Smith chart there are some small notations with arrows that show us that for "wavelengths toward generator" (which is what the question asks about) we go around the scale clockwise. So we start at the far left end of the resistance scale – zero ohms – and follow around to 0.125 (1/8) wavelength. 1/8 wavelength is ¼ of the way clockwise around the chart – remember that and you'll crush any of these questions.

Notice, that impedance point is in the upper half of the circle – and that means it has an **inductive**, +j reactance.

E9F11 (C) What impedance does a 1/8-wavelength transmission line present to a generator when the line is open at the far end?
    A. The same as the characteristic impedance of the line
    B. An inductive reactance
    C. **A capacitive reactance**
    D. Infinite

A 1/8 wavelength long transmission line with an open end presents a **capacitive reactance**.

Yep, we can do this with the Smith chart, too. See Figure 56.3. This time we have an infinite resistance number, so rather than starting at the far left of the resistance line, we start on the far right. We trace clockwise around that reactance axis until we reach a spot opposite 0.125 wavelength ...., ¼ way around the circle, going clockwise ...

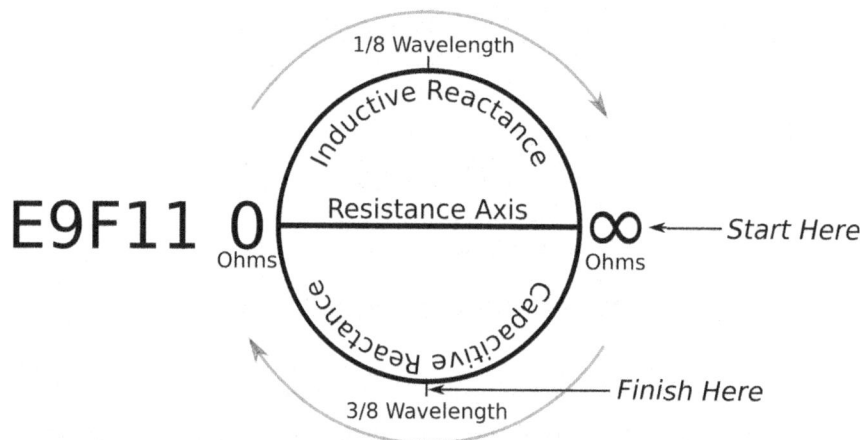

Figure 56.3: Solution for E9F11

That plotted impedance is in the lower half of the circle, so it is an impedance with **capacitive reactance**.

E9F12 (D) What impedance does a 1/4-wavelength transmission line present to a generator when the line is open at the far end?

    A. The same as the characteristic impedance of the line

    B. The same as the input impedance to the generator

    C. Very high impedance

    D. **Very low impedance**

Give me a Smith chart! Hold my beer!

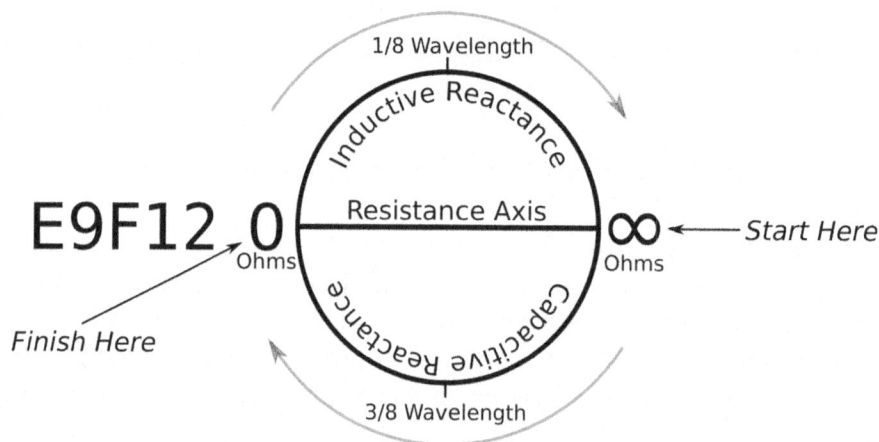

Figure 56.4: Solution for E9F12

We have an open-ended transmission line – infinite resistance. As seen in Figure 56.4, we'll start at the infinite resistance point of the resistance axis – the far right end, then we'll follow the reactance axis around the chart to a point that says 0.25 wavelength on the wavelength scale. (It's at the far left-hand side of the chart – straight across from where we started.) ½ a trip around the circle, always clockwise for these problems, is ¼ wavelength.

We end up with an "impedance" that has zero resistance and zero reactance – that's a **very low impedance** indeed!

378

E9F13 (A) What impedance does a 1/4-wavelength transmission line present to a generator when the line is shorted at the far end?

    A. **Very high impedance**

    B. Very low impedance

    C. The same as the characteristic impedance of the transmission line

    D. The same as the generator output impedance

You can guess this one, right?

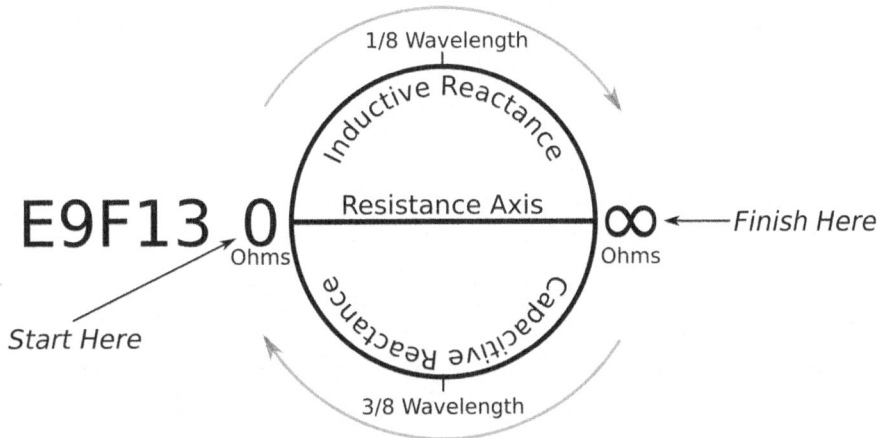

Figure 56.5: Solution for E9F13

We'll start at the far left hand side of the Smith chart, follow the reactance axis halfway around and end up with an impedance with infinite resistance (and still no reactance.) That's one **very high impedance**.

E9F04 (B) What impedance does a 1/2-wavelength transmission line present to a generator when the line is shorted at the far end?

    A. Very high impedance

    B. **Very low impedance**

    C. The same as the characteristic impedance of the line

    D. The same as the output impedance of the generator

In the case of a 1/2 wavelength transmission line that is shorted at one end, we start at the zero ohm mark on the far left, go all the way around the circle, and end up at an impedance that looks just like the 1/4 wave open end coax – mighty close to zero. **Very low impedance**.

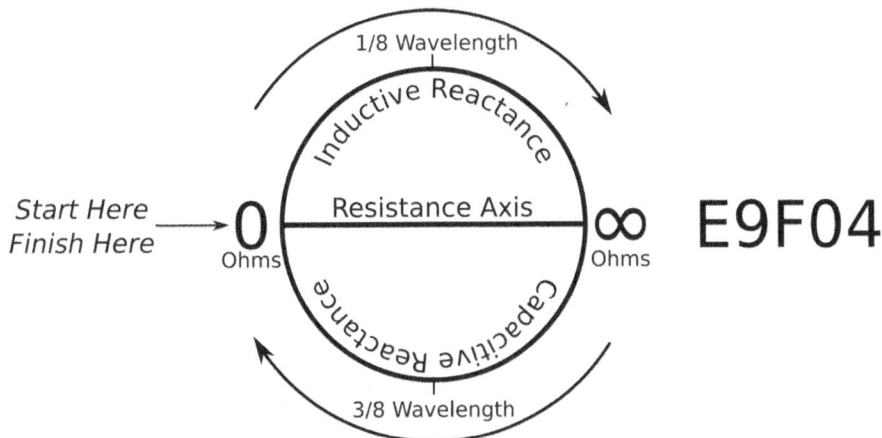

Figure 56.6: Solution for E9F04

I have to tell you; a friend of mine decided to get his Extra and he was totally frustrated because he couldn't memorize the answers to these five questions. I said, "Why are you trying to memorize it? Why don't you do it with the Smith chart?"

"Oh, that looks really hard."

No it's really not. If you'll just try it a few times, it is so simple – you'll have these completely handled and you won't be beating your brains out trying to memorize five completely arbitrary answers.

Key Concepts in This Chapter

- The velocity factor of a transmission line is the velocity of the wave in the transmission line divided by the velocity of light in a vacuum.

- What has the biggest effect on the velocity factor of a transmission line is the dielectric materials used in the line.

- The physical length of a coaxial cable transmission line is shorter than its electrical length because electrical signals move more slowly in a coaxial cable than in air.

- The approximate physical length of a solid polyethylene dielectric coaxial transmission line that is electrically one-quarter wavelength long at 14.10 MHz is 3.5 meters.

- The approximate physical length of a foam polyethylene coaxial transmission line that is electrically 1/4 wavelength long at 7.2 MHz is 8.3 meters.

- The approximate physical length of an air-insulated, parallel conductor transmission line that is electrically 1/2 wavelength long at 14.10 MHz is 10.6 meters.

- There are several significant differences between foam dielectric coaxial cable and solid dielectric cable. Foam dielectric has lower safe operating voltage limits, foam dielectric has lower loss per unit of length, and foam dielectric has a higher velocity factor.

- At 50 MHz ladder line has lower loss than small-diameter coaxial cable such as RG-58.

- A 1/8 wavelength transmission line presents an inductive reactance to a generator when the line is shorted at the far end. If that same transmission line is open at the far end, it presents a capacitive reactance.

- A 1/4 wavelength transmission line presents a very low impedance to a generator when the line is open at the far end. If the line is shorted at the far end it presents a very high impedance.

- A 1/2 wavelength transmission line that is shorted at one end presents a very low impedance.

Chapter 57

Receiving Antennas

General Purpose Receiving Antennas

E9H01 (D) When constructing a Beverage antenna, which of the following factors should be included in the design to achieve good performance at the desired frequency?

    A. Its overall length must not exceed 1/4 wavelength

    B. It must be mounted more than 1 wavelength above ground

    C. It should be configured as a four-sided loop

    D. **It should be one or more wavelengths long**

I've joked before that the Beverage antenna is seen far more often on ham exams than in real life, but I have uncovered evidence of at least a few hams with a lot of land and a serious interest in DX who have built them. At least one has built a multi-directional array of them, with switching controlled from the shack. (Google W8JI if you're interested.)

The Beverage is a non-resonant long-wire (*really* long) receiving antenna. **It should be one or more wavelengths long**, and the largest on record was an old AT&T installation that had a wire 9 *miles* long. With the Beverage, the longer the better.

The Beverage is extraordinarily directional; it's so directional there are a few military installations around the world that use an array of Beverage antennas for critical direction-finding tasks.

It doesn't have to be hoisted high in the air – in fact, it works better low to the ground.

E9H09 (B) What is a Pennant antenna?

    A. A four-element, high-gain vertical array invented by George Pennant

    B. **A small, vertically oriented receiving antenna consisting of a triangular loop terminated in approximately 900 ohms**

    C. A form of rhombic antenna terminated in a variable capacitor to provide frequency diversity

    D. A stealth antenna built to look like a flagpole

A Pennant antenna is **a small, vertically oriented receiving antenna consisting of a triangular loop terminated in approximately 900 ohms.** "Small" is a relative term, as you can see from Figure 57.1, but compared to an effective Beverage antenna the pennant is positively dainty.

The Pennant was designed to be a directional receiving antenna for the 160-meter band, but is reported to work well on 80 and 40 meters as well.

E9H02 (A) Which is generally true for low band (160-meter and 80-meter) receiving antennas?

    A. **Atmospheric noise is so high that gain over a dipole is not important**

    B. They must be erected at least 1/2 wavelength above the ground to attain good directivity

    C. Low loss coax transmission line is essential for good performance

    D. All these choices are correct

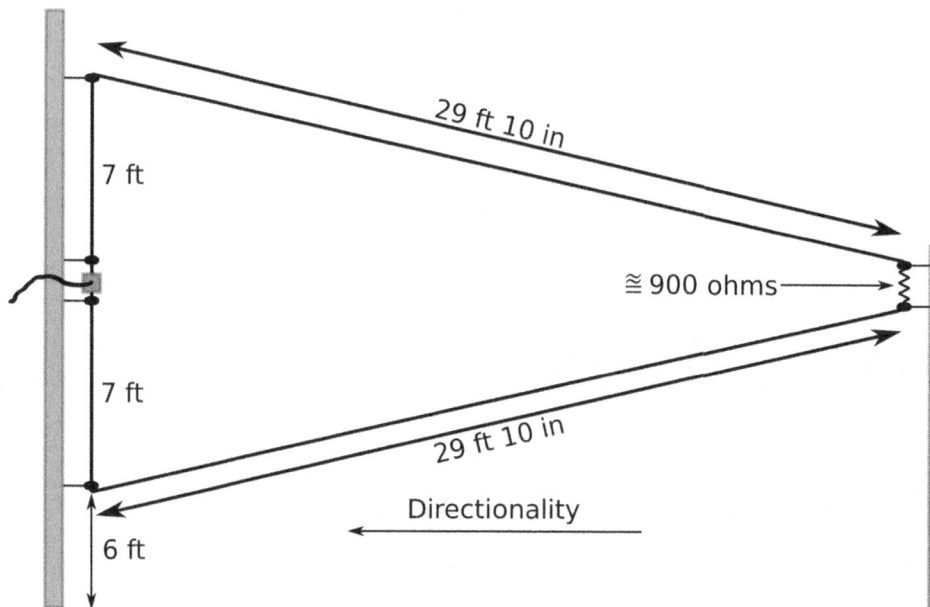

Figure 57.1: Pennant Antenna

Let's kill off the wrong answers on this one first. We know low band receiving antennas don't necessarily have to be ½ wavelength above the ground to attain good directivity – the Beverage antenna can ramble along at a few feet off the ground and is very directive.

You might think low loss coax transmission line is *essential* for good receiving performance, but it really isn't. Have you ever noticed that throughout your ham studies the question of impedance matching your receiving antenna to your receiver never came up? Unlike with transmitters, it really isn't essential for the receiving side. The reason is, our receivers have such huge gain, we can amplify RF signals far beyond any amount we might lose to at least a mild SWR loss. And if we do have high SWR from the receiving antenna to the receiver, so what? It's not like the antenna is going to burn up from all the reflected power. So low loss coax is on the "nice to have" list, not the "essential" list.

That leaves us with **atmospheric noise is so high that gain over a dipole is not important**. Atmospheric noise is a constant – there are about a hundred lightning strikes around the globe every second. That's 15 million volts and up to 500,000 amps (!) of electricity arcing and creating radio waves in the process. Just like our lower frequency signals, the lower frequencies created by each lightning strike have a good chance of covering a lot of distance either by ground wave or by skywave, and in turn have a good chance of appearing at our 160-meter or 80-meter antenna from all directions. They're loud, too; 50 or 60 dB above the background noise. If you have a very high gain antenna – say 20 dB over a dipole – you're still trying to hear over 30 to 40 dB of noise coming in from the sides, never mind the full 50 or 60 coming in the front of the antenna. It's like cupping your hands next to your ears to try to hear someone in a noisy factory – it's still going to sound very noisy.

Direction Finding Antennas

E9H04 (B) What is an advantage of placing a grounded electrostatic shield around a small loop direction-finding antenna?

    A. It adds capacitive loading, increasing the bandwidth of the antenna

    B. **It eliminates unbalanced capacitive coupling to the surroundings, improving the nulls**

    C. It eliminates tracking errors caused by strong out-of-band signals

    D. It increases signal strength by providing a better match to the feed line

Figure 57.2 shows a somewhat famous shielded loop antenna in the hands of someone a good deal more famous.

Figure 57.2: Amelia Earhart with Loop Antenna

That's Amelia Earhart with a radio direction finding antenna, probably the one on the Lockheed Electra that disappeared in 1937. Earhart aficionados have bandied about one theory that her inexperience with the system may have contributed to – well, to whatever happened to her.

The shielded loop consists of a conductor inside a "shield" and it is highly directional, making it useful for radio direction finding applications like fox hunts or tracking down that annoying source of RF interference you can't seem to find.

In order to work properly a shielded loop antenna must be constructed so that it is electrostatically balanced *relative* to ground.

When a loop antenna is directly facing the signal source – so you're looking through the loop at the signal source – the two sides of the circle are equally energized by the signal. The left side is positive at the same moment and to the same extent that the right side is positive, in other words. If everything is the same between the points that get energized and the input to the receiver, we should get absolutely nothing from the antenna – because the two sides cancel each other out. So simple, what could possibly go wrong? Impedance could go wrong, that's what.

Everything even a little bit conductive in the immediate environment forms a capacitance with the two sides of the antenna. If our antenna was in "free space" everything would be fine, but we're never in free space, and there's never a perfect balance of miscellaneous stuff around the antenna. It would be a rare occasion indeed when the stuff on the left side of the antenna exactly balanced the stuff on the right side – so the world constantly tries to unbalance our loop antenna, and when the sides are unbalanced, we don't get a good strong null for our direction finding project.

The shield breaks up this unequal capacitance effect, and that's how **it eliminates unbalanced capacitive coupling to the surroundings, improving the nulls**.

E9H05 (A) What is the main drawback of a small wire-loop antenna for direction finding?

    A. **It has a bidirectional pattern**

    B. It has no clearly defined null

    C. It is practical for use only on VHF and higher bands

    D. All these choices are correct

The loop antenna above has nodes around each side. In other words, **it has a bidirectional**

pattern. There are two nulls, each perpendicular to the loop. That means if you find a signal you are hunting with it, you have a 50/50 chance of knowing the direction to walk to get closer to the source of the signal.

E9H10 (D) How can the output voltage of a multiple-turn receiving loop antenna be increased?
    A. By reducing the permeability of the loop shield
    B. By utilizing high impedance wire for the coupling loop
    C. By winding adjacent turns in opposing directions
    D. **By increasing the number of turns and/or the area**

Most loop antennas are non-resonant antennas, so we're not worried about making our antenna "too long", we're worried about getting enough voltage to have a useful received signal. We can accomplish that **by increasing either the number of wire turns in the loop or the area of the loop structure, or both**.

E9H06 (C) What is the triangulation method of direction finding?
    A. The geometric angles of sky waves from the source are used to determine its position
    B. A fixed receiving station plots three headings to the signal source
    C. **Antenna headings from several different receiving locations are used to locate the signal source**
    D. A fixed receiving station uses three different antennas to plot the location of the signal source

In the triangulation method of direction finding **antenna headings from several different receiving locations are used to locate the signal source**.

    Let's say Ham Radio Operator A and Ham Radio Operator B are trying to find a sooper-sekrit clandestine radio station. Ham Radio Operator A hooks up her wire loop antenna to her receiver and twists and turns it a bit until she nulls out the signal. She also has a GPS with her to get her precise position. She gets on her 2-meter HT and reports her position and that she has the signal on a bearing of "due north or due south." (See Figure 57.3.)

Figure 57.3: "Triangulation" With One Receiver

    Well, should we march north or south? We don't know. But, happily, A's friend B also has a direction-finding setup and he is somewhere up to the north of A. B fires up his radio and finds the clandestine station is somewhere on a line running from SW to NE. (See Figure 57.4.)

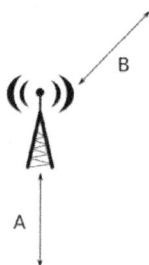

Figure 57.4: Triangulation With Two Receivers

Ham B reports his position and the bearing of the signal from his position. They plot the lines on their maps and they've at least approximately located the station – that station has been *triangulated*.

If they can get Ham Radio Operator C to report in from a different position, they can get an even more precise fix on that station.

E9H07 (D) Why is RF attenuation used when direction finding?
    A. To narrow the receiver bandwidth
    B.To compensate for isotropic directivity and the antenna effect of feed lines
    C. To increase receiver sensitivity
    D. **To prevent receiver overload which reduces pattern nulls**

When we're direction finding, we're trying to "null out" the signal we're seeking. It's easy to hear a null point – the signal vanishes. We could also look for a peak, but that's harder. Either way, as we get closer to the target, the signal gets stronger and either the null direction gets very skinny (or non-existent) or the peak direction gets wider and wider. By attenuating the signal, we get back to a more useful situation. Attenuation is used **to prevent receiver overload which reduces pattern nulls**.

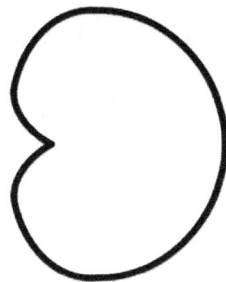

Figure 57.5: Cardioid Pattern

E9H08 (A) What is the function of a sense antenna?
    A. **It modifies the pattern of a DF antenna array to provide a null in one direction**
    B. It increases the sensitivity of a DF antenna array
    C. It allows DF antennas to receive signals at different vertical angles
    D. It provides diversity reception that cancels multipath signals

"DF" is Direction Finding. When you've looked at a small airplane – probably an older one – you might have noticed some sort of wire running from near the top of the tail to a spot near the cockpit. That wasn't a guy wire to hold the tail on, it was the sense antenna for the airplane's radio DF, for direction finding, system.

A sense antenna is an "extra" antenna. It can be a dipole or even a vertical whip. When its signal is added to a loop antenna's signal, it makes the loop more directional. Specifically, it creates a cardioid pattern. **It modifies the pattern of a DF antenna array to provide a null in one direction**, as in Figure 57.5.

You can see that looking for a peak with that pattern won't do much good – it's fairly equal across the "front" – but there's a strong null and it's only in one direction. Now we know which direction to march! The sense antenna **modifies the pattern of a DF antenna array to provide a null in one direction.**

E9H11 (B) What feature of a cardioid pattern antenna makes it useful for direction finding?
    A. A very sharp peak
    B. **A very sharp single null**
    C. Broadband response
    D. High radiation angle

It's that **very sharp single null** that makes a cardioid antenna so useful for direction finding.

Key Concepts in This Chapter

- When constructing a Beverage antenna, to achieve good performance at the desired frequency the antenna should be one or more wavelengths long.

- A Pennant antenna is a small, vertically oriented receiving antenna consisting of a triangular loop terminated in approximately 900 ohms.

- What's generally true for low band (160-meter and 80-meter) receiving antennas is that atmospheric noise is so high that gain over a dipole is not important.

- An advantage of using a shielded loop antenna for direction finding is that it eliminates unbalanced capacitive coupling to the surroundings, improving the nulls.

- The main drawback of a wire-loop antenna for direction finding is that it has a bidirectional pattern.

- The output voltage of a multiple turn receiving loop antenna can be increased by increasing the number of wire turns and/or the area.

- The triangulation method of direction finding can be described as, "antenna headings from several different receiving locations are used to locate the signal source."

- RF attenuation is used when direction finding to prevent receiver overload which reduces pattern nulls.

- The function of a sense antenna is to modify the pattern of a DF antenna array to provide a null in one direction.

- The characteristic of a cardioid pattern antenna that makes it useful for direction finding is a very sharp single null.

Chapter 58

Safety

Can you believe it? We're coming up on the final chapter. Don't be sad.

Let me say up front, I am not a safety instructor, and the topics covered on the exam fall far short of being a complete course in proper safety procedures for our hobby. If you're up to something potentially hazardous, get complete safety instruction and proper equipment for the job.

Station Grounding

E0A01 (B) What is the primary function of an external earth connection or ground rod?
   A. Reduce received noise
   B. **Lightning protection**
   C. Reduce RF current flow between pieces of equipment
   D. Reduce RFI to telephones and home entertainment systems

The primary function of our *external* earth connections or ground rods is **lightning protection**. The grounds from inside the shack to the ground rod(s) are for RF grounding, and they're important, too, but we *really* want to diffuse that lightning to ground before it has a chance to do damage.

RF Exposure

E0A02 (B) When evaluating RF exposure levels from your station at a neighbor's home, what must you do?
   A. Ensure signals from your station are less than the controlled Maximum Permitted Exposure (MPE) limits
   B. **Ensure signals from your station are less than the uncontrolled Maximum Permitted Exposure (MPE) limits**
   C. Ensure signals from your station are less than the controlled Maximum Permitted Emission (MPE) limits
   D. Ensure signals from your station are less than the uncontrolled Maximum Permitted Emission (MPE) limits

Since you're getting your Extra license, you're pretty committed to the hobby, and that commitment might include spending the dough on a high-power rig. (Or not – there's always QRP to explore.)

If you go off in that high-power direction, this might serve as a timely reminder that you need to **make sure signals from your station are less than the <u>uncontrolled</u> MPE limits** over there in your neighbor's house. MPE is "Maximum Permissible Exposure." For us, those are spelled out in FCC Bulletin 65 and Supplement B, both of which you can download for free from https://www.fcc.gov/general/oet-bulletins-line.

There are MPE limits on controlled environments – environments where the occupants know RF radiation is present – and lower MPE limits for uncontrolled environments, where the general public is at risk of exposure.

Until recently, RF Exposure Evaluation was a simple process for most hams. We consulted a table of "what power at which frequency" and if we were under the power listed, we were in good shape.

New regulations recently came into effect that complicated things considerably, but your RF Exposure Evaluation is still easily accomplished. There are online tools to get it done in minutes. We evaluated our mobile and home rigs at the following site, and it took less than two minutes.

http://hintlink.com/power_density.htm

E0A03 (C) Over what range of frequencies are the FCC human body RF exposure limits most restrictive?
A. 300 kHz to 3 MHz
B. 3 to 30 MHz
C. **30 to 300 MHz**
D. 300 to 3000 MHz

Take a look at the chart of maximum power levels vs. wavelength in Table **??**. From 160 meters – 1.7 MHz – up through 40 meters, the limit before an evaluation is required is 500 watts. Then the limits take a rather sudden nosedive starting at 30 meters, until they're down to 50 watts for everything from 30 to 300 MHz. The reason is those are the frequencies to which the human body is most vulnerable.

E0A04 (C) When evaluating a site with multiple transmitters operating at the same time, the operators and licensees of which transmitters are responsible for mitigating over-exposure situations?
A. Only the most powerful transmitter
B. Only commercial transmitters
C. **Each transmitter that produces 5 percent or more of its MPE limit in areas where the total MPE limit is exceeded**
D. Each transmitter operating with a duty-cycle greater than 50 percent

Interesting rule, isn't it? If you are at a site with multiple transmitters operating *at the same time* and your transmitter is at just **5% of its MPE limit**, you're on the hook for mitigating over-exposure situations.

The only common situations I can think of where this might come into play are some repeater installations and some Field Day setups.

E0A05 (B) What is one of the potential hazards of operating in the amateur radio microwave bands?
A. Microwaves are ionizing radiation
B. **The high gain antennas commonly used can result in high exposure levels**
C. Microwaves often travel long distances by ionospheric reflection
D. The extremely high frequency energy can damage the joints of antenna structures

One of the potential hazards of using microwaves in the amateur radio bands is linked to the fact that it's relatively easy to create very high gain antennas in the microwave range.

Legend has it that the microwave oven was invented when a gentleman named Percy Spencer was working on a magnetron for a radar installation and noticed the candy bar in his pocket had melted. He and some colleagues threw together a makeshift oven to see if they could heat other food. Wouldn't you know it, the first thing they microwaved was – popcorn.

So if high-gain microwave antennas can pop popcorn – think what they could do to your innards. 'nuff said?

E0A11 (C) Which of the following injuries can result from using high-power UHF or microwave transmitters?

    A. Hearing loss caused by high voltage corona discharge

    B. Blood clotting from the intense magnetic field

    C. **Localized heating of the body from RF exposure in excess of the MPE limits**

    D. Ingestion of ozone gas from the cooling system

If the popcorn story didn't get the point across, the injuries that can result from using high-power UHF or microwave transmitters are **localized heating of the body from RF exposure in excess of the MPE limits.**

E0A06 (D) Why are there separate electric (E) and magnetic (H) field MPE limits?

    A. The body reacts to electromagnetic radiation from both the E and H fields

    B. Ground reflections and scattering make the field strength vary with location

    C. E field and H field radiation intensity peaks can occur at different locations

    D. **All these choices are correct**

You know radio waves are electromagnetic waves, and probably remember that they consist of electric waves and magnetic waves. The formal physics names for these are the E field and the H field.

    The body reacts to electromagnetic radiation from both the E and H fields.

    In part because they the H field and E field are polarized at 90° from each other, **ground reflections and scattering make the field impedance vary with location** and **E field and H field radiation intensity peaks can occur at different locations.**

    (Don't worry – all this stuff gets calculated by the RF exposure modeling program.)

E0A08 (C) What does SAR measure?

    A. Synthetic Aperture Ratio of the human body

    B. Signal Amplification Rating

    C. **The rate at which RF energy is absorbed by the body**

    D. The rate of RF energy reflected from stationary terrain

The MPE figures are derived from the "SAR" figures. SAR is Specific Absorption Rate. It's **the rate at which RF energy is absorbed by the body.**

    The MPE converts those SAR numbers into something much more useful for us, which is, basically, "here's how much power how close at what frequency for how long before you have a problem."

Carbon Monoxide

E0A07 (B) How may dangerous levels of carbon monoxide from an emergency generator be detected?

    A. By the odor

    B. **Only with a carbon monoxide detector**

    C. Any ordinary smoke detector can be used

    D. By the yellowish appearance of the gas

Carbon monoxide is deadly, odorless, colorless, and undetectable by ordinary smoke detectors. Did I mention it's deadly? It can be detected **only with a carbon monoxide detector**.

Hazardous Materials

E0A09 (C) Which insulating material commonly used as a thermal conductor for some types of electronic devices is extremely toxic if broken or crushed and the particles are accidentally inhaled?

    A. Mica

    B. Zinc oxide

    C. **Beryllium Oxide**

    D. Uranium Hexafluoride

If you find a chunk of ceramic under something like a magnetron or a high-power transistor, don't break it up with a hammer and inhale the particles. You'll get real sick and probably die. In fact, handle it with a lot of care.

Beryllium oxide gets made into a ceramic for such applications, and it's an excellent thermal conductor. It's quite stable in that form, but poisonous as a powder.

It's not the stuff in that paste you put between your computer's CPU and the heat sink, so no worries there – but still, don't eat or inhale that either.

Uranium hexafluoride, mentioned as a wrong answer, isn't good for you either, but it is not an insulating material commonly used as a thermal conductor; it's a chemical used in uranium enrichment.

E0A10 (A) What toxic material may be present in some electronic components such as high voltage capacitors and transformers?

    A. **Polychlorinated biphenyls**

    B. Polyethylene

    C. Polytetrafluoroethylene

    D. Polymorphic silicon

Polychlorinated biphenyls – better known as PCB's – haven't been made for more than 40 years, because the stuff turned out to be viciously carcinogenic. It used to be in every power pole transformer in the country and lots of other high voltage electric and electronic parts. A lot of those parts might still be out there somewhere, especially in vintage ham radio gear.

You don't want to be exposed to that stuff.

Key Concepts in This Chapter

- The primary function of an external earth connection or ground rod is lightning protection.

- When evaluating RF exposure levels from your station at a neighbor's home you must make sure signals from your station are less than the uncontrolled Maximum Permissible Exposure limits.

- The range of frequencies where the FCC human body RF exposure limits are most restrictive is 30 to 300 MHz.

- When evaluating a site with multiple transmitters operating at the same time, the operators and licensees of each transmitter that produces 5 percent or more of its MPE limit in areas where the total MPE limit is exceeded are responsible for mitigating over-exposure situations.

- One of the potential hazards of using microwaves in the amateur radio bands is that the high gain antennas commonly used can result in high exposure levels.

- Injuries from localized heating of the body from RF exposure in excess of the MPE limits can result from using high-power UHF or microwave transmitters.

- There are several reasons why there are separate electric (E) and magnetic (H) field MPE limits. The body reacts to electromagnetic radiation from both the E and H fields; ground reflections and scattering make the field impedance vary with location, and; E field and H field radiation intensity peaks can occur at different locations.

- SAR measures the rate at which RF energy is absorbed by the body.

- Dangerous levels of carbon monoxide from an emergency generator can only be detected with a carbon monoxide detector.

- The insulating material commonly used as a thermal conductor for some types of electronic devices that is extremely toxic if broken or crushed and the particles are accidentally inhaled is beryllium oxide.

- The toxic material that may be present in some electronic components such as high voltage capacitors and transformers is called polychlorinated biphenyls.

If you have been following the *Fast Track* study plan, you're almost ready to start taking Final Practice Exams (available on fasttrackham.com) covering all subelements and groups. We'd suggest you take Progress Check #12 at least once, just as a final check on any group-by-group work you might need to do. When you're consistently scoring 85% or better on the Final Practice Exam, go get that upgrade. Congratulations!

Progress Check #12

Extra Class Final Practice Exam

Chapter 59

QRT

It's about time for sign-off. I hope you've mastered the material you needed to master and have even enjoyed learning some new stuff about our hobby. I congratulate you for daring to challenge your mind by exploring new territory. Now it's time to go take the exam. If you have taken a ham exam recently, you are familiar with the way a testing session goes, but if it has been a while, here's a review.

First, get a good night's sleep before the exam. Research shows that aside from proper preparation, the single most important contributor to good test performance is being well rested.

When you arrive at the testing location, it's pretty certain you'll be greeted by one of the Volunteer Examiners, and directed to a seat. You'll have some paperwork to fill out, and the VE's will verify you are who you say you are. Here's what you'll need for the exam:

- **A form of legal ID**. Your driver's license or US passport will suffice.

- A copy of **your current FCC issued ham license**, if you have one. (Some people choose to take the Technician, General and Extra exams on the same day.) Don't just bring that card you probably carry in your wallet or purse, make a copy of the original license the examiners can keep. It's also acceptable to bring a "Reference Copy" of your license, which you can get through the FCC's Universal Licensing System web site.

- If you do not already have an FCC license, bring your **Social Security Number**. This becomes your temporary FCC registration number. Once you have your license, you'll be issued what's called an FRN – an FCC Registration Number – so your Social Security number will never show up on the FCC public database. However, if you don't have an FRN and you don't want to be handing out your Social Security Number, that's perfectly fine, but you need to get on the FCC website ahead of time and register there for an FRN, then bring that to the testing session.

- **Pencils and a pen.** Pencils for the test, a pen to fill out your paperwork.

- **A calculator**. It's "optional" but, come on, this is the Extra exam. The VE's will ask you to clear the memory of your calculator before you start the exam, so know how to do that. (That's to prevent you from storing any handy formulas or answers in your calculator.) On the TI-30XS it's 2nd -> Reset/0 -> 2. You will not be allowed to use your cell phone calculator and programmable calculators are barred from the exam room.

- **A relaxed attitude**. This is optional, too, but it will boost your results. Remember, those "scary" Volunteer Examiners have been through this experience themselves and they're on your side. They're your neighbors and your fellow hams. They're Extra Class hams and had to take yet another test to be accredited as a VE. They will take their jobs seriously, and you want them to. If there are irregularities in the testing procedure, the FCC has the right to invalidate everyone's test at the session (yes, yours, too), and invite everyone to retest. A clean test procedure benefits you. But under that serious attitude, trust me, they want you to succeed.

- **The test fee** – currently about $15 but subject to change annually.

If you have special needs, such as a visual disability, hearing disability, mobility challenges, dyslexia, or any other disability that would hinder your ability to take and pass the test, the FCC requires that appropriate accommodations be made. This might include someone reading the test to you, providing access to the testing site, offering an exam without the illustration-based questions, or other measures. If you want some accommodation, your part is to let the VE team know ahead of time so they can make the proper arrangements.

At the test site you'll fill out an application for your upgrade. It's called a Form 605, and it asks you, among other things, to certify that you have "read and WILL COMPLY with Section 97.13(c) of the Commission's Rules regarding RADIOFREQUENCY (RF) RADIATION SAFETY and the amateur service section of OST/OET Bulletin Number 65." If you would like to read that Bulletin before you swear that you did, here's a place you can do so:

https://www.arrl.org/files/file/RFSAFIN2.pdf

As of September 2017, there's a "new" question on Form 605, which is actually an old question that was, somehow, left off the next-to-latest 605 when it was created. Each applicant now must state whether they have ever been convicted of or plead guilty to a felony. If the answer is yes, that doesn't necessarily disqualify them from holding an Amateur license, but it does mean they must "submit as an exhibit a statement explaining the circumstances and a statement giving the reasons why the Applicant believes that grant of the application would be in the public interest notwithstanding the actual or alleged misconduct." That statement is to be sent by the applicant to the FCC, not given to the VE's. The FCC will review the statement and decide whether or not to grant the license. They have not made any public statement about what standards they will apply in such a case.

There's also the pressing question about whether or not you want a shiny new Amateur Extra Class call sign, or if you want to keep your current call sign. It's your choice. I opted for the change because my old call sign was KG7DVV and on phone nobody seemed to be able to get it straight – it would come back as KG7DBB, DVD, DBD, BVD, or almost anything that rhymed with "E" except KG7DVV, so I rolled the dice and got AF7KB. (Years later, a friend pointed out it was a pretty good call sign for me, since "KB" is an abbreviation for Knowledge Base in some circles.) Other folks are very fond of their current call sign, and if that's you, you don't have to give it up.

You can always upgrade later through the vanity call system, but you'll probably pay for the privilege, since there is currently talk at the FCC of reinstating fees for the service. Upgrading your call sign when you take the exam is free. If you check the box that says "CHANGE my station call sign systematically" and initial to confirm, you'll get the next available 1 x 2 or 2 x 2 call sign in the stack when they process your new license. Otherwise, your call sign will stay the same.

Once the paperwork is complete and checked, they'll hand out the exams and the answer sheets, and you'll take your exam. The exam is not timed. Take your time, read each question carefully, double check each answer and be sure you are entering the answer in the right spot on the answer sheet. You won't be troubling anyone by taking your time. Those VE's are expecting the session to last at least a couple of hours, and they're committed to staying the whole time. In fact, by law they can't leave the room while testing is going on.

During the exam, if you really blank out on a question and just can't come up with an answer, I suggest you leave the answer sheet blank for that one and go on to the next. Do be sure you get the next answer in the right spot on the answer sheet, or every answer from then on will be "one off" and you'll miss them all. Your brain might come up with the answer later in the test, or perhaps you'll get a hint from one of the other questions. Be sure to come back to the question and answer it with your best guess. There is no penalty for guessing, only rewards for correct answers, and a blank answer is automatically not a right answer.

During the test, if you are given a direction by a VE, do it. As you know, FCC law requires that your exam be terminated immediately if you do not follow the directions of the VE's.

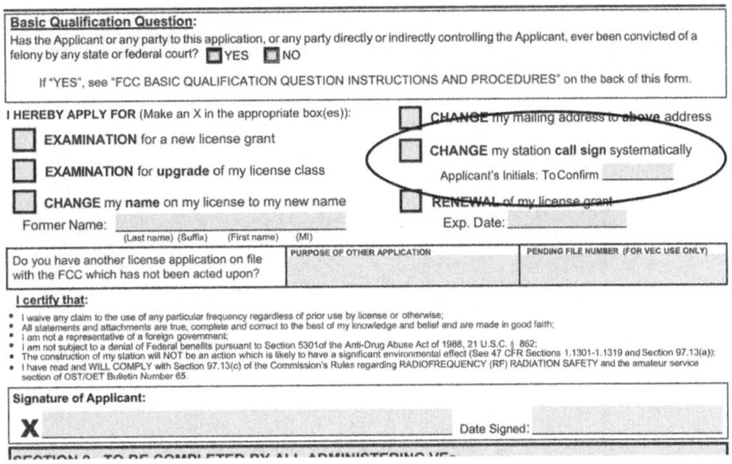
Basic Qualification Question:
Has the Applicant or any party to this application, or any party directly or indirectly controlling the Applicant, ever been convicted of a felony by any state or federal court? YES NO

If "YES", see "FCC BASIC QUALIFICATION QUESTION INSTRUCTIONS AND PROCEDURES" on the back of this form.

I HEREBY APPLY FOR (Make an X in the appropriate box(es)):
- **EXAMINATION** for a new license grant
- **EXAMINATION** for **upgrade** of my license class
- **CHANGE** my **name** on my license to my new name

Former Name: (Last name) (Suffix) (First name) (MI)

- ~~**CHANGE** my mailing address to above~~ address
- **CHANGE** my station **call sign** systematically

Applicant's Initials: To Confirm

- ~~**RENEWAL** of my license grant~~

Exp. Date:

Do you have another license application on file with the FCC which has not been acted upon?	PURPOSE OF OTHER APPLICATION	PENDING FILE NUMBER (FOR VEC USE ONLY)

I certify that:
- I waive any claim to the use of any particular frequency regardless of prior use by license or otherwise;
- All statements and attachments are true, complete and correct to the best of my knowledge and belief and are made in good faith;
- I am not a representative of a foreign government;
- I am not subject to a denial of Federal benefits pursuant to Section 5301of the Anti-Drug Abuse Act of 1988, 21 U.S.C. § 862;
- The construction of my station will NOT be an action which is likely to have a significant environmental effect (See 47 CFR Sections 1.1301-1.1319 and Section 97.13(a));
- I have read and WILL COMPLY with Section 97.13(c) of the Commission's Rules regarding RADIOFREQUENCY (RF) RADIATION SAFETY and the amateur service section of OST/OET Bulletin Number 65.

Signature of Applicant:
X Date Signed:

Figure 59.1: Part of Form 605

When your test is complete, you'll hand it in and it will be graded during the session. Each exam must be graded independently by three VE's, so it takes a bit of time. Then, someone will come over and – I fully expect – congratulate you on passing the exam. In the wildly unlikely event you don't pass your exam, they'll have a quiet – probably outside the room – conversation with you. VE's are directed by their training to take pains to avoid embarrassing anyone.

What the VE's almost certainly will *not* do is tell you which, if any, questions you missed. They're not even required to tell you your score. Just *pass* or *fail*.

Even though luck should play no part in it, I'll still wish you good luck with your exam. It can't hurt.

Thank you so much for participating in the *Fast Track* program. There are more licensed amateur radio operators today than ever before in history, and we're really happy to have been a small part of that. Thanks to your support, whether through just buying the programs, or leaving nice reviews, or telling your friends about it, I'm not on top of Snoqualmie Pass throwing snow chains onto a big rig today, and for that alone I am tremendously grateful.

Finally, remember; the main privilege of your ham license is that it is a license to keep learning. There's always something new to dive into in ham radio, partly because it relates to so many of the sciences, partly because there are so many different hams and ham activities, and partly because there are so many ways hams can contribute positively to their communities.

We wish you a long and happy ham hobby.

73 DE AF7KB & KC7YL

Education is not the filling of a pail, but the lighting of a fire.
- William Butler Yeats; Poet, Nobel Prize Winner

Appendix: Important Formulas

dB to Power Ratio

$$Power\,Ratio = 10^{(dB \div 10)}$$

Power Ratio to dB

$$dB = 10_{\log}(Power\,Ratio)$$

Intermodulation frequency

$$f_{imd} = 2f_1 - f_2$$

and

$$f_{imd} = 2f_2 - f_1$$

For the problem on the exam:

$$f_2 = 2f_1 - f_{imd}$$

and

$$f_2 = \frac{f_{imd} + f_1}{2}$$

Resonant Frequency of a Circuit

$$f = \frac{1}{2\pi\sqrt{L \times C}}$$

Inductive Reactance

$$X_L = 2\pi f L$$

Capacitive Reactance

$$X_C = \frac{1}{2\pi f C}$$

Impedance

$$Z = \sqrt{R^2 + X^2}$$

Q of a Series Resonant Circuit

$$Q = \frac{X}{R}$$

Q of a Parallel Resonant Circuit

$$Q = \frac{R}{X}$$

Half power bandwidth

$$half\,power\,bandwidth = \frac{f_{resonant}}{Q}$$

Time Constant of a Capacitor

$$\tau = RC$$

Time Constant of an Inductor

$$\tau = \frac{L}{R}$$

Susceptance

$$Susceptance = \frac{1}{Reactance}$$

Phase Angle

$$Phase\,Angle = \tan^{-1}\left(\frac{X_L - X_c}{R}\right)$$

Power Factor

$$Power\,Factor = \cos(Phase\,Angle)$$

Turns on a Coil for a Desired Inductance

$$N = \sqrt{\frac{L}{A_L}}$$

Power Dissipation of a Series Connected Linear Voltage Regulator

$$(E_{input} - E_{output}) \times I_{output}$$

Gain of an Op-amp

$$Gain = \frac{R_F}{R_1}$$

Modulation Index (Deviation ratio)

$$Modulation\,Index = \frac{\Delta f}{f_{max}}$$

Bandwidth from WPM or Baud rate

For CW : $bandwidth = WPM \times 4$

For frequency shifted carriers: $bandwidth = (K \times shift) + B$

(For AM: $bandwidth = B \times K$)

Antenna Efficiency

$$Antenna\,Efficiency = (radiation\,resistance / total\,resistance) \times 100\,per\,cent$$

Impedance Matching Transformer

$$Z_{Transformer} = \sqrt{Z_1 \times Z_2}$$

Index

E1D09, 35
E1D10, 35
E1D11, 35
E1D12, 34
E1E01, 32
E1E02, 29
E1E03, 29
E1E04, 29
E1E05, 30
E1E06, 30
E1E07, 30
E1E08, 30
E1E09, 30
E1E10, 31
E1E11, 31
E1E12, 31
E1F01, 13
E1F02, 13
E1F03, 14
E1F04, 15
E1F05, 15
E1F06, 16
E1F07, 16
E1F08, 16
E1F09, 13
E1F10, 17
E1F11, 14
E2A01, 37
E2A02, 39
E2A03, 40
E2A04, 37
E2A05, 37
E2A06, 42
E2A07, 38
E2A08, 38
E2A09, 38
E2A10, 42
E2A11, 41
E2A12, 42
E2A13, 43
E2B01, 45
E2B02, 46
E2B03, 47
E2B04, 49
E2B05, 48
E2B06, 47
E2B07, 48
E2B08, 48
E2B09, 50
E2B10, 49
E2B11, 49
E2B12, 50
E2C01, 25
E2C02, 53
E2C03, 53
E2C04, 68
E2C05, 55
E2C06, 53
E2C07, 54
E2C08, 55

E2C09, 68
E2C10, 54
E2C11, 55
E2C12, 68
E2D01, 60
E2D02, 61
E2D03, 64
E2D04, 65
E2D05, 65
E2D06, 65
E2D07, 67
E2D08, 67
E2D09, 65
E2D10, 66
E2D11, 67
E2E01, 71
E2E02, 313
E2E03, 74
E2E04, 71
E2E05, 72
E2E06, 72
E2E07, 74
E2E08, 72
E2E09, 73
E2E10, 73
E2E11, 73
E2E12, 73
E2E13, 72
E3A01, 63
E3A02, 64
E3A03, 64
E3A04, 58
E3A05, 57
E3A06, 54
E3A07, 58
E3A08, 59
E3A09, 60
E3A10, 57
E3A11, 58
E3A12, 58
E3A13, 59
E3A14, 77
E3B01, 80
E3B02, 80
E3B03, 80
E3B04, 81
E3B05, 82
E3B06, 82
E3B07, 81
E3B09, 82
E3B10, 83
E3B11, 82
E3B12, 83
E3C01, 84
E3C02, 87
E3C03, 88
E3C04, 88
E3C05, 88
E3C06, 78
E3C07, 89

E3C08, 90
E3C09, 90
E3C10, 87
E3C11, 85
E3C12, 79
E3C13, 79
E3C14, 78
E3C15, 90
E4A01, 93
E4A02, 94
E4A03, 95
E4A04, 94
E4A05, 100
E4A06, 94
E4A07, 96
E4A08, 96
E4A09, 94
E4A10, 97
E4A11, 97
E4B01, 103
E4B02, 104
E4B03, 99
E4B04, 99
E4B05, 100
E4B06, 103
E4B07, 98
E4B08, 146
E4B09, 104
E4B10, 96
E4B11, 98
E4C01, 107
E4C02, 111
E4C03, 112
E4C04, 115
E4C05, 116
E4C06, 116
E4C07, 118
E4C08, 119
E4C09, 110
E4C10, 117
E4C11, 119
E4C12, 119
E4C13, 117
E4C14, 109
E4C15, 108
E4D01, 121
E4D02, 122
E4D03, 125
E4D04, 127
E4D05, 126
E4D06, 123
E4D07, 122
E4D08, 124
E4D09, 111
E4D10, 125
E4D11, 125
E4D12, 122
E4E01, 132
E4E02, 131
E4E03, 129

E7D05, 255	E7H13, 299	E9A13, 327
E7D06, 257	E7H14, 301	E9B01, 331
E7D07, 257	E7H15, 302	E9B02, 332
E7D08, 256	E8A01, 305	E9B03, 333
E7D09, 259	E8A02, 276	E9B04, 334
E7D10, 258	E8A03, 306	E9B05, 334
E7D11, 255	E8A04, 284	E9B06, 334
E7D12, 258	E8A05, 306	E9B07, 325
E7D13, 255	E8A06, 307	E9B08, 333
E7D14, 258	E8A07, 307	E9B09, 335
E7D15, 259	E8A08, 274	E9B10, 335
E7E01, 261	E8A09, 275	E9B11, 336
E7E02, 262	E8A10, 282	E9C01, 337
E7E03, 265	E8A11, 276	E9C02, 339
E7E04, 263	E8B01, 267	E9C03, 338
E7E05, 262	E8B02, 268	E9C04, 341
E7E06, 262	E8B03, 268	E9C05, 342
E7E07, 261	E8B04, 269	E9C06, 347
E7E08, 263	E8B05, 269	E9C07, 343
E7E09, 263	E8B06, 269	E9C08, 342
E7E10, 264	E8B07, 270	E9C09, 343
E7E11, 265	E8B08, 270	E9C10, 341
E7F01, 273	E8B09, 269	E9C11, 344
E7F02, 279	E8B10, 270	E9C12, 341
E7F03, 277	E8B11, 271	E9C13, 344
E7F04, 282	E8C01, 314	E9C14, 344
E7F05, 274	E8C02, 309	E9D01, 348
E7F06, 275	E8C03, 309	E9D02, 349
E7F07, 279	E8C04, 310	E9D03, 350
E7F08, 281	E8C05, 311	E9D04, 351
E7F09, 281	E8C06, 313	E9D05, 349
E7F10, 275	E8C07, 313	E9D06, 351
E7F11, 275	E8C08, 314	E9D07, 351
E7F12, 284	E8C09, 315	E9D08, 327
E7F13, 283	E8C10, 310	E9D09, 351
E7F14, 284	E8C11, 309	E9D10, 326
E7G01, 288	E8C12, 312	E9D11, 352
E7G02, 290	E8D01, 317	E9D12, 352
E7G03, 289	E8D02, 318	E9E01, 355
E7G04, 289	E8D03, 318	E9E02, 357
E7G05, 291	E8D04, 318	E9E03, 358
E7G06, 289	E8D05, 319	E9E04, 357
E7G07, 291	E8D06, 320	E9E05, 360
E7G08, 289	E8D07, 319	E9E06, 359
E7G09, 292	E8D08, 319	E9E07, 361
E7G10, 292	E8D09, 320	E9E08, 360
E7G11, 292	E8D10, 320	E9E09, 358
E7G12, 288	E8D11, 320	E9E10, 359
E7H01, 295	E9A01, 323	E9E11, 361
E7H02, 297	E9A02, 328	E9F01, 373
E7H03, 295	E9A03, 324	E9F02, 373
E7H04, 295	E9A04, 326	E9F03, 374
E7H05, 296	E9A05, 324	E9F04, 379
E7H06, 297	E9A06, 328	E9F05, 374
E7H07, 298	E9A07, 329	E9F06, 375
E7H08, 298	E9A08, 327	E9F07, 376
E7H09, 300	E9A09, 324	E9F08, 376
E7H10, 300	E9A10, 325	E9F09, 375
E7H11, 300	E9A11, 326	E9F10, 376
E7H12, 298	E9A12, 323	E9F11, 377

About the Authors

Michael Burnette, AF7KB, started playing with radios at age 8.

As a commercial broadcaster for 25 years, he did a bit of everything from being a DJ to serving as a vice president and general manager with Westinghouse Broadcasting (now CBS/Infinity.)

In 1992, Burnette left the radio business behind, and took to traveling the world designing and delivering experiential learning seminars on leadership, management, communications, and building relationships. He has trained people across the US and in Indonesia, Hong Kong, China, Taiwan, Mexico, Finland, Greece, Austria, Spain, Italy, and Russia.

In addition to his public and corporate trainings, he has been a National Ski Patroller, a Certified Professional Ski Instructor, a Certified In-Line Skating Instructor, a Certified NLP Master Practitioner, a big-rig driving instructor, and a Certified Firewalking Instructor. Kerry Burnette, KC7YL, came later in life to the world of radio, but embraced it enthusiastically and serves as net control for our local club's weekly net as well as leading the club's weekly YL net.

If you want to communicate with Michael, his e-mail is AF7KB@fasttrackham.com. The *Fast Track* web site is

http://fasttrackham.com.

There's a *Fast Track* Facebook page at

http://facebook.com/AF7KB

There's also a YouTube AF7KB *Fast Track* video channel.

He is still playing with radios.

Kerry was the original inspiration for the first *Fast Track* license course, which started as a set of notes for her to use preparing for her Technician exam.

A graduate of the University of Washington, Kerry spent a decade teaching high school math and science and holds a Masters of Education degree.

Michael and Kerry often appear at major hamfests, where Michael is a sought-after speaker. Together, they have created

The Fast Track to Getting Started in Ham Radio
The Fast Track to Your Technician Class Ham Radio License
The Fast Track to Mastering Technician Class Ham Radio Math
The Fast Track to Your General Class Ham Radio License
The Fast Track to Mastering General Class Ham Radio Math
The Fast Track to Your Extra Class Ham Radio License
The Fast Track to Mastering Extra Class Ham Radio Math
The Fast Track to Understanding Ham Radio Propagation
The Fast Track Ham Radio Facts Book

Most *Fast Track* programs are available in paperback, e-book, and audio formats from major online retailers.

Image Credits

- APRS Radio Setup, courtesy of Fred and Anita Kemmerer's excellent "Our Ham Station" blog,
https://stationproject.wordpress.com/

- Ferrite Toroid Core Coil: By Peripitus (Own work) GFDL
(http://www.gnu.org/copyleft/fdl.html) or CC BY-SA 4.0-3.0-2.5-2.0-1.0
(http://creativecommons.org/licenses/by-sa/4.0-3.0-2.5-2.0-1.0), via Wikimedia Commons

- 555 Timers: By Swift.Hg (Own work) CC BY-SA 3.0 (http://creativecommons.org/licenses/by-sa/3.0), via Wikimedia Commons

- DIP Switch: By Winxpcn (Own work) CC BY 3.0 (http://creativecommons.org/licenses/by/3.0), via Wikimedia Commons

- Solar flare photo by NASA.

- All MFJ equipment photos courtesy of MFJ Enterprises.

- All other illustrations either the work of the author or Public Domain

- Typeset in TexStudio LaTeX

Made in the USA
Las Vegas, NV
13 December 2022

62347166R00236